Biomolecular Catalysis

ACS SYMPOSIUM SERIES **986**

Biomolecular Catalysis

Nanoscale Science and Technology

Jungbae Kim, Editor
Pacific Northwest National Laboratory

Seong H. Kim, Editor
The Pennsylvania State University

Ping Wang, Editor
The University of Akron

Sponsored by the
ACS Division of Colloid and Surface Chemistry

American Chemical Society, Washington, DC

Library of Congress Cataloging-in-Publication Data

Biomolecular catalysis : nanoscale science and technology / Jungbae Kim, Seong H. Kim, Ping Wang, editors.

p. cm.—(ACS symposium series ; 986)

"A symposium...was organized at the 230th conference of American Chemical Society during August 28th to September 1st, 2005 in Washington, D.C.

Includes bibliographical references and index.

ISBN : 978–0–8412–7415–0 (alk. paper)

1. Nanotechnology—Congresses. 2. Biotechnology—Congresses. 3. Immobolized enzymes—Congresses.

I. Kim, Jungbae. II. Kim, Seong H. III. Wang, Ping. IV. American Chemical Society. Meeting (230th : 2005 : Washington, D.C..)

TP248.25.N35B563 2008
660.6—dc22

2007060589

The paper used in this publication meets the minimum requirements of American National Standard for Information Sciences—Permanence of Paper for Printed Library Materials, ANSI Z39.48–1984.

Distributed by Oxford University Press

PRINTED IN THE UNITED STATES OF AMERICA

Foreword

The ACS Symposium Series was first published in 1974 to provide a mechanism for publishing symposia quickly in book form. The purpose of the series is to publish timely, comprehensive books developed from ACS sponsored symposia based on current scientific research. Occasionally, books are developed from symposia sponsored by other organizations when the topic is of keen interest to the chemistry audience.

Before agreeing to publish a book, the proposed table of contents is reviewed for appropriate and comprehensive coverage and for interest to the audience. Some papers may be excluded to better focus the book; others may be added to provide comprehensiveness. When appropriate, overview or introductory chapters are added. Drafts of chapters are peer-reviewed prior to final acceptance or rejection, and manuscripts are prepared in camera-ready format.

As a rule, only original research papers and original review papers are included in the volumes. Verbatim reproductions of previously published papers are not accepted.

ACS Books Department

Contents

I: Nanostructures for Biomolecular Catalysis: I-3: Nanobiocomposites

II: Applications of Nanoscale Biocatalysts

Indexes

Preface

One might argue that the greatest advances in the frontier research in recent years have been made in the nanoscale science and technology realm. New nanomaterials and systems were developed and these developments are opening new opportunities in electronics, sensors, energy harvest and storage, etc. These influences are now extended to the area of biomolecular catalysis systems by many research groups around the world. The marriage between nanomaterials and biomolecular catalysis provides unprecedented prospects, such as high surface area to volume ratios that allow high biomolecular loadings, wide range of surface chemistry to regulate biomolecules, unique three dimensional structures allowing efficient mass transports, and flexible reactor designs. Various nanomaterials—such as nanoparticles, nanofibers, nanotubes and nanoporous matrices—have demonstrated promising potentials in revolutionizing the preparation and use of biocatalysts.

In order to coordinate and facilitate the synergy between the nanoscale materials and biomolecular catalysis, a symposium of "Nano-scale Science and Technology in Biomolecular Catalysis" was organized at the 230th national conference of the American Chemical Society in Washington D.C., August 28 to September 1, 2005,. From the presentation and discussions, several main themes and needs were identified. These include the lack of understanding of the fundamentals that interface nano and bio, the needs of new tools and methodologies to design and realize nano-scale structures and materials that are specifically suited to biomolecular applications, and the development of unique applications of nano–bio hybrid systems. To summarize and document these issues and perspectives for future development, this book was prepared by selecting presentations that were well received at the meeting along with several contributions that were not presented at the symposium but reflect certain important aspects of the current art of nanobiocatalysis. These additional contributions were invited from presentations made at a nanobiocatalysis-related symposia in national conferences of the American Chemical Society and the American Institute for Chemical Engineers in the following year.

The chapters presented in this book are grouped into two parts, although the editors found it difficult indeed to draw a clear dividing line between them. The first part of the book presents several chapters describing the use of various nanostructures, including nanoporous materials, nanotubes and nanofibers, and composite materials for biocatalysis. Although these chapters may not present all the reported and potential structures available for biocatalysis, they do demonstrate the current frontier exploration with these probably most popular structures for nanobiocatalysis. The second part of the book illustrates potential applications of nanoscale biocatalysts in heterogeneous catalysis, energy conversion, reaction engineering, and nanotransport.

We hope that this book reflects the nature of the current research activities in nanobiocatalysis and provides a reference or guide for future advances in the field. We are grateful to all the authors who have contributed to this book, the ACS Division of Colloid and Surface Chemistry and ACS Petroleum Research Foundation program who sponsored the symposium in the 230th ACS National Meeting, and the ACS publishing department who encouraged us to organize this book. We also acknowledge the efforts of Kathy Peters at Penn State and Xueyan Zhao at the University of Minnesota, who helped process the manuscripts and reviews.

Jungbae Kim, Ph.D.
Senior Research Scientist
Pacific Northwest National Laboratory
Richland, WA 99352
Current address:
Associate Professor
Department of Chemical & Biological Engineering
Korea University
Seoul 136-701, Korea
email: jbkim3@korea.ac.kr

Seong H. Kim, Ph.D.
Associate Professor
Department of Chemical Engineering
The Pennsylvania State University
University Park, PA 16802
email: shkim@engr.psu.edu

Ping Wang, PhD
Associate Professor
Department of Bioproducts and Biosystems Engineering
Biotechnology Institute
University of Minnesota
St Paul, MN 55018
email: ping@umn.edu

Biomolecular Catalysis

I: Nanostructures for Biomolecular Catalysis

I–1: Nanoporous Materials

Chapter 1

General Synthesis of Ordered Nonsiliceous Mesoporous Materials

Ying Wan[1,2], Haifeng Yang[2], and Dongyuan Zhao[2,3,*]

[1]Department of Chemistry, Shanghai Normal University, Shanghai 200234, People's Republic of China
[2]Department of Chemistry, Shanghai Key Laboratory of Molecular Catalysis and Innovative Materials, Advanced Materials Laboratory, Fudan University, Shanghai 200433, People's Republic of China
[3]Key Laboratory of Molecular Engineering of Polymers and Engineering Center of Coating of Chinese Ministry of Education, Fudan University, Shanghai 200433, People's Republic of China

The design of ordered nonsiliceous mesoporous materials is illustrated based on the surfactant assembly and confined-space growth. These materials include mesoporous metal oxides, polymers, and carbons with open framework structures, and single-crystal metal, metal oxide and carbon nanoarrays with replicated mesostructures. A generalized "acid–base pair" concept, which self-adjusts the acidity and homogeneity of the inorganic precursor, is proposed to prepare highly ordered mesoporous metal oxides, phosphates and borates, as well as mixed metal oxides and phosphates, with diverse structures. Mesoscopically ordered polymer frameworks with uniformly large pore sizes are derived from the self-assembly from organic templates and organic precursors. Heating these materials transforms them to homologous carbon frameworks. Microwave digested method is an efficient way to enhance the adsorption property of hard silica templates in fabricating ordered nonsiliceous single-crystal nanoarrays. The coordination of organic surfactants with metal ions is also a method to increase the interaction between hard templates and precursors. A one-step impregnation process is used to fabricate ordered silica monoliths with various metal oxide nanocrystals. Improving the interactions between the precursors themselves plays an essential role in replicating ordered mesoporous CdS, SiC and graphitized carbons. Preliminary applications in bone-forming materials, biosensors and electrodes are presented as well.

1. Introduction

Periodic mesoporous materials represent a new class of solids, which possess uniform pore sizes (1.5 – 40 nm), large surface areas (up to 2000 m^2/g) and tunable structures (*1-5*). This kind of materials, without a doubt, contribute one of the most promising candidates for the nanoreactors of large molecules. They are becoming a commercial necessity and will continue to grow (*6*) in significance as the demand for oil outpacing productions and bio-applications (*7, 8*). Mesoporous silicates, since the first report in 1992, have attracted more and more attention owing to numerous applications in catalysis, adsorption, separation, bio-molecule purification, drug delivery and even nanomaterials (*8-13*).

Diverse silicate mesoscopically periodic pore structures have been synthesized by cationic quaternary ammonium surfactants, for example, two-dimensional (2D) hexagonal (*p6m*), 3D cubic ($Pm\bar{3}n$), bicontinuous cubic ($Ia\bar{3}d$), and 3D hexagonal ($P6_3/mmc$) structures (*1,2,14-18*). Designed multi-head quaternary ammonium surfactants like bolaform surfactants with different chain lengths and rigid benzene ring groups as structure-directing agents (SDAs), 3D cubic ($Pm\bar{3}n$) and face-centered cubic ($Fd\bar{3}m$) mesoporous silicates were synthesized (*19*). Lately, the low symmetrical mesostructures with tetragonal and orthorhombic pores were also obtained (*20*). It has been clearly demonstrated that the derived mesostructures are remarkably influenced by the rational control of inorganic–organic interactions and cooperative assembly of the silica species and the surfactants. Compared with that of the cationic surfactants, the repulsive interaction between anionic surfactant and silicate species fails to organize ordered mesostructure. Concerning the charge matching effect, Che *et al.* (*21-23*) demonstrated a synthetic route to create a family of mesoporous silica structures under a basic condition by employing anionic surfactants as the SDAs and aminosilanes or quaternized aminosilanes as the co-SDAs. Most strikingly, even helical mesoporous silica materials can be templated by the anionic surfactants containing chiral units (*24*). In addition, nonionic alkyl poly(ethylene oxide) (PEO) oligomeric surfactants and poly(alkylene oxide) block copolymers can also direct the mesoscopic assembly. These nonionic surfacants become more and more popular and powerful in the synthesis of ordered mesoporous silica materials due to their diverse structural characteristics, rich phase behaviors, low cost, non-toxic degradation and many other merits. Except for 2D (*p*6*m*) large pore mesoporous silica SBA-15 (*3,25*), 3D mesostructures ($Im\bar{3}m$, $Ia\bar{3}d$, $Fm\bar{3}m$, *etc.*) have also been prepared under acidic media by using poly(ethylene oxide)–*block*–poly(propylene oxide)–*block*–poly(ethylene oxide) triblock copolymers (PEO–PPO–PEO) as the templates (*26-32*). Mesotunnels can be generated on the pore walls of SBA-15 (*33*). The window size and the pore size of FDU-12 can be enlarged to larger than 20 nm and to 27 nm (*34*), respectively. Morphologies can also be controlled as of thin films, spheres, rods, single-crystals, monoliths, fibers, *etc* (*27,35-39*).

Considering the intrinsic chemical and physical properties of nonsiliceous solids, in particular crystallites, together with the confined-space effects of their nanopores, nonsiliceous mesoporous materials have been developed. It includes important design features and manufacturing processes that are vital to a long-term success in the mesoporous solids technology. These materials are capable of opening up inestimable potential applications in optics, magnetics, electronics, mechanics, microdevices and quantum dots (*40-45*). From the viewpoint of synthesis, there are two challenging questions for material scientists. (1) Are there generalized assembly approaches for producing ordered nonsiliceous mesoporous materials, especially in the case of multicomponent and 3D channels? (2) Are there versatile methodologies to generate crystalline mesostructures?

Two classes of materials have often been integrated as components in this mesoporous family, *i.e.* mesoporous molecular sieves with open framework structures and mesoporous replicas constructed by nanowire arrays, *etc*. The former is normally prepared *via* surfactant self-assembly, and the latter is fabricated by confined-space assembly or nanocasting based on a hard-template approach, as shown in Figure 1 (*46,47*). Surfactant self-assembly is a construction of atoms, ions, molecules and/or nanoclusters into ordered arrays and patterns by weak non-covalent bonds such as hydrogen bonds, van der Waals forces and electrovalent bonds. Instead of a simple superposition of the weak interaction, an integrated and complicated synergistic reaction facilitates the process. Cooperative assembly between organic surfactant templates and precursors is involved. In the process of confined-space growth, hard-template materials possess mesoscopically topological structures. and the assembly of the guest is restricted to the confined space in hard templates. A true replication of the guest from hard templates depends upon either the interactions between them or the outside forces (*e.g.* capillary force, electrical force, *etc.*). A large number of studies have been explored at least on the following three aspects: (1) rationally synthesizing mesoporous materials with desired components and structures, (2) investigating morphology, pore structures and surface chemistry, and (3) implementing the physical and chemical properties into applications.

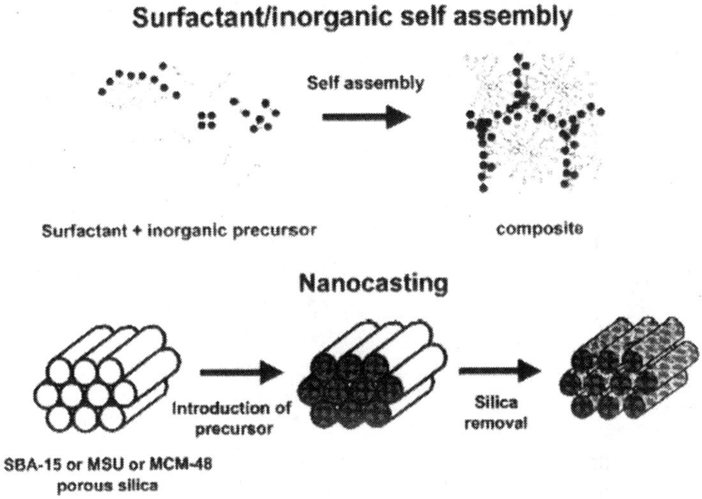

Figure 1. Schematic representation of general pathways leading to mesostructured and mesoporous nonsiliceous materials. Reprinted from reference 46. Copyright 2001, ACS.

This chapter presents the recent developments in the synthesis of ordered nonsiliceous mesoporous materials mainly from our group. We summarize in the first section the current status of mesoporous nonsiliceous molecular sieves with open frameworks *via* surfactant self-assembly approach. This part includes a description of: (1) highly ordered metal oxides, phosphates and borates, as well as mixed metal oxides and phosphates obtained by using "acid–base pair" route from inorganic–organic self-assembly, and (2) highly ordered polymers and carbons with diverse mesostructures from organic–organic self-assembly. In the second section, we will focus on the key factors for synthesizing ordered mesoporous replicas in the hard-template approach. They include improving the interactions between template and precursor, and between precursors themselves.

These materials provide further motivations for the development of applications in biochemistry, electronics, *etc.*

2. Surfactant self-assembly approach

Supramolecular self-assembly provides routes to a range of materials with diverse components and structures. Inorganic–organic assembly and organic–organic assembly are the two major branches according to the general chemistry, to get inorganic and organic frameworks, respectively.

(1) Inorganic–organic self-assembly

The classical products from inorganic–organic self-assembly, which is based on the cooperative organization of an organic surfactant and an inorganic precursor, are mesoporous silicates (*1-4*). As generally accepted, the organic–inorganic interaction between organic templates and inorganic precursor species is the main driving force to form ordered silicate mesostructures (*48,49*). Elaborate investigations on mesoporous materials have been focused on understanding and utilizing the organic–inorganic interactions. Stucky and coworkers proposed four general synthetic routes which are S^+I^-, S^-I^+, $S^+X^-I^+$ and $S^-X^+I^-$ (S^+ = surfactant cation, S^- = surfactant anion, I^+ = precursor cation, I^- = precursor anion, X^+ = counterion cation and X^- = counterion anion), respectively, as shown in Figure 2 (*48*). A series of ordered nonsiliceous mesostructured solids (*e.g*, tungstate and molybdate) were prepared. However, unlike the silicate or aluminate precursors, most inorganic precursors can not hydrolyze or hydrolyze in a quite high speed. 3D interlinked frameworks of these components are difficult to be generated with stable long-range periodic structures (*50,51*).

To synthesize nonsiliceous ordered mesoporous materials, the important success appeared when the versatile evaporation induced self-assembly (EISA) strategy under the non-aqueous system was introduced to synthesize mesoporous metal oxides (*50,53*). It was initially used in the preparation of mesoporous silica thin films by Brinker and coworkers (*52*). The EISA method is beginning with a solution in which the initial concentration of structural directing agent is lower than the critical micelle concentration (CMC) value. Upon solvent evaporation, the inorganic–organic composite liquid crystal mesophases are induced. The inorganic species are further polymerized and condensed, gradually forming the mesostructures (Figure 3) (*54-56*). Semicrystalline frameworks were obtained where nanocrystallites nucleated within the amorphous inorganic matrices (*50*). However, the compositions of ordered mesoporous oxides directly obtained from the EISA approach are still limited. Most oxides are disordered or hybrid-mesostructure that are more distinct for multicomponent mesoporous solids (*50*).

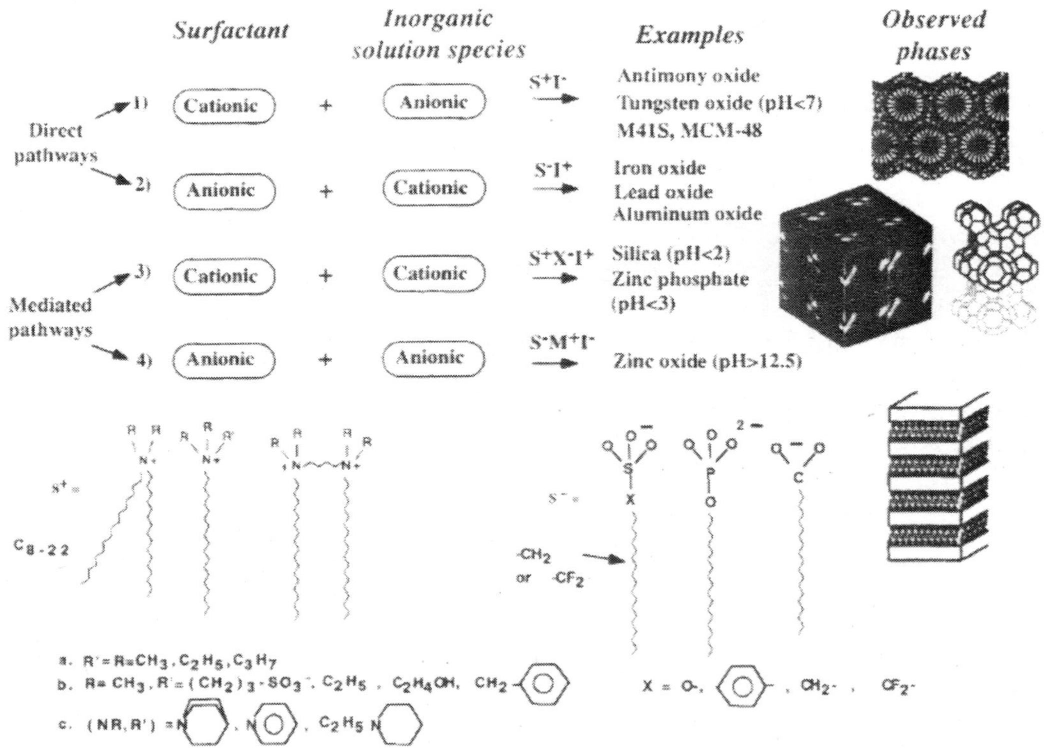

Figure 2. A general scheme for the self-assembly of different surfactants and inorganic species. Reproduced with permission from reference 48. Copyright 1994, Nature.

The reasons may arise from the complexity of reactive properties and chemical stability intrinsic in the desired constituents, as well as the difficult control of hydrolysis, polymerization, cross-linkage and inorganic–organic assembly. It is not general although several methods have been applied into controlling the hydrolysis of inorganic precursors, for example, relying on metal chlorides and coordination (*50,54*). Templates removal is also very difficult since the unstable frameworks collapse during calcination owing to the crystallization of metal oxides at relative low temperatures (*57,58*).

A better understanding of the interactions between precursors themselves is, therefore, beneficial for careful regulation of synthesis. Recently, on the basis of

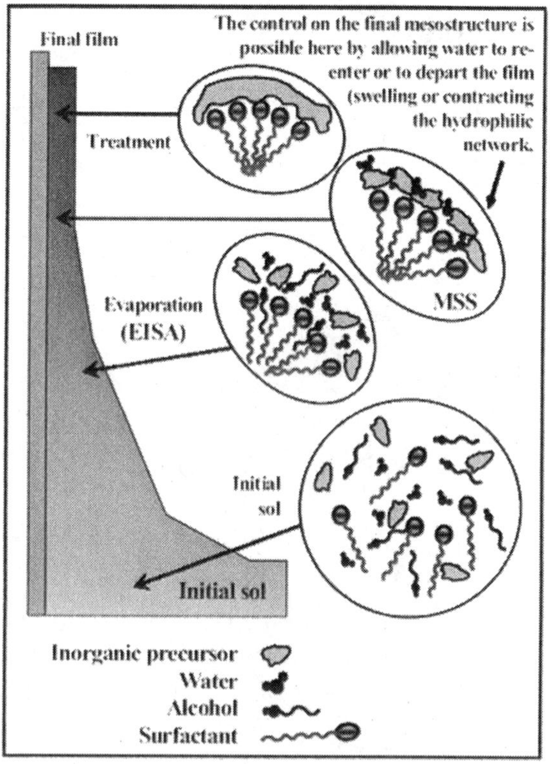

Figure 3. Mesostructured thin-film formation by dip-coating. Step1: the isotropic initial sol where the condensation is optimally slowed down. Step 2: the evaporation proceeds and micelles start to form above the CMC. Step 3: the evaporation is complete; the film equilibrates with its environment and the final mesostructure is selected by adjusting the relative humidity before further inorganic condensation. Step 4: the inorganic network is condensed; the hybrid mesostructure is stabilized. Reproduced from reference 55. Copyright 2004, Wiley.

compatibility of acid and base precursors, an "acid–base pair" concept was proposed to guide the selection of precursors in fabricating mesoporous metal oxides with various components under the non-aqueous systems (*59,60*). The simple neutralization concept in the "acid–base" chemistry principle and the appropriate acidity self-adjusted by the precursor pairs are introduced. This concept, together with the increased understanding on EISA strategy, sol–gel chemistry and organic–inorganic interaction, which are interdependent of each other, will pave the way for preparing ordered mesoporous nonsiliceous materials.

Firstly, the inorganic precursors are divided into "acid" and "base" according to their alcoholysis (here, alcohol is used as the solvent) behaviors. Inorganic metallic or nonmetallic chlorides are considered as strong "acid" since a large amount of acid is generated during the alcoholysis process. Hydrate metallic salts and inorganic acid (Brönsted acid) are attributed to middle acidic precursors. Metallic alkoxides and nonmetallic alkoxides (*e.g.* phosphatides) are assigned to base because acid substances are seldom generated.

As depicted in Figure 4, an "acid" mineral precursor is designed to couple with a "base" counterpart, forming the "acid–base" pair according to their relative acidity and alkalinity on solvation. The pair not only generates a proper acidic medium by tuning the ratio of "acid" to "base" precursors for both the inorganic–organic assembly and the gelation of inorganic precursors, but also is crucial for the homogenous mineral composition within the whole framework. For assembling ordered mesostructures, normally the "acid–base" pairs formed from strong "acid" and strong "base", or strong "acid (base)" and medium "base (acid)" in non-aqueous media are required. These pairs can be applied in the formation of homogenous multi-component inorganic precursors. Generally, the larger the acidity or alkalinity difference between the metallic and/or nonmetallic sources is, the more effectively the pairs will form and function.

Five fundamental (A to E) acid–base pair connections between metallic sources and/or nonmetallic sources are described in Figure 4. For convenience, nonmetallic alkoxides such as $Si(OR)_4$, $Ge(OR)_4$ (where R is a short-chain alkyl such as CH_3, C_2H_5, C_3H_7, C_4H_9) are listed in the metallic alkoxides column. Routes A, B and E are reminiscent of the well-known non-hydrolytic sol–gel process which produces molecular homogenous oxides (*61*). Routes C and D are similar to the synthesis of mesoporous silicates (*25*) or mesostructured metal chalcogenides (*62*) and microporous aluminophosphates (*63,64*). While routes F and G are experimentally proven to be less effective with poor function. Non-aqueous synthetic media are used to maximize the utility of this method and to promote inorganic–inorganic polymerization for assembling ordered mesostructured materials. Polar organic solvents, such as C_2H_5OH or CH_3OH, are recommended for their oxygen donating property to improve the proton transferring within the synthetic system. The various mesostructured phases can be synthesized by tuning ratios of inorganic species to surfactants, or by using

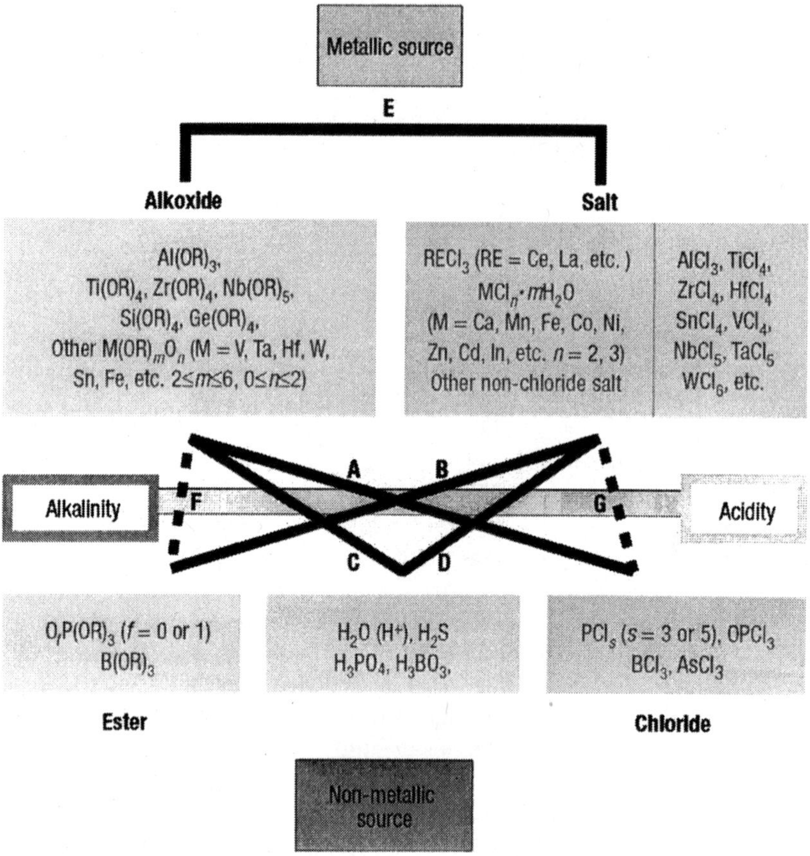

Figure 4. General scheme of the "acid–base pair" concept and guided synthetic routes for mesoporous minerals. Reproduced from reference 59. Copyright 2003, Nature.

different surfactants, which are in agreement with previous synthesis pathways (*48,50*).

Figure 5 shows the examples of synthesizing various mesoporous metal phosphates through fundamental routes A to E and their derivatives. Route C is not effective in our synthetic system. It is probably due to the less-controlled inorganic precipitation. However, route C will work with the addition of some strong "acid", either metal chloride or phosphorus chloride (namely routes H and K). Route E is suited for the preparation of metal oxides or mixed metal oxides. Routes I and J derive from route A + B and route B + D, respectively. To

	Metallic source (I_1)	Phosphorous source (I_2)	Explanation	Examples
A	$M(OR)_n$	PCl_s	$I_1 I_2 >> I_1 I_1$ (suppressed), $I_2 I_2$.	TiPO, ZrPO, AlPO, etc.
B	$MCl_n (mH_2O)$	$O_fP(OR)_3$	$I_1 I_2 >> I_1 I_1, I_2 I_2$,	TiPO, ZrPO, NbPO, WPO, CePO, etc.
C	$M(OR)_n$	H_3PO_4	$I_1 I_2$ or $I_1 I_1$ is too strong	None
D	$MCl_n (mH_2O)$	H_3PO_4	$I_1 I_2 >> I_1 I_1, I_2 I_2$,	AlPO, TiPO, NbPO, CePO, VPO, etc.
E	$MCl_n (mH_2O) + M'(OR)_n$		$I_1 I_2 >> I_1 I_1, I_2 I_2$	VO_x, TiO_2, $Sn_xTi_{1-x}O_2$, ZrW_2O_8, etc.
F	$MCl_n (mH_2O)$	PCl_s	$I_1 I_2$ is too weak	None
G	$M(OR)_n$	$O_fP(OR)_3$	$I_1 I_1 >> I_1 I_2$	None
H	$M(OR)_n$ and MCl_m	H_3PO_4	$I_1 I_2 >> I_1 I_1$ (suppressed), $I_2 I_2$,	TiPO, AlPO, etc.
I	$M(OR)_n$ and MCl_m	PCl_s and $O_fP(OR)_3$	$I_1 I_2 >> I_1 I_1, I_2 I_2$,	TiPO, ZrPO, etc.
J	$MCl_n (mH_2O)$	H_3PO_4 and $O_fP(OR)_3$	$I_1 I_2 >> I_1 I_1, I_2 I_2$,	CePO, TiPO, InPO, CaPO, ZnPO, etc.
K	$M(OR)_n$	PCl_s and H_3PO_4	$I_1 I_2 >> I_1 I_1, I_2 I_2$.	TiPO, etc.

$MCl_n (mH_2O)$ means MCl_n or $MCl_n \cdot mH_2O$.

The subscript *m*, *n*, *x*, *f* and *s* are designated as the number of the atoms.

Figure 5. Syntheses of various mesoporous metal phosphates through fundamental routes A to E and their derivatives. Adapted from Tian et al.(59) Copyright 2003, Nature.

assemble a multi-component (I_1I_2) composite, it is necessary to match acid–base interactions of the various species presented during nucleation of the mesostructured phases in the order of the interaction $I_1I_2 >> I_1I_1$, I_2I_2, and $O(I_1I_2) >> OI_1$, OI_2 (Figure 5). For example, in the synthesis of TiPO, the starting precursors can be $TiCl_4$, $Ti(OC_3H_7)_4$ and $P(OC_2H_5)_3$, PCl_3. The acidity can be well managed by changing the $TiCl_4/P(OC_2H_5)_3$ or $PCl_3/Ti(OC_3H_7)_4$ ratio. Beginning with a PCl_3–$Ti(OC_3H_7)_4$ pair as the precursors and triblock copolymer P123 (EO_{20}–PO_{70}–EO_{20}) as a template, the interactions between Ti and P oxide species are stronger than those between Ti–Ti and P–P species, resulting in a 2D hexagonal mesostructure (Figure 6). N_2 sorption measurements show that the ordered mesoporous TiPO displays a narrow pore size distribution with 5.4 nm and a BET surface area of around 220 m^2/g. By tuning the P123/(Ti + P) molar ratio, bicontinuous cubic $Ia\bar{3}d$ and unstable lamellar mesostructures can be modulated. Body-centered cubic mesoprous TiPO ($Im\bar{3}m$) can also be obtained by switching the template to F108 ($EO_{132}PO_{50}EO_{132}$). Narrow pore size distributions are observed at the mean values of 6.4 – 9.8 nm.

Interestingly, ordered mesoporous crystalline metal oxides can be obtained by using $Ti(OC_3H_7)_3$ and $PO(C_2H_5)_3$ as the precursor and triblock copolymer P123 as a SDA, followed by the controlled *in situ* crystallization of nanocomposites, in which nanocrystals can precipitate within an amorphous matrix (Figure 7) (*42*). The target structure of a designed mesoporous nanocomposite is shown in Figure 7c. The designed nanocomposite is made up of a large number of functional nanocrystals and small quantity of the multi-component glass phase. It is the glass phase that helps in forming and maintaining the network, controls *in situ* crystallization of materials on a nanoscale and "glues" the nanocrystals together. The mesostructure can then be retained even after crystallization.

It should be noted that the "acid–base pair" can either be two or more kinds of metal or nonmetal precursors, *e.g.* $TiCl_4$, $ZrCl_4$ and $P(OR)_3$, or the inorganic precursors with the same metal, *e.g.* $Ti(OR)_4$ and $TiCl_4$. The latter which was used in the synthesis of ordered mesoporous titania is a good example to explain the efficiency of "acid–base pair" concept. Titanium oxides are one of the largest research areas in semiconductor photocatalysts, which offer the advantages of cleaner and quieter power generation while utilizing solar energy (*65,66*). Some interesting features related to titania are the surface, quantum and macroscopic quantum tunnel effects with the nanosize (1 – 100 nm), imparting them extraordinary photoemission and photocatalysis properties (*67*). Ying and coworker firstly used tetradecylphosphate as a template to prepare ordered mesostructured titanium oxides relying on the strong affinity between phosphorus and titania framework. Amorphous phosphorus modified titania were prepared (*68*). The article by Yang *et al.*(*50*) covered the synthesis of mesoporous titania solids with semicrystalline wall using titanium tetrachlorides

($TiCl_4$) as a precursor and triblock copolymer Pluronic P123 as a SDA *via* an EISA strategy. The thick inorganic pore wall improves the thermal stability, and thus inhibits the structure collapse during the template removal. However, the products are less ordered compared to their silica counterparts. Detailed and complete studies on EISA-based synthesis of titania thin films have been investigated by Sanchez and coworkers (*54,55,69*). Reproducible preparation and rational design of mesostructured thin films, as well as better understanding of the EISA mechanism become possible, owing to their critical and informative discussions. We believe that the acidity of precursor solution is an important factor in the entire reaction process of preparing mesoporous titania under the non-aqueous system (*60*).

A mixture of titanium alkoxide (ethoxide, isopropoxide or *n*-butoxide) and $TiCl_4$ were used as the precursors. As expected, this method was proved to be very effective, fast and easy to synthesize highly ordered, pore size tunable mesoporous titania with both hexagonal and cubic structures. In this case, titanium alkoxide was used as the main titanium source and titanium chloride served as the pH "adjustor" and hydrolysis–condensation "controller". Compared with the synthesis from a single titanium source, namely $TiCl_4$, the acidity of precursor solution is significantly and controllably reduced by the addition of titanium alkoxide which decreases the amount of $TiCl_4$ and neutralizes the acid. The added metal alkoxide is also an extra oxygen donor. Therefore, the cross-linkage and gelation of inorganic precursor molecules may be easier and better. It shortens the synthetic period, improves the thermal stability and hence, maintains the mesostructure and facilitates the crystallization. Later on, successes were acquired in the cases of mesoporous alumina and zirconia from the similar route.

On the basis of "acid–base" route, aluminum organophosphonates (AOPs) were synthesized by using $(HO)_2OPCH_2PO(OH)_2$ and $AlCl_3$ with the initial molar Al:2P ratio of 1: 0.75 in the presence of oligomeric surfactants and triblock copolymers. 2D hexagonal mesoporous AOPs can be obtained with variable pore diameters of 3.3 – 9.2 nm (Figure 8) (*70*). Using the phosphorus trichloride–zirconium propoxide pair as inorganic precursors, and triblock copolymers or alkyltrimethylammonium bromide as organic templates, mesoporous zirconium phosphates were prepared with surface areas between 78 and 177 m^2/g and controlled pore sizes between 2 and 4 nm (*71*).

Thus far, a wide variety of mesoscopically ordered, large-pore, homogeneous and stable metal oxides, mixed metal oxides, metal phosphates, metal borates, *etc* have been successfully synthesized *(59)*. Most of these mesoporous products possess semicrystalline pore walls and relatively high thermal stability. These materials show high surface areas, uniform pore sizes and tunable periodic structures, which may cause fascinating chemical and physical properties.

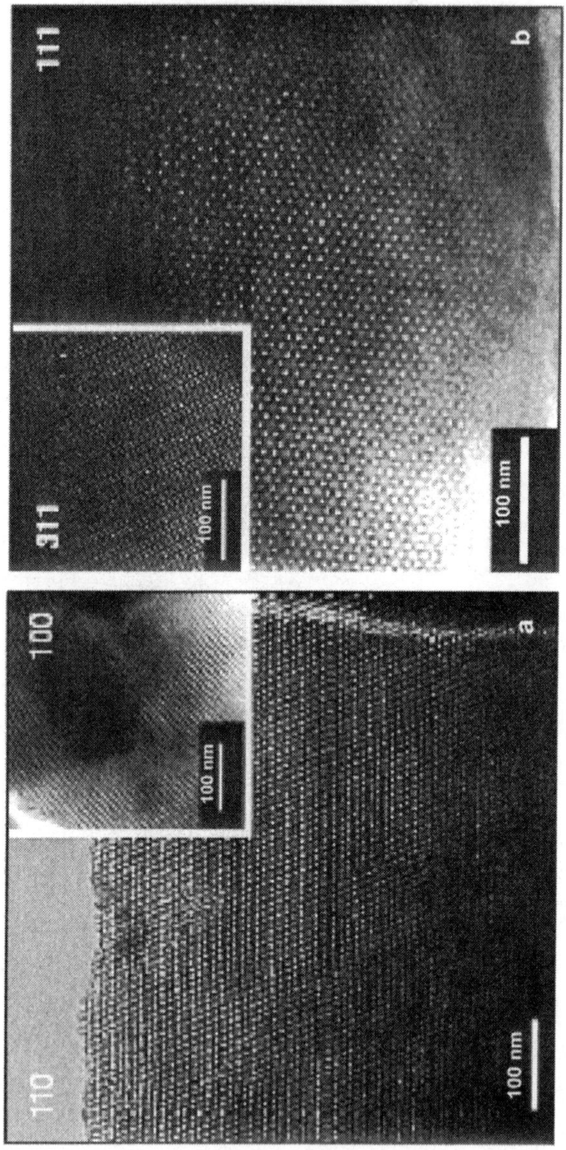
110
100 nm
100
100 nm
a
311
100 nm
111
100 nm
b

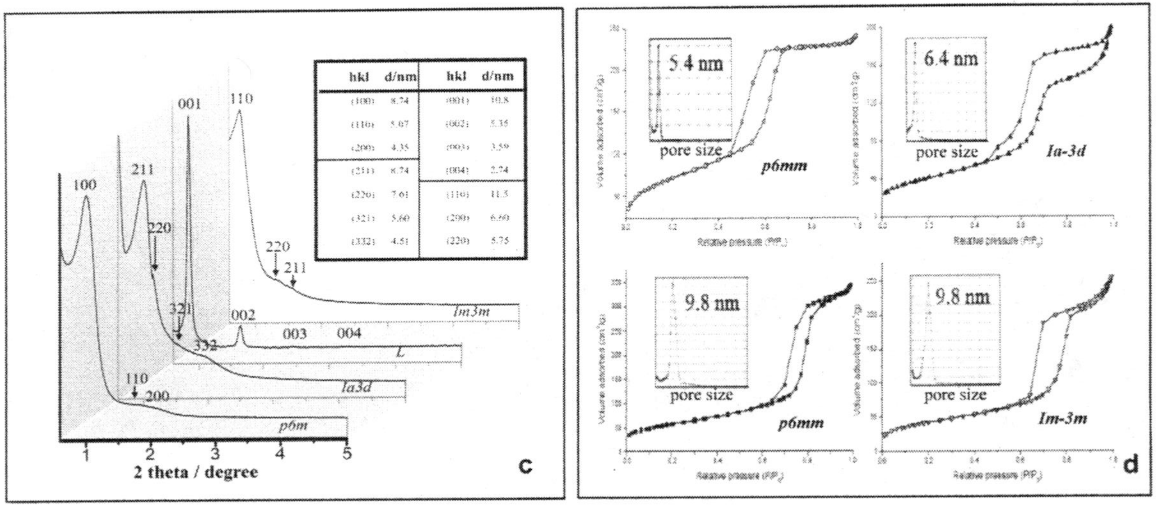

Figure 6. a) TEM images of calcinated hexagonal (p6m) TiPO recorded along [110] and [100] directions. b) cubic bicontinuous (Ia ω d) TiPO along [111] and [311] directions. c) XRD patterns and d) N_2 adsorption isotherms of mesoporous TiPO with diverse structures. Inset (d) are the pore size distribution curves. Reproduced from reference 59. Copyright 2003, Nature.

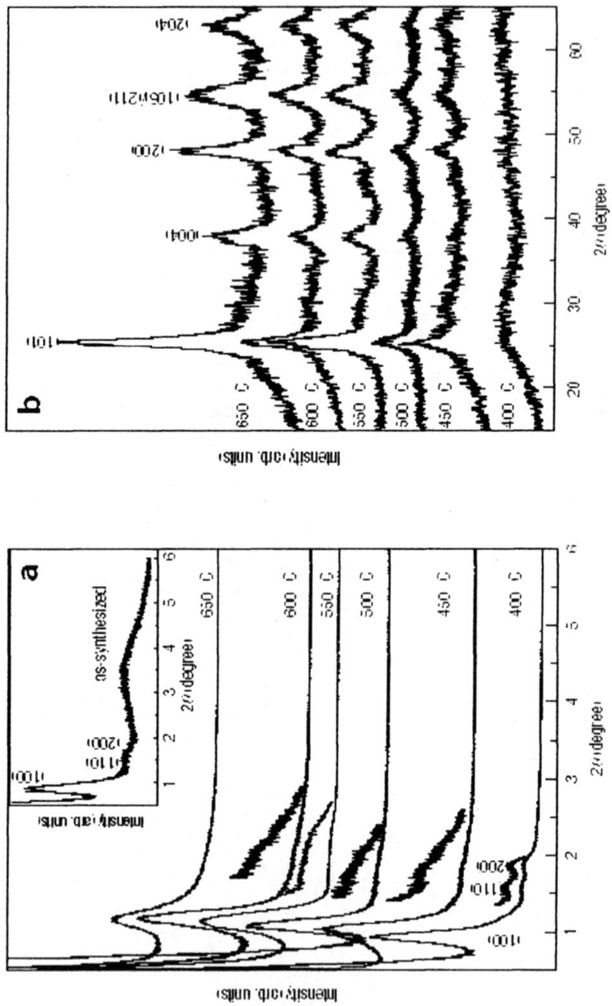
a
as-synthesized
Intensity (arb. units)
2θ (degree)
(100)
(110)
(200)
650 °C
600 °C
550 °C
500 °C
450 °C
400 °C
b
(101)
(004)
(200)
(105)(211)
(204)

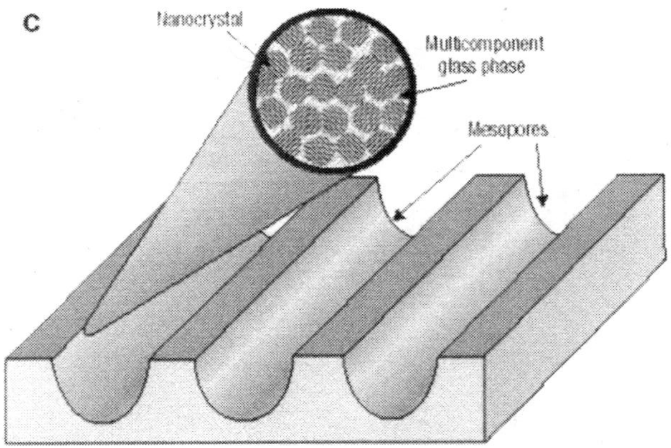

Figure 7. a) Powder small-angle and b) wide-angle X-ray diffraction patterns of $75TiO_2$–$25P_2O_5$ materials (in molar ratio of 75:25) after heat treatment at 400, 450, 500, 550 and 650 °C. Inset a is small-angle X-ray diffraction pattern of as-synthesized material. c) Target structure of designed mesoporous nanocomposites. Reproduced from reference 42. Copyright 2004, Nature.

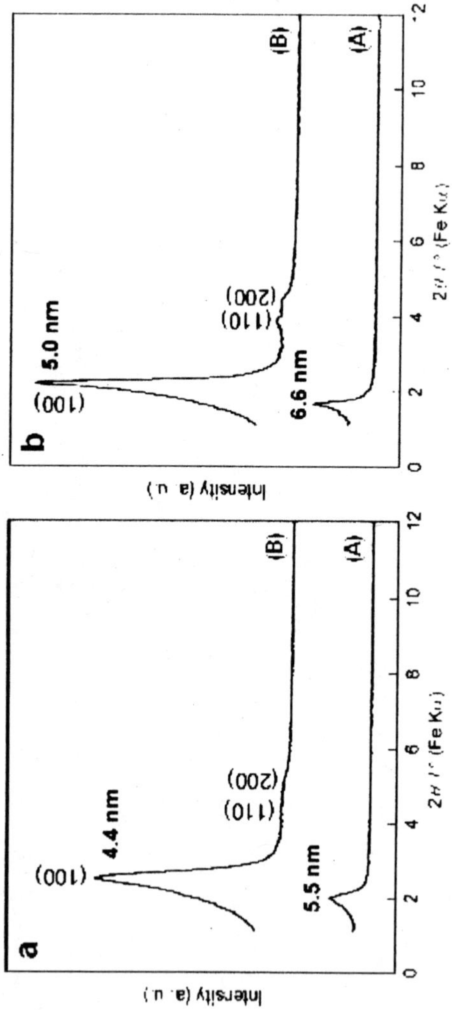
a
b
Intensity (a. u.)
(100)
(001)
4.4 nm
5.5 nm
5.0 nm
6.6 nm
(110)
(200)
(B)
(A)
0
2
4
6
8
10
12

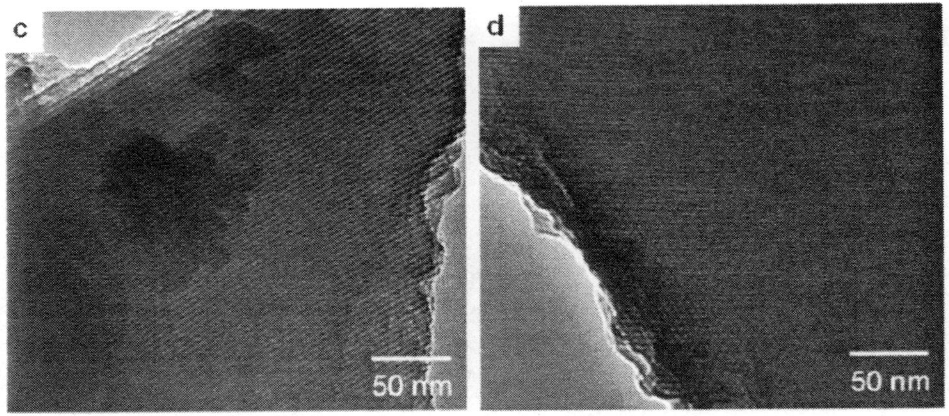

Figure 8. XRD patterns of (A) as-synthesized and (B) calcinated (at 400 ℃) aluminum methylenediphosphonates prepared in the presence of $C_{16}EO_{10}$ (a) and $C_{16}EO_{20}$ (b). TEM images of (B) prepared in the presence of $C_{16}EO_{10}$ (c) and $C_{16}EO_{20}$ (d). Reproduced from reference 70. Copyright 2003, ACS.

2) Organic–organic self-assembly

As shown above and elsewhere, the inorganic–organic self-assembly has been extended to synthesize ordered mesoporous metal oxides, metal sulfides and even metals (*63,72*). However, the preparation of mesoporous carbon materials with ordered open pore structures is extremely difficult in solution and remains challenging because of the high formation energy of C–C bonds. Significant research activity has been undertaken to develop mesoporous carbons due to their enormous potential for many high-tech applications such as hydrogen storage, catalysis, adsorption, separations and electrochemical double capacitors, even semiconductor and space technologies (*73,74*). Most of the studies were focused on mesoporous carbon replicas prepared by a mesoporous silica hard-templating method. It is the carbonization of sucrose-filled ordered mesoporous silicates using sulfuric acid as a catalyst, and then removal of the silica frameworks by HF or NaOH (*75*). The replicas obtained from the time-consuming and costly procedure are constructed as ordered nanowire, nanorod or nanotube arrays, replicating the ordered structures of mesoporous silicates. It is an obviously elaborate, high-cost, and thus industrially unfeasible method. Therefore, a soft surfactant self-assembly approach towards ordered mesoporous carbons with open frameworks is desirable.

It is noticed that disordered porous carbon networks can be transformed by pyrolysis of organic aerogels (*76*). The open frameworks and tunable porosity give carbon aerogels accessibility to metal ions and good contact with reactants which are extremely important in the fields of catalysis, electronic devices, gas separation, *etc* (*76*). An attractive question therefore arises. Can the ordered mesoporous organic framework directly transform to its carbon counterpart? This strategy is important because it can simultaneously get the organic and carbon frameworks.

In general, ordered organic materials can be fabricated from organic–organic phases through several routes including phase separation (*77*), cross-linked lyotropic liquid crystal (LLC) assemblies (*78*), and hard templating approach by colloidal particles or porous inorganic materials (*79*). In the phase separation, the pore structures can be formed after etching or degrading one block (A) from the assembled block copolymer (A–B) (*77*). Perhaps due to the thermal unstable framework deriving from weak van der Waals forces between polymer chains instead of covalent bonds, the ordered mesoporous polymer channels with large surface areas and porosity have not been obtained. Gin and coworkers(*78*) prepared ordered organic mesostructures by utilizing cross-linked LLC obtained from amphiphilic monomer molecules containing polymerizable head groups. Unfortunately, the polymerization of liquid crystals occurs at only their nearest neighbor head groups of amphiphilic surfactant molecules around the water core. The pore channels are occupied with water molecules, resulting in no porosity.

The successful syntheses of ordered mesoporous organic materials from organic–organic self-assembly are however, rare, by reason of the intrinsic characters of organic molecules. An excellent example for polymer nanostructure from organic–organic self-assembly was given by Hillmyer and coworkers (*80*). They used low-molecular-weight thermosetting epoxy as an organic precursor and diblock copolymer poly (ethylene oxide)–poly (ethylethylene) (PEO–PET) as an amphiphilic template to synthesize highly ordered polymer nanostructures. The interaction between epoxy resin precursor and template is stronger than that inside the resin precursor itself. Hence, highly ordered polymer nanostructures could be obtained *via* the organic–organic self-assembly, and subsequent cross-linkage of epoxy matrices did not destroy the nanostructures. Unfortunately, the removal of templates and the evidence for porosity in the products were not reported. Other instances are mesoporous polyacryonitrile structures from the self-assembly of triblock copolymer poly (propylene oxide)–poly (ethylene oxide)–poly (propylene oxide) ($PO_{19}EO_{33}PO_{19}$), and nanoporous carbon prepared by assembling cetyltrimthylammonium bromide (CTAB) with resorcinol/ formaldehyde (*81,82*). Their structures, however, are disordered. The main reason may be the extremely weak interaction between organic polymer frameworks and amphiphilic surfactants. It results in the reduction of miscibility between organic frameworks and surfactants after polymerization, and consequent macroscopic phase separation.

More recently, Dai and coworkers (*83*) successfully prepared ordered mesoporous carbon films with open pore structures by using resorcinol and formaldehyde as carbon sources and diblock copolymer poly(styrene)–*block*–poly(4-vinylpyridine) (PS–P4VP) as a template. Electron microscopy images of the carbon film are shown in Figure 9. Similar results were reported by Tanaka *et al.* (*84*), in which the same carbon precursor and Pluronic F127 ($EO_{106}PO_{70}EO_{106}$) were used for organic–organic self-assembly (Figure 10). Triethyl orthoacetate (EOA) was added as a carbon co-precursor. It may decrease the polymerization rate of resorcinol and formaldehyde under strong acid conditions and enhance the interaction between them and the surfactant templates. Consequently, the interactions between both organic precursors and templates and the organic precursors themselves should be considered in an carbons with diversified structures. A unique amphiphilic surfactant- templating organic assembly approach was therefore applied to create organic mesoporous materials and homologous carbon with open frameworks *via* a well-controlled polymerization of the proper precursor.

We independently demonstrated a reproducible synthesis of highly ordered mesoporous polymers and carbons from a solvent evaporation induced assembly of low-molecular-weight and water-soluble phenolic resins (resols) and low-cost, commercial triblock copolymers PEO–PPO–PEO (*85*). The schematic presentation of synthesis is shown in Figure 11. The preparation procedure

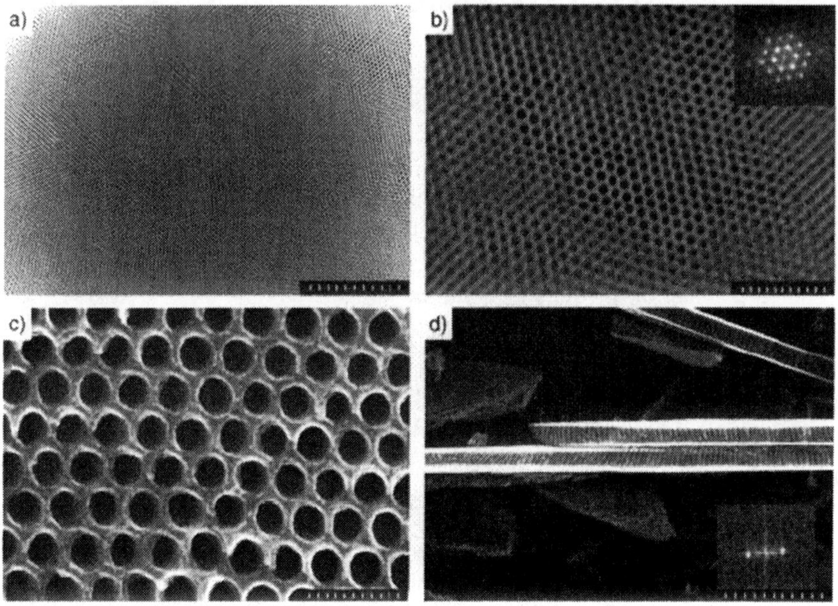

Figure 9. Electron microscopy images of the carbon film. a) Z-contrast image of the large-scale homogeneous carbon film in a 4H3 mm area. The scale bar is 1 mm. b) Z-contrast image showing details of the highly ordered carbon mesostructure. Inset b is the Fourier transform (FT) of the image. The scale bar is 300 nm. c) High-resolution SEM image of the surface of the carbon film. The scale bar is 100 nm. d) SEM image of the carbon film cross section. The scale bar is 100 nm. The inset is the FT of the cross section image. Reproduced from reference 83. Copyright 2004, Wiley.

mainly includes five steps, that are resol precursor preparation, organic–orgnanic hybrid mesophase formation from EISA, thermopolymerization and drying process, template removal, carbonization.

The choice of resols as the organic precursor is the characteristic of this work and is essential for the EISA of organic–organic templating process. The resols are polymerized by phenol and formaldehyde under alkaline conditions. Base-catalyzed processes can result in phenolic resins with 3D network structures with benzene rings as three or four cross-linking sites. This kind of cross-linked structure is similar to that of covalently-bonded silicate zeolite frameworks, in which one silicon atom is linked to four other silicon atoms through Si–O bonds. Simultaneously, basic condition facilitates the formation of resol with plenty of hydroxyl groups (–OH), which can strongly interact with the

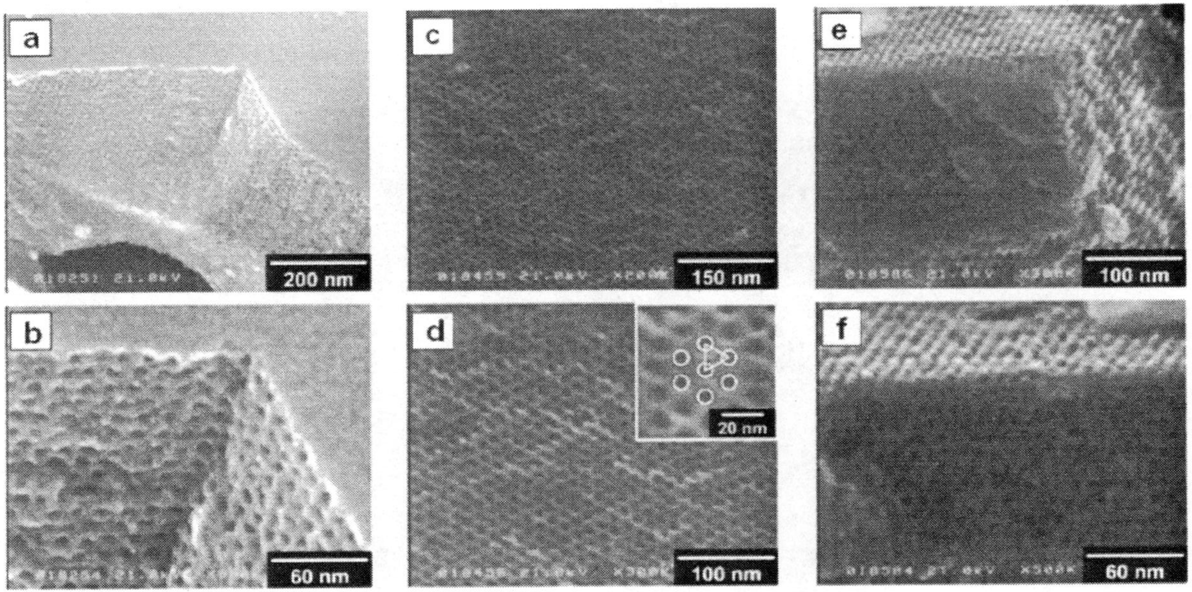

Figure 10. FE-SEM images of mesoporous carbon materials. The carbonization temperatures were as follows: (a, b) 400 ℃, (c, d) 600 ℃ and (e, f) 800 ℃. Reproduced from reference 84. Copyright 2005, RSC.

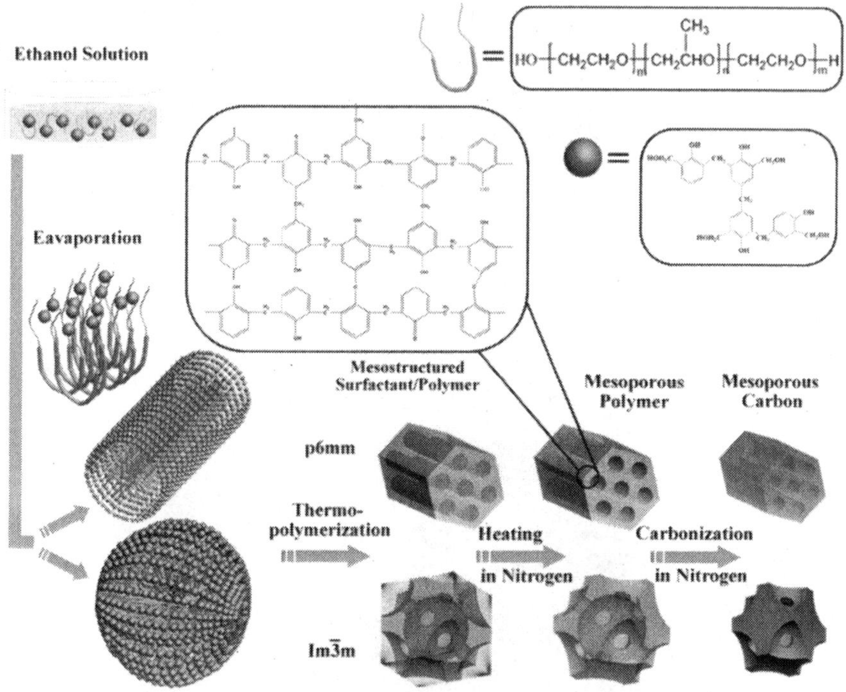

Figure 11. Schematic diagram of the synthetic procedure for mesoporous polymer and homologous carbon frameworks. Adapted from Meng et al.(85) Copyright 2005, Wiley.

PEO blocks of triblock copolymer templates *via* hydrogen bonds. The phenol has fewer reactive sites which are more inert than those in resorcinol. Therefore, the polymerization rate of resols is more easily controlled. Compared with the resorcinol/formaldehyde system, the resols decreases the interaction between the polymer resin precursors themselves. As a consequence, the assembly of phenolic resins and copolymer templates occurs more readily to form ordered mesostructures without macrophase separation.

Owing to the successful syntheses of mesoporous silica films (*52*), the EISA method is engaged to prepare ordered mesoporous polymer and carbon materials. The beginning homogeneous solution is prepared by dissolving the triblock copolymer and resol precursor in ethanol. The preferential evaporation of ethanol progressively enriches the concentration of the copolymer and drives the organization of resol–copolymer composites into ordered liquid-crystalline mesophase. Furthermore, the ordered mesophase is solidified by the cross-linkage of resols, which can be easily induced by thermopolymerization.

The thermopolymerization step involves cross-linkage and polymerization of the resol resins, further solidification of the ordered mesophases generated by organic–organic assembly, and finally formation of integral 3D resin frameworks. This step is rather important for the stability of the mesoporous products, because the covalent bonds constructing the framework are essentially formed during this step. A fast rate of resol polymerization would lead to the deformation of preformed mesostructures and even macroscopic phase separation. Based on the consideration of the mesophase behaviors of PEO–PPO–PEO copolymers, a relatively low temperature (100°C) is adopted for thermopolymerization to deliberately decrease the cross-linking rate of the resols. To ensure complete cross-linkage, a long time (> 24 h) thermopolymerization is utilized.

At least two distinct advantages can then be obtained. The big difference in the thermal stability or glassy transformation temperature and the chemical behavior between the resin frameworks and triblock copolymers contributes to thermally decompose the soft templates. It gives large mesoporosity of polymer resins by calcination under an inert atmosphere. In addition, the polymer resin frameworks are stable enough to resist the deformation caused by the removal of templates. Further increasing the heating temperature leads to a framework transformation to carbon with ordered homologous mesostructures.

A family of highly ordered mesoporous polymer resins and carbons have been synthesized, including mesostructures with lamellar, 2D hexagonal *p6m* and body-centered cubic $Im\bar{3}m$ symmetries, which are obtained by simply varying the EO/PO ratios in the templates. Partial results are shown in Figure 12. The reason may be that the resols can interact with the hydrophilic PEO blocks in copolymers. It causes a change of hydrophilic/hydrophobic ratio in the resol–surfactant mesophase, and hence a difference in the interfacial curvature. Higher ratios favor the formation of mesostructures with higher curvatures.

The mesoporous polymer resin materials exhibit highly ordered structures, high surface areas (~ 670 m^2/g), large pore volumes (~ 0.65 cm^3/g) and uniform, large pore sizes (~ 7.0 nm). Further heating to a temperature above 600°C transforms the mesoporous polymers to the homologous carbon frameworks. Ordered mesoporous carbon products show high surface areas (~ 1500 m^2/g), large pore volumes (~ 0.85 cm^3/g) and uniform, large pore sizes (~ 4.9 nm), as well as very thick pore walls (6 ~ 8 nm). The open carbon frameworks with covalently bonded construction and thick pore walls exhibit high thermal stability (> 1400°C). Compared with conventional mesoporous silicates, the new families of ordered mesoporous organic polymers and inorganic carbon solids offer great opportunities with applications emerging in adsorption, catalysis, drug delivery, electrodes, and bioengineering.

After we recall the manufacture of ordered mesoporous silicates, the assembly of organic surfactant and inorganic precursor in a dilute aqueous

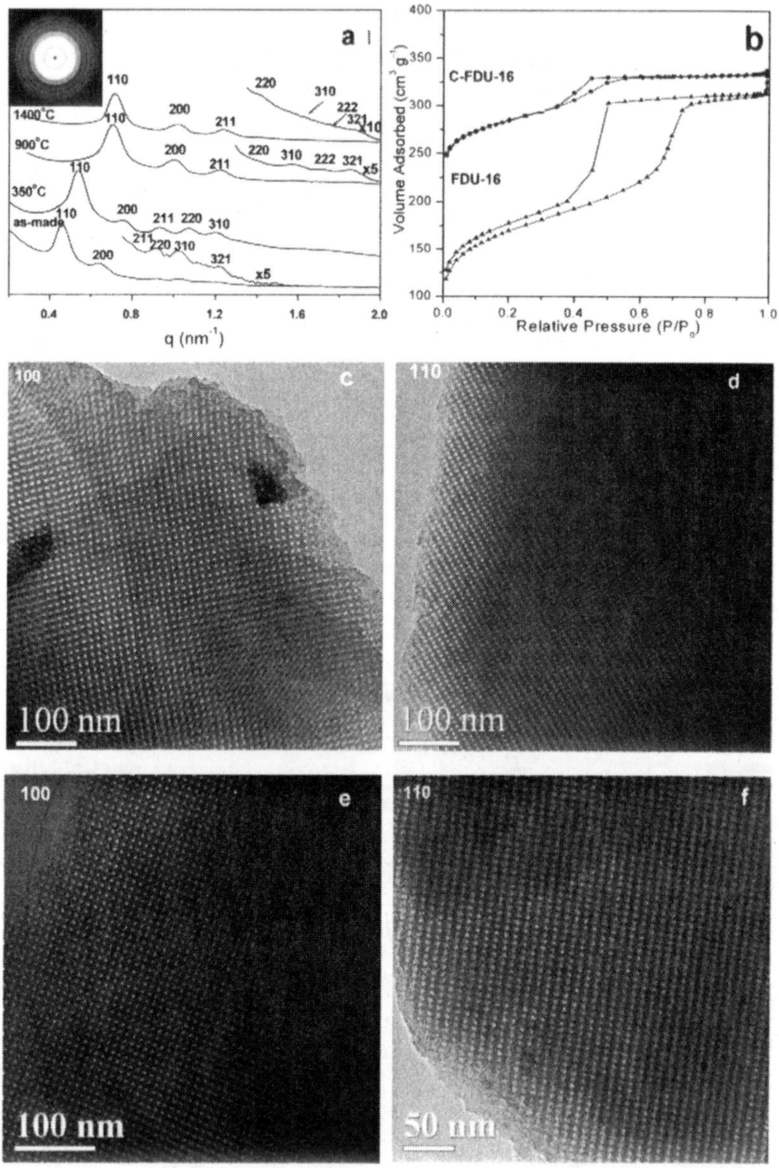

Figure 12. a) SAXS patterns of mesoporous polymer FDU-16 and mesoporous carbon C-FDU-16. The inset is the 2D SAXS pattern of FDU-16. b) N_2 adsorption isotherms of FDU-16 and C-FDU-16. c) and d) TEM images of FDU-16 recorded along [100] and [110] directions. e) and f) TEM images of C-FDU-16 viewed along the same directions. Reproduced from reference 85. Copyright 2005, Wiley.

solution is now considered to be a facile, low-cost and versatile approach which has been demonstrated to be feasible in the industrial production. Can the organic mesoporous materials and homologous carbons with open frameworks be prepared in aqueous media? The example of bicontinuous cubic mesoporous polymer (FDU-14) and carbon (C-FDU-14) with $Ia\bar{3}d$ symmetry can illustrate it (Figure 13) (*86*). The dilute aqueous route resembles to that used in preparing mesoporous silica (*e.g.* MCM-41 and SBA-15). The preformed resols first cooperatively assemble with the triblock copolymer P123, leading to a block–copolymer–resin mesophase in the dilute solution at a pH value of about 8.5. Further polymerization and carbonization direct the mesoporous polymer and homologous carbon molecular sieve to a desired mesostructure.

3. Confined-space assembly

Confined-space assembly is the growth of a precursor is restricted to the surface, cages, channels and substrates of a hard template (*46*). Perfectly duplicating topology of the hard template derives the replica. In particularly, the Au–silica mesostructured superlattice thin film is a model of the hard-template approach which is templated by gold nanoparticles (*87*). Frequently used inorganic hard templates are anode porous alumina (*88*), mesoporous silica (*75*), *etc.* The latter is advantageous because of the flexibly modulated structures, textures and morphology. A large variety of silicate periodic mesostructures have been synthesized by using various surfactants as the SDAs, adding organic swelling substances and inorganic salts, as well as adjusting synthetic parameters.

Using ordered mesoporous silica materials as hard templates, a nanocasting process can be carried out to obtain replicated mesostructures constructed by nanorod, nanowire or nanotube arrays (*46*). The most fascinating success is the replication of mesoporous carbon materials (*75*). However, other inorganic replicas are not as successful, that normally consist of disordered nanowires or nanoparticles without long-range periodicity owing to the partial loadings of precursors inside the mesopores and the fast growth of precursor molecules outside the mesopores (*89*). How to promote the filling of precursors into the hard templates is therefore an urgent task. Recently, we summarize the work about the synthesis of the replicated mesostructures prepared by the hard-template approach of mesoporous silica (*89*). Evidently, improving the interactions between template and precursor and between precursors themselves are essential in the hard-template approach.

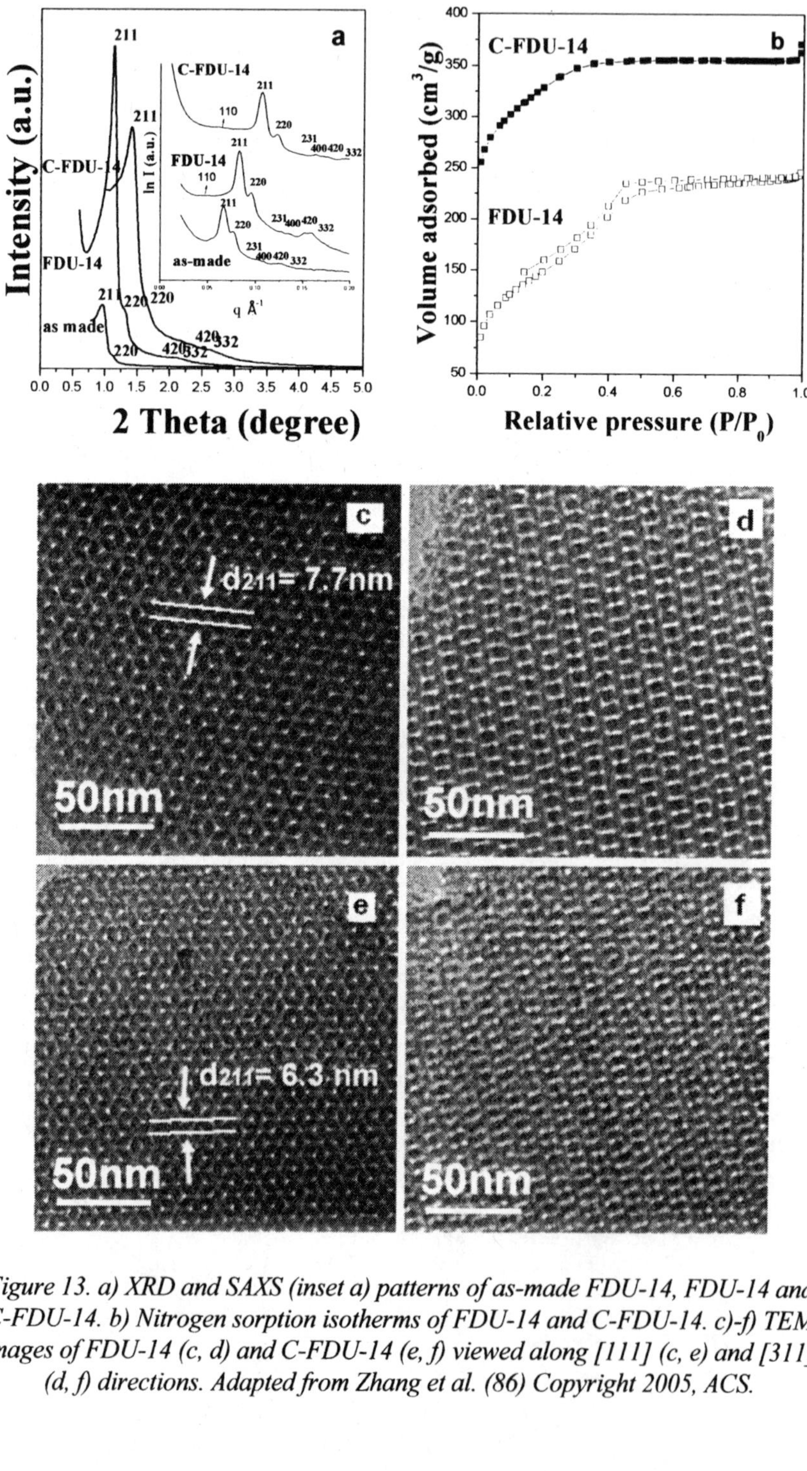

Figure 13. a) XRD and SAXS (inset a) patterns of as-made FDU-14, FDU-14 and C-FDU-14. b) Nitrogen sorption isotherms of FDU-14 and C-FDU-14. c)-f) TEM images of FDU-14 (c, d) and C-FDU-14 (e, f) viewed along [111] (c, e) and [311] (d, f) directions. Adapted from Zhang et al. (86) Copyright 2005, ACS.

1) Improving the interaction between template and precursor

Filling the guest molecules inside the channels of the hard template is important for nanocasting. The main driving force of a precursor entering into a hard template comes from the capillary force (*46*). However, the interaction of the inorganic precursor with the mesopore surface is also a fundamental point. If the interaction is too weak, a partial loading is the result, thus producing disordered nanowires or nanoparticles. On the contrary, orifice growth occurs in the case of extraordinary interplay, which blocks further precursor transportation. Which functional groups in the inorganic templates can have proper interactions with the precursor molecules? As reported previously, both the silanols located on hollowed out mesopore walls and the copolymers in surfactant/silica hybrid materials can coordinate with the guest molecules (*75,89*).

Silanols (≡Si–OH) located on the pore walls of mesoporous silicates are important in attracting guest molecules, which participate in the formation of hydrogen bonds with H atoms and the electron-negative atoms (*e.g.* oxygen atoms) of the precursors, as well as the chelation with metal ions in the precursors. The more silanols on the pore walls are retained, the stronger interaction between the hard template and guest molecules may occur.

Here a microwave-digested (MWD) technique is used to remove surfactants and to protect surface hydroxyl groups. Taking SBA-15 for an example, a procedure is that puts as-synthesized SBA-15 and an appropriate amount of HNO_3 and H_2O_2 into a reactor. The instantaneous high temperature (~ 200°C) and pressure (~ 1.3 MPa) generated by microwave radiation facilitates the oxidation of surfactants by HNO_3 and H_2O_2. Surfactants in the mesopores can be totally eliminated. Less than detectable amount (0.05 wt%) of organic species left. In contrast to calcination, MWD technique is facile and effective in removing the surfactant without sacrificing the silanols on pore walls and distinct framework shrinkage (*90*). It is reasonable that the microwave digested silicates with large pore volumes, high surface areas, and the most important of all, abundant silanols serve as the ideal hard templates for nanocasting replicas (*90,91*).

After various surfactant/silica hybrid materials are prepared, four steps are required to get the final mesostructured replicas. Firstly, the hollow-out inorganic templates with abundant silanols are acquired upon removal of the SDAs *via* MWD method. The silanols can strongly interact with the precursors to improve their transportation. Introducing of the precursors into the pores in templates is the second step *via* enhanced interactions between hard templates and precursors, such as capillary force, chelation bond, and hydrophilic affinity. A following crystallite process upon heating forms mestostructured patterns within the pore channels. The last step is etching the hard templates by NaOH or

HF solution to obtain the self-supported ordered nanoarrays with replicated mesostructures.

The combination of hollowed out mesopores and abundant silanol groups enables the general synthesis of mesostructured metal oxides and sulfides within the channels and cavities of mesoporous silicates. The template–precursor incorporation is easily obtained by mixing the mesoporous silica powders with the alcohol solutions of conventional metal salts. After the solvents are naturally evaporated and the silica templates are dissolved, highly ordered metal oxide nanowire arrays can be replicated with various components including Cr_2O_3, Mn_xO_y, Fe_2O_3, Co_3O_4, NiO, CuO, WO_2, CeO_2 and In_2O_3 (Figure 14 and 15). The second-generation mesostructures can also vary from hexagonally packed nanowire arrays (*p6m*), 3D bicontinuous nanowire arrays ($Ia\bar{3}d$), to nanosphere arrays ($Im\bar{3}m$) derived from SBA-15, FDU-5 or KIT-6, and SBA-16, respectively. More interestingly, they show doubly ordered structures on both meso- and atomic scales (*91, 92*). For example, small-angle XRD patterns (Figure 14) and TEM images (Figure 15) display that replicated NiO and Mn_xO_y nanowires have highly ordered hexagonally mesostructured packing. Wide-angle XRD patterns (inset Figure 14) indicate well-crystallized inorganic frameworks. N_2 sorption measurements show that these mesostructured metal oxide nanoarrays have BET surface areas of 50 – 140 m^2/g, total pore volumes of 0.20 – 0.45 cm^3/g, and uniform pore (intercrystalline voids) sizes of 3 – 7 nm.

MWD-mesoporous silicates are also efficient in replicating mesostructured metal sulfides. By employing metal salts as precursors and thiourea as a sulfur source, the pore channels of mesoporous silica templates can be completely

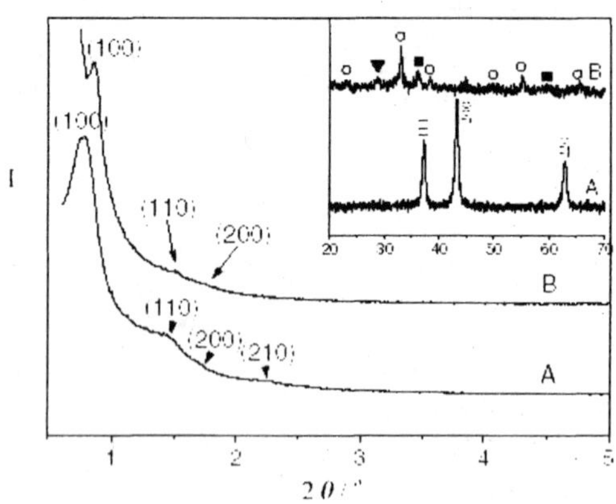

Figure 14. Small-angle XRD patterns of A) NiO and B) Mn_xO_y nanowire arrays templated by MWD-SBA-15. Inset: wide-angle XRD patterns of A) NiO nanowire arrays and B) Mn_xO_y nanowire arrays (○, ▼, and ■ represent typical diffraction peaks of Mn_2O_3, MnO_2, and Mn_3O_4 phases, respectively). Reproduced from reference 91. Copyright 2003, Wiley.

filled with a one-step impregnation process. Subsequent heating treatment at 120°C and dissolution of silicates by NaOH solution result in ordered mesostructured metal sulfide nanowires with high surface areas, such as CdS, ZnS, In_2S_3 and CuS (*93*).

Normally mesoporous silicates with small sizes of 1D channels and caged windows, like MCM-41 and SBA-16, respectively, are thought to be unsuitable for preparing mesoporous carbon. It is probably because of the lack of complementary micropores within the silicate walls in contrast to SBA-15 (*94*). However, mesoporous carbon, constructed of ordered hexagonal packed carbon nanowire arrays with a small diameter around 2.2 nm, can be replicated from a MWD MCM-41 hard template (Figure 16) (*90*). It can be attributed to the unique microwave radiation in the strongly acidic and oxidative system under a high pressure, where the siliceous framework may undergo local structural evolution, *e.g.* recondensation and reconstruction. The process causes some disordered meso/microtunnels on the pore walls, which are important for the assembly of self-supported ordered carbon nanowires. As evidence, the adjacent carbon nanowires are connected by irregular rods. Besides that, the preserved rich silanol groups on the mesopore walls offer better inclusion and contact between the hard templates and guest molecules, which provide strong binding sites for sucrose and facilitate the transcription of the silica structural details, for example, connecting tunnels, to carbon. Similarly, mesoporous carbon arrays replicates the topological structure of cubic mesoporous silica SBA-16 whose cage and window size are enlarged by the MWD method.

Considering the chelation between metal ions and —O— groups in the block copolymer, a one-step nanocasting process is designed to fabricate the ordered mesoporous silica monolith with nanocrystals inside the channels (*95*). The mesophase of the copolymer can be substantially maintained in a certain concentration range of the metal ions and the resulting complex can assist in the assembly.

It is interesting that a mutual effect can occur between the hard template and precursor. Their roles during the assembly can be exchanged. A metal complex formed in the micelles of the block copolymer primarily serves as the soft template to direct the mesoporous silicate through cooperative assembly and assists in dispersing and transferring the inorganic metal salt precursor. By employing a liquid paraffin protected heating treatment, highly ordered mesostructured regularity and monolithic morphology is retained (*95*). The following pyrolysis of inorganic salts to metal oxide nanocrystals can be assigned to the guest growth in the mesopores of silicate upon the removal of surfactant by calcination. The assembly in the confined space of mesoporous silica channels leads to the metal oxide nanocrystals with a uniform size.

This method works well for the low-melting-point metals. For example, indium metal clusters can be generated by the pyrolysis of indium nitrate and the reduction from copolymer decomposition. The low-melting-point indium clusters

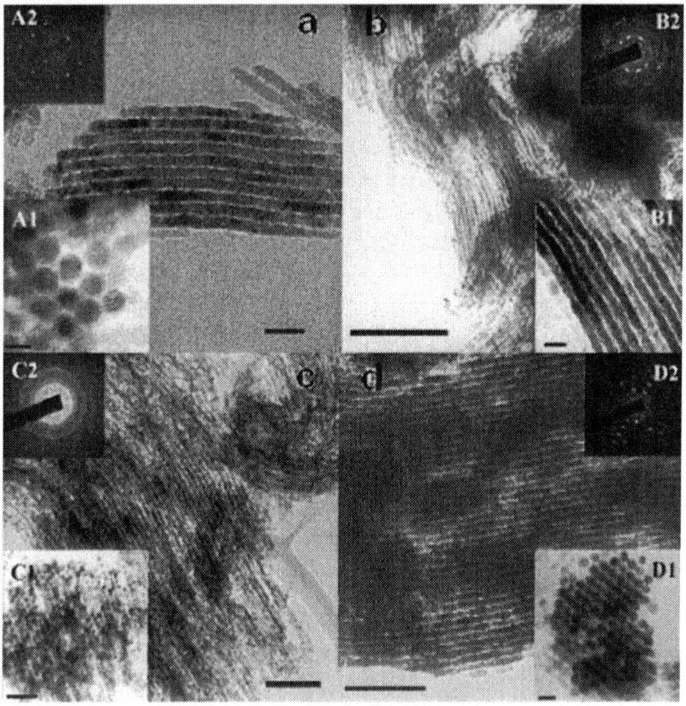

Figure 15. TEM images of ordered metal-oxide nanowires obtained from MWD-SBA-15 with crystallized mesostructured frameworks. (a) Cr_2O_3 nanowires depicted on the (100) plane (with the incident electron beam perpendicular to the c-direction of the 2D hexagonal mesochannels) and the (001) plane (inset A1); inset A2 is the SAED of a. The bars are 25 nm and 10 nm for a and A1, respectively. (b) Mn_xO_y nanowires depicted on the (100) plane; inset B1 showing the (100) plane with large magnification; inset B2 is the SAED of B1. The bars are 200 nm and 20 nm for B and B1, respectively. (c) Fe_2O_3 nanowires depicted on the (100) plane and the (001) plane (inset C1); SAED of c is shown in inset C2. The bars are 100 nm and 50 nm for c and C1, respectively. (d) Co_3O_4 nanowires, (100) and (001) planes; SEAD of d is shown in inset D2. The bars are 100 nm and 20 nm for d and D1, respectively. (e) NiO nanowires, depicted on the (001)plane and (100) plane (inset E1); inset E2 is the SAED of E1. The bars are 20 nm for both e and E1. (f) In_2O_3 nanowires, represented as the (001) plane and the (100) plane (inset F1); SAED of f is shown in inset F2. The bars are 50 nm and 100 nm for f and F1, respectively. Reproduced from reference 91. Copyright 2003, Wiley.

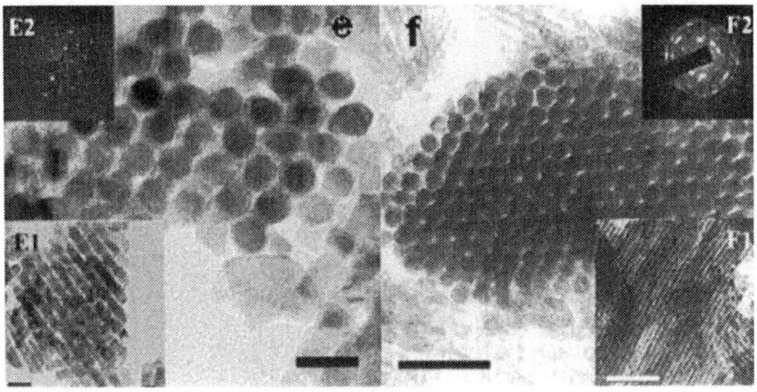

Figure 15. Continued.

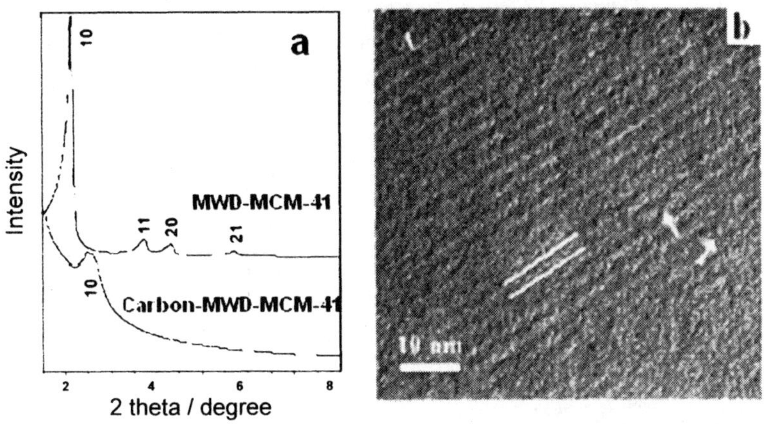

Figure 16. a) XRD patterns of MWD-MCM-41 and its carbon replica, and b) TEM image of carbon nanowires replicated from MWD-MCM-41. Adapted from Tian et al.(90) Copyright 2003, RSC.

(lower than 160°C) can easily diffuse inside the 1D channels of mesoporous silicate monolith and then aggregate to indium nanowires. Once the temperature rises to 400 – 500°C, the oxidation occurs, and the resultant mesostructured material is highly ordered with In_2O_3 single-crystal nanowires as building blocks. Ordered mesostructure and atomic periodicity of the single crystal exist simultaneously.

2) Enhancing the interaction between precursors themselves

Reminiscent to the soft-template self-assembly, the interactions between the precursors themselves should be fully considered. The greater tendency toward migration inside the mesopore channels is found in the precursors to van der Waals force or non-covalent bonds such as hydrogen and coordination bonds (*96, 97*). The interaction plays a further role in inter-linking the stable framework during the annealing stage (*98*).

Mesoporous graphitized carbon is the first example of building the guest units. It is well known that the electrical and adsorption properties of carbon materials are closely related to their graphite nature. However, mesoporous carbons with graphitized structures have remained an attractive topic for researchers and facile synthetic routes are anticipated. A method of one-step melt impregnating mesophase pitches (MPs) inside the mesopores of silica template is demonstrated to replicate the mesoporous carbon with a graphitized structure at a low temperature (*96*). MPs are derived from polyaromatic systems and generally form polydomain liquid crystals constructed by well-stacking layers of carbon rings. The π–π stacking between the molecular units of MPs keeps the incorporating process in a continuous fashion, leading to a perfect packing of MPs in the pores (*96, 99, 100*). Etching the silica results in a true replica of the 3D bicontinuous cubic mesostructure ($Ia\bar{3}d$) with the complicated graphitized orientation (Figure 17). Hexagonal mesoporous carbon can be replicated from a mesoporous silica SBA-15 template. The guest is composed of graphitized domains with a (002) crystallographic plane perpendicular to the long axis of carbon nanorods. Compared with the amorphous mesoporous carbon, the graphitized materials are proved to be anisotropic and present low surface areas (~ 350 m^2/g) and pore volumes (~ 0.4 cm^3/g) lacking of micropores for the crystalline nature.

The precursors with internal hydrogen bonds are suitable for nanocasting. The replication of mesoporous carbon from a sucrose/silica composite is a good example (*75*). A complete filling of precursor molecules inside the mesochannels depends on the internal hydrogen bonds between the –OH groups of sucrose and the silanols (Si–OH) on inorganic pore walls. $[Cd_{10}S_{16}]^{4+}$ is a soluble cadmium sulfide cluster with ample –OH groups, which is structurally comparative with

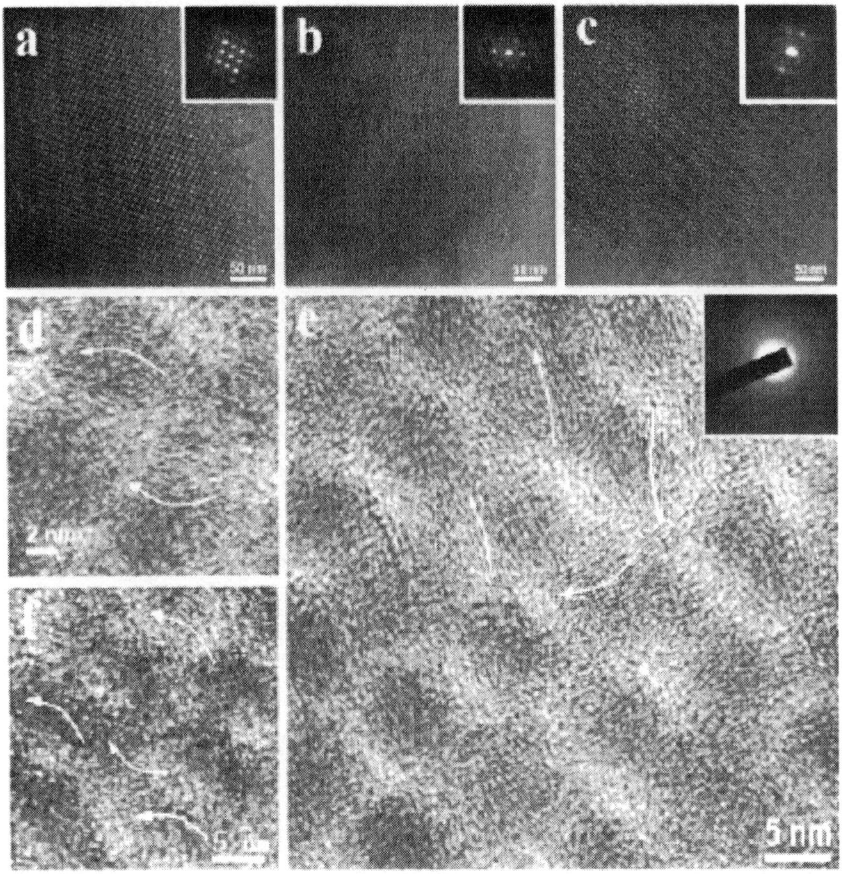

Figure 17. TEM images of ordered mesoporous graphitized carbon with bicontinuous cubic $Ia\bar{3}d$ symmetry along the [110] (a), [111] (d), and [311] (c) directions; the insets of panels a, b, and c are the corresponding SAED patterns; HRTEM images of the sample along the [110] (d), [111] (e), and [311] (f) directions. The inset of panel e is the corresponding SAED pattern of the observed area. The white arrows are the observed orientation of the graphitized lattices. Reproduced from reference 96. Copyright 2004, ACS.

sucrose to some extent. Analogous inter-linking function to that in sucrose is expected to attain a large loading rate of $[Cd_{10}S_{16}]^{4+}$, and in turn, a continuous mesoporous sulfide framework (*101*). Indeed, ordered CdS replica mesostructure can be obtained following a two-step impregnation procedure which is similar to that for casting the mesoporous carbon from sucrose (*97*). The resultant copied from a SBA-15 template is constituted by inter-connected polycrystalline CdS nanowires and the BET surface area is as high as 150 m^2/g.

The active interaction during the annealing process occurs in the nanocasting procedure of mesoporous silicon carbides (*98*). Silicon carbides are a kind of non-oxide ceramic with unique mechanical and functional characteristics. However, because of the high formation temperature, ordered porous SiC solids are very difficult to be obtained (*102*). Recently, highly ordered mesoporous silicon carbides with high specific surface areas of about 700 m^2g^{-1}, large pore volumes of around 0.8 cm^3/g and extraordinary thermal stability up to 1400°C under nitrogen atmosphere have been successfully prepared (*98*). After a polycarbosilane solution is impregnated in a mesoporous silica (with the 2D hexagonal (*p6m*) or 3D bicontinuous ($Ia\bar{3}d$) mesostructure), the nanocasting process is proceeded at 300 °C for 5 hours so that the rough surface of silica can be entirely replicated to the polycarbosilane. Concretionary amorphous SiC solid is produced during the following pyrolysis of polymer precursor, fully copying the shape of channels in the mesoporous silica template (Figure 18 and Figure 19). A higher total ceramic yield after calcination is found at a lower heating rate which can be assigned to the stronger inter-linkage interactions among polycarbosilane molecules. The step is therefore carried out up to 700°C at a rate of 0.5°C min^{-1}. Subsequent crystallization is achieved by the local region Si–C bonds reassembly. A rough surface of the SiC nanowires can be obtained after removal of the silica matrix, resulting in an unusually high surface area.

4. Applications

1) Bio-Application

Numerous applications of the mesoporous materials emerge in biology owing to the mesopore sizes matching well with the dimensions of proteins. Although silica mesostructures manifest outstanding prospects in the fields of protein separation, fixation, catalysis and identification (*12, 103, 104*), nonsiliceous mesoporous materials possess more predominant optical, electric and magnetic properties. These characters, therefore, pave the way for valuable applications in biosensors, bone-forming materials, electrodes, photochromic materials, photo-inducing substrates and even microdevices (*105, 106*).

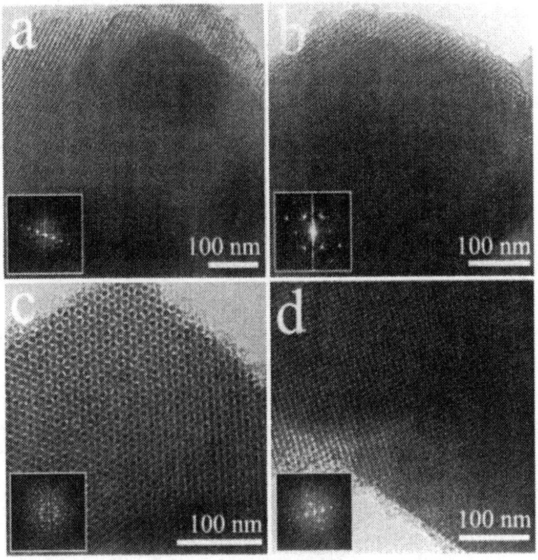

Figure 18. TEM images of mesoporous silicon carbides SiC-SBA15-1200 taken along [110] (a)and [100] (b) directions, and SiC-KIT6-1400 along [111] (c) and [531] (d) directions. Insets show their corresponding Fourier diffractograms. Reproduced from reference 98. Copyright 2006, Wiley.

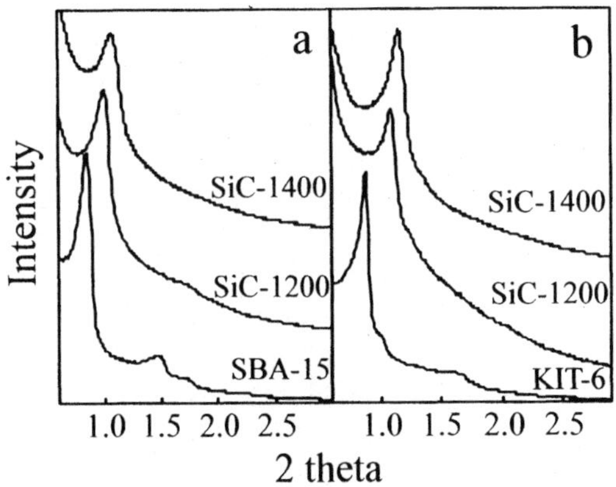

Figure 19. Small-angle XRD patterns of mesoporous silicon carbide replicas and their "parent" mesoporous silica materials with 2D hexagonal (p6m) structure (a) and 3D cubic ($Ia\bar{3}d$) structure (b). Reproduced from reference 98. Copyright 2006, Wiley.

Glass ceramics constituted by $CaO–P_2O_5–SiO_2–MO$ (M = Na, Mg, *etc.*) have been widely studied and used in clinical applications owing to the chemical bonds with living bones (*107*). Does mesostructure have any predominance in bone forming? Mesoporous bioactive glasses (MBGs, $CaO–P_2O_5–SiO_2$) have been synthesized by a simple EISA process in non-aqueous systems (*105*). The structures and components of these materials can be easily tuned. The bone-forming activities of MBGs *in vitro* were tested in simulated body fluid (SBF) to monitor the formation of hydroxycarbonate apatite (HCA) on the surfaces of MBGs as a function of time. The formation of a rod-like morphology, which is similar to that of HCA in human bones, is an evaluation to access the biomimetic chemistry (Figure 20). The growth of HCA nanoparticles less than 100 nm in diameter is observed after soaking the MBG 80S15C (80S and 15C represents the molar fraction of Si, 80% and Ca, 15%, respectively) sample for 4 hours. Rod-like morphology is shaped after another 4 hours (Figure 20c,d). However, such a biomimetic morphology hasn't been molded even after 3 days for the sol–gel-derived bioactive glass (BG) sample. The better bone-forming bioactivity *in vitro* can be attributed to the higher surface area, larger pore volume and more accessible mesopore surface of the mesostructure than the conventional BG material. Animal tests were carried out on the legs of mice. Compared with the case from commercial BG, the shorter bone-forming periods and higher yields in the stocky mice bones with less macroporous voids give further evidences on the advantage of mesostructure.

Nb_2O_5 has been reported as an electron acceptor and host for a family of electron-donating guest species and is thus applicable to a wide variety of reactions involving electron transfer (*108*). It is therefore reasonable to design mesoporous niobium oxide electrodes with high surface areas, tunable pore sizes to match the dimensions of diverse biomolecules, favorable bio-compatibility and electrochemical activities by the "acid–base pair" strategy (*106*).

Figure 21 reveals that assembling cytochrome *c* guest molecules onto an ordered mesoporous Nb_2O_5 host with a pore size of 6 nm effectively promotes the direct electron transfer of redox proteins. The rapid catalytic response to hydrogen peroxide indicates that the high activity of immobilized biomolecules is retained. Horseradish peroxidaze can also be immobilized on the mesoporous Nb_2O_5 host whose high porosity matrix offers a good environment for enzyme loading and substrate diffusion causing high sensitivity and long-term stability. Another virtual electrode is the mesoporous carbon replicated from a mesostructured silica template. Anchored glucose oxidase in the channels of mesoporous carbons, the reversible electron transfer of the proteins can be directly detected. Biosensors based on this kind of electrodes exhibit a fast response to the glucose, which have potential applications in biocatalysis for glucose oxidation.

Ordered hexagonal mesoporous $WO_3–TiO_2$ prepared according to the "acid–base pair" concept was used as a matrix in matrix-assisted laser

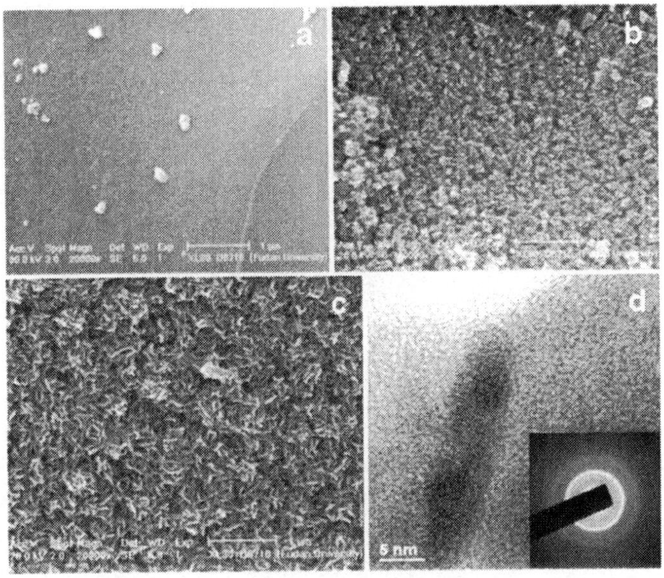

Figure 20. SEM images of mesoporous bioglass, MBG 80S15C after immersing in SBF for 0 (a), 4 (b), and 8 h (c). HRTEM image of the above case c (d). Inset d is the ED pattern of HCA deposited on MBG sample. Reproduced from reference 105. Copyright 2004, Wiley.

desorption/ionization mass spectrometry (MALDI-MS) (*109*). The matrix can trap the analytic molecules, such as peptides and large biomolecules, and act as an energy receptacle for laser irradiation, owing to their prominent absorption abilities in ultraviolet light range and excellent competences in laser desorption/ionization systems. The comparatively higher ionic yield in the analysis of peptide Gramicidin S than that of the nonporous counterpart indicates the privilege of mesostructure. In addition, the measurement of myoglobin digests and their derived fragments in MALDI-MS provides the complete detection of protein sequence and accurate protein identification. On the combination with the benefits of facile preparation, exceptional sensitivity, outstanding reproducibility, as well as high chemical and physical stability, ordered mesoporous metal oxides are proved to be a kind of potential matrices for protein identification in the field of proteomic.

2) Optical and electronic applications

Nanocomposites obtained from the hard-template approach combine the single-crystalline components with large surface areas and uniformly nanoporous

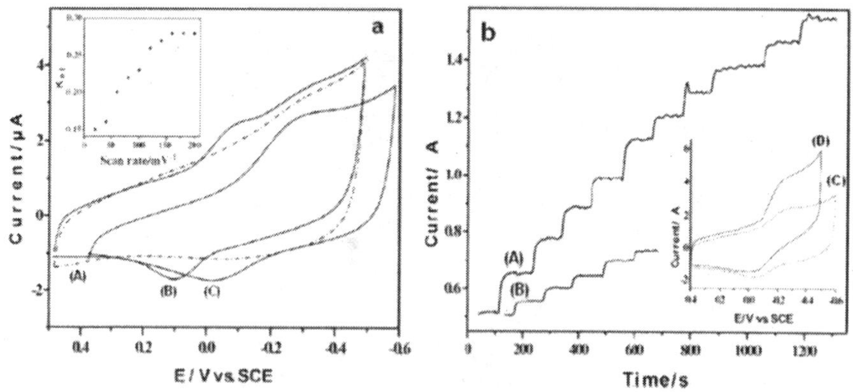

Figure 21. a) CVs of (A) the bare Nb_2O_5 film in blank buffer solution compared with the immobilized Cyt-c/Nb_2O_5 film in a protein-free buffer, (B) before and (C) after immersion overnight: scan rate =100 mV/s. The inset shows the electron transfer rate constant versus the scan rate for adsorbed Cyt-c molecules on such films. b) Dynamic response of the (A) Cyt-c/Nb_2O_5 and (B) Nb_2O_5 electrode to the successive addition of H_2O_2 in pH 7.0 buffer solution at an applied potential of -0.25V. The inset shows CVs of the Cyt-c/Nb_2O_5 electrode in (C) the absence and (D) the presence of H_2O_2, at a scan rate of 50 mV/s. Reproduced from reference 106. Copyright 2003, Wiley.

confinement which may derive unexpected characters (*110*). Mesoporous tin-based oxide/carbon composite (ONTC) prepared from a mesoporous carbon CMK-3 template offers a promising negative-electrode material in lithium-ion battery (*111*). In comparison with the nano-sized tin oxides and tin-based oxides (MTBO), the mesostrtured tin-based oxide/carbon composite displays a superior cycling performance and stability cycled with a wide voltage range of 0.02 – 2.0 V (Figure 22). It can be concluded that 3D interconnected mesoporous carbon framework provides the confined-growth nanospace of tin-based oxide, prevents the particle–to–particle aggregation of tin species, and ensures the good electrical contact in tin composite.

The unique function of ordered mesoporous nonsiliceous oxides can also be applied in optical devices. 3D caged mesoporous WO_3–TiO_2 composite exhibits an extraordinary electrochromic property and a stable contrast ratio in a wavelength ranging from 400 to 800 nm. The advanced applications can be achieved in “smart windows” (*112*).

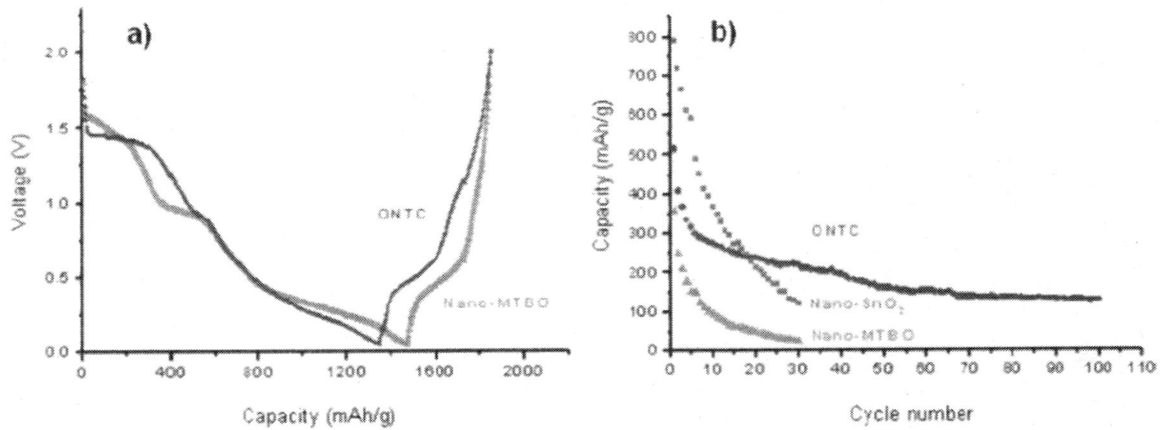

Figure 22. a) First-cycle charge and discharge curves of Li/ONTC (500°C/in air), and Li/nano-MTBO cells. b) Capacity versus cycle number for the cell, with samples ONTC (500°C/in air), nano SnO2 and nano-MTBO, as indicated. Potential limits, 0.02 – 2.0 V vs. Li, current density of 0.5 mA/cm^2. Reproduced from reference 111. Copyright 2004, Wiley.

5. Summary and Outlook

The field of ordered mesoporous materials represents a great opportunity for materials to have an impart on modern technology. Substantial progress has been made in the inorganic–organic and organic–organic interactions that have been reflected in mesoporous silicates from inorganic–organic self-assembly. Considerations on modulating the surfactants with various types and hydrophobic/hydrophilic nature and controlling the hydrogen bonds and Coulomb forces between the templates and inorganic molecules deduce a large number of highly ordered mesostructures. This chapter demonstrates that the precise control of the interactions inside the inorganic or organic species and between organic template/precursor species is very important for understanding the surfactant self-assembly approach in the fabrication of ordered mesoporous nonsiliceous materials. In addition, improving the interactions either between the hard template and precursor or between the precursors themselves favors the confined-space growth. Valuable applications emerge in the fields of biology, photoelectronic materials, sensors and electrodes.

The EISA strategy skillfully avoids the true inorganic–organic interaction in the cooperative assembly process, which facilitates the soft-template approach.*(113)* A generalized "acid–base pair" concept and an organic–organic self-assembly strategy are therefore established, which give rise to a large number of highly ordered mesoporous metal oxides, mixed metal oxides, metal phosphates, metal borates, polymers and carbon molecular sieves. In the former case, the organization of inorganic–organic mesophase is dissociated with the densification of inorganic framework. While in the preparation of ordered polymers and carbons from organic–organic self-assembly, the cross-linking and thermopolymerization processes of the resols are separated from the assembly. Therefore, carefully choosing the reactants and the conditions may enhance and balance those interactions, and result in high-quality products. In the "acid–base" route, it is necessary to match acid–base interactions of the various species presented during nucleation of the mesostructured phases. The key issue for the successful organization of organic–organic mesostructures is the preformed polymer guest, which has plenty of hydroxyl groups and strongly interacts with the amphiphilic triblock copolymer *via* hydrogen bonds. Furthermore, three- and four-connected benzene rings can be generated *via* covalent bonds, the same as the silicate in zeolites or/and mesoporous silicate molecular sieves.

Current improvements are also essential for the synthesis by a hard-template pathway. It materializes in enhancing the interactions between the hard template and precursor and the precursors themselves to a substantial extent. A microwave-digested method, which can remove the surfactant, retain the majority surface hydroxyl and generate connected micropores on pore walls, is used to produce a wide variety of mescopically ordered nanoarray replicas with single-crystalline structures. By utilizing the coordination of organic surfactant with metal ions, a one-step impregnation process is used to fabricate ordered silica monoliths with various metal oxides nanocrystals. $[Cd_{10}S_{16}]^{4+}$ clusters with plenty of hydroxyl groups and mesophase pitches are excellent candidates for nanocasting ordered mesoporous CdS and graphitized carbons, respectively. Concretionary amorphous SiC solid is produced during the continuous pyrolysis of polycarbosilane precursor, fully copying the shape of channels in the mesoporous silica template.

Although the number of research in nonsiliceous mesoporous materials has increased because of the high potential impact on practical applications, there is still a significant need to design, synthesize, and evaluate new materials. A method for regulating the interaction between guest molecules and/or hard template has not yet been determined, but such regulation could produce novel, convenient, economic, and reproducible strategies. Bulk production of nonsiliceous mesoporous materials are also imperative under the situation. The present syntheses are limited to the qualitative description. The quantitative illustration is the future target from the viewpoint of theoretical and configurable computation in order to develop generalized synthetic pathways and rationally design the structures, compositions and properties. Once this is achieved, it will be possible to fabricate ordered mesoporous metal oxides and sulfides with crystallite walls, high melting point semiconductors, as well as mesoporous Si, SiC, metal carbides and nitrides with open frameworks, together with achieving direct organic functional walls or channels of the mesoporous materials. The challenges are inevitably great in exploiting their functions in catalysis, sensors, microelectrodes and dielectric materials.

Acknowledgments

This work was supported by NNSF of China (20233030, 20373013, 20421303, 20407014 and 20521140450), State Key Basic Research Program of China (2001CB610506), Shanghai Science & Technology and Educational Committee (04JC14087, 03527001, T0402 and 03QF14037), Shanghai HuaYi Chemical Group and Unilever research institute of China.

References

1. Kresge, C. T.; Leonowicz, M. E.; Roth, W. J.; Vartuli, J. C.; Beck, J. S. *Nature* **1992**, *359*, 710-712.
2. Beck, J. S.; Vartuli, J. C.; Roth, W. J.; Leonowicz, M. E.; Kresge, C. T.; Schmitt, K. D.; Chu, C. T. W.; Olson, D. H.; Sheppard, E. W.; McCullen, S.

B.; Higgins, J. B.; Schlenker, J. L. *J. Am. Chem. Soc.* **1992**, *114*, 10834-10843.
3. Zhao, D. Y.; Feng, J. L.; Huo, Q. S.; Melosh, N.; Fredrickson, G. H.; Chmelka, B. F.; Stucky, G. D. *Science* **1998**, *279*, 548-552.
4. Ying, J. Y.; Mehnert, C. P.; Wong, M. S. *Angew. Chem. Int. Ed.* **1999**, *38*, 56-77.
5. Hatton, B.; Landskron, K.; Whitnall, W.; Perovic, D.; Ozin, G. A. *Acc. Chem. Res.* **2005**, *38*, 305-312.
6. Stein, A.; Melde, B. J.; Schroden, R. C. *Adv. Mater.* **2000**, *12*, 1403-1419.
7. Davis, M. E. *Nature* **2002**, *417*, 813-821.
8. Corma, A. *Chem. Rev.* **1997**, *97*, 2373-2419.
9. He, X.; Antonelli, D. *Angew. Chem. Int. Ed.* **2001**, *41*, 214-229.
10. Winkler, H.; Birkner, A.; Hagen, V.; Wolf, I.; Schmechel, R.; von Seggern, H.; Fischer, R. A. *Adv. Mater.* **1999**, *11*, 1444-1448.
11. Chang, J. H.; Shim, C. H.; Kim, B. J.; Shin, Y.; Exarhos, G. J.; Kim, K. J. *Adv. Mater.* **2005**, *17*, 634-637.
12. Fan, J.; Shui, W. Q.; Yang, P. Y.; Wang, X. Y.; Xu, Y. M.; Wang, H. H.; Chen, X.; Zhao, D. Y. *Chem. A Eur. J.* **2005**, *11*, 5391-5396.
13. Schuth, F.; Schmidt, W. *Adv. Mater.* **2002**, *14*, 629-638.
14. Huo, Q. S.; Margolese, D. I.; Stucky, G. D. *Chem. Mater.* **1996**, *8*, 1147-1160.
15. Kim, J. M.; Kim, S. K.; Ryoo, R. *Chem. Commun.* **1998**, 259-260.
16. Chen, F. X.; Huang, L. M.; Li, Q. Z. *Chem. Mater.* **1997**, *9*, 2685-2686.
17. Huo, Q. S.; Leon, R.; Petroff, P. M.; Stucky, G. D. *Science* **1995**, *268*, 1324-1327.
18. Che, S.; Sakamoto, Y.; Terasaki, O.; Tatsumi, T. *Chem. Mater.* **2001**, *13*, 2237-2239.
19. Shen, S. D.; Li, Y. Q.; Zhang, Z. D.; Fan, J.; Tu, B.; Zhou, W. Z.; Zhao, D. Y. *Chem. Commun.* **2002**, 2212-2213.
20. Shen, S. D.; Garcia-Bennett, A. E.; Liu, Z.; Lu, Q. Y.; Shi, Y. F.; Yan, Y.; Yu, C. Z.; Liu, W. C.; Cai, Y.; Terasaki, O.; Zhao, D. Y. *J. Am. Chem. Soc.* **2005**, *127*, 6780-6787.
21. Che, S.; Garcia-Bennett, A. E.; Yokoi, T.; Sakamoto, K.; Kunieda, H.; Terasaki, O.; Tatsumi, T. *Nature Mater.* **2003**, *2*, 801-805.
22. Garcia-Bennett, A. E.; Kupferschmidt, N.; Sakamoto, Y.; Che, S.; Terasaki, O. *Angew. Chem. Int. Ed.* **2005**, *44*, 5317-5322.
23. Garcia-Bennett, A. E.; Terasaki, O.; Che, S.; Tatsumi, T. *Chem. Mater.* **2004**, *16*, 813-821.
24. Che, S.; Liu, Z.; Ohsuna, T.; Sakamoto, K.; Terasaki, O.; Tatsumi, T. *Nature* **2004**, *429*, 281-284.
25. Zhao, D. Y.; Huo, Q. S.; Feng, J. L.; Chmelka, B. F.; Stucky, G. D. *J. Am. Chem. Soc.* **1998**, *120*, 6024-6036.
26. Yu, C. Z.; Yu, Y. H.; Zhao, D. Y. *Chem. Commun.* **2000**, 575-576.

27. El-Safty, S. A.; Hanaoka, T. *Chem. Mater.* **2004**, *16*, 84-400.
28. Wiesner, U., In *Nanotechnology In Mesostructured Materials*, 2003; Vol. 146, pp 407-411.
29. Finnefrock, A. C.; Ulrich, R.; Toombes, G. E. S.; Gruner, S. M.; Wiesner, U. *J. Am. Chem. Soc.* **2003**, *125*, 13084-13093.
30. Liu, X. Y.; Tian, B. Z.; Yu, C. Z.; Gao, F.; Xie, S. H.; Tu, B.; Che, R. C.; Peng, L. M.; Zhao, D. Y. *Angew. Chem. Int. Ed.* **2002**, *41*, 3876-3878.
31. Kim, T. W.; Kleitz, F.; Paul, B.; Ryoo, R. *J. Am. Chem. Soc.* **2005**, *127*, 7601-7610.
32. Wang, Y. Q.; Yang, C. M.; Zibrowius, B.; Spliethoff, B.; Linden, M.; Schuth, F. *Chem. Mater.* **2003**, *15*, 5029-5035.
33. Fan, J.; Yu, C. Z.; Wang, L. M.; Tu, B.; Zhao, D. Y.; Sakamoto, Y.; Terasaki, O. *J. Am. Chem. Soc.* **2001**, *123*, 12113-12114.
34. Fan, J.; Yu, C. Z.; Lei, J.; Zhang, Q.; Li, T. C.; Tu, B.; Zhou, W. Z.; Zhao, D. Y. *J. Am. Chem. Soc.* **2005**, *127*, 10794-10795.
35. Zhao, D. Y.; Sun, J. Y.; Li, Q. Z.; Stucky, G. D. *Chem. Mater.* **2000**, *12*, 275-279.
36. Yu, C. Z.; Fan, J.; Tian, B. Z.; Zhao, D. Y.; Stucky, G. D. *Adv. Mater.* **2002**, *14*, 1742-1745.
37. El-Safty, S. A.; Hanaoka, T.; Mizukami, F. *Chem. Mater.* **2005**, 17, 3137-3145.
38. Wang, J. F.; Tsung, C. K.; Hong, W. B.; Wu, Y. Y.; Tang, J.; Stucky, G. D. *Chem. Mater.* **2004**, *16*, 5169-5181.
39. Miyata, H.; Suzuki, T.; Fukuoka, A.; Sawada, T.; Watanabe, M.; Noma, T.; Takada, K.; Mukaide, T.; Kuroda, K. *Nature Mater.* **2004**, *3*, 651-656.
40. Innocenzi, P.; Brusatin, G. *Chem. Mater.* **2001**, *13*, 3126-3139.
41. Sanchez, C.; Lebeau, B.; Ribot, F.; In, M. *J. Sol-Gel Sci. Tech.* **2000**, *19*, 31-38.
42. Li, D. L.; Zhou, H. S.; Honma, I. *Nature Mater.* **2004**, *3*, 65-72.
43. Gratzel, M. *Inorg. Chem.* **2005**, *44*, 6841-6851.
44. Bartl, M. H.; Puls, S. P.; Tang, J.; Lichtenegger, H. C.; Stucky, G. D. *Angew. Chem. Int. Ed.* **2004**, *43*, 3037-3040.
45. Sakatani, Y.; Grosso, D.; Nicole, L.; Boissiere, C.; Soler-Illia, G.; Sanchez, C. *J. Mater. Chem.* **2006**, *16*, 77-82.
46. Schuth, F. *Chem. Mater.* **2001**, *13*, 3184-3195.
47. Schuth, F. *Angew. Chem. Int. Ed.* **2003**, *42*, 3604-3622.
48. Huo, Q. S.; Margolese, D. I.; Ciesla, U.; Feng, P. Y.; Gier, T. E.; Sieger, P.; Leon, R.; Petroff, P. M.; Schuth, F.; Stucky, G. D. *Nature* **1994**, *368*, 317-321.
49. Huo, Q. S.; Margolese, D. I.; Ciesla, U.; Demuth, D. G.; Feng, P. Y.; Gier, T. E.; Sieger, P.; Firouzi, A.; Chmelka, B. F.; Schuth, F.; Stucky, G. D. *Chem. Mater.* **1994**, *6*, 1176-1191.

50. Yang, P. D.; Zhao, D. Y.; Margolese, D. I.; Chmelka, B. F.; Stucky, G. D. *Nature* **1998**, *396*, (6707), 152-155.
51. Grosso, D.; Soler-Illia, G.; Babonneau, F.; Sanchez, C.; Albouy, P. A.; Brunet-Bruneau, A.; Balkenende, A. R. *Adv. Mater.* **2001**, *13*, 1085-1090.
52. Lu, Y. F.; Ganguli, R.; Drewien, C. A.; Anderson, M. T.; Brinker, C. J.; Gong, W. L.; Guo, Y. X.; Soyez, H.; Dunn, B.; Huang, M. H.; Zink, J. I. *Nature* **1997**, *389*, 364-368.
53. Yang, P. D.; Zhao, D. Y.; Margolese, D. I.; Chmelka, B. F.; Stucky, G. D. *Chem. Mater.* **1999**, *11*, 2813-2826.
54. Soler-illia, G. J. D.; Sanchez, C.; Lebeau, B.; Patarin, J. *Chem. Rev.* **2002**, *102*, 4093-4138.
55. Grosso, D.; Cagnol, F.; Soler-Illia, G.; Crepaldi, E. L.; Amenitsch, H.; Brunet-Bruneau, A.; Bourgeois, A.; Sanchez, C. *Adv. Funct. Mater.* **2004**, *14*, 309-322.
56. Crepaldi, E. L.; Soler-Illia, G.; Grosso, D.; Cagnol, F.; Ribot, F.; Sanchez, C. *J. Am. Chem. Soc.* **2003**, *125*, 9770-9786.
57. Soler-Illia, G.; Louis, A.; Sanchez, C. *Chem. Mater.* **2002**, *14*, 750-759.
58. Yun, H. S.; Miyazawa, K.; Zhou, H. S.; Honma, I.; Kuwabara, M. *Adv. Mater.* **2001**, *13*, 1377-1380.
59. Tian, B. Z.; Liu, X. Y.; Tu, B.; Yu, C. Z.; Fan, J.; Wang, L. M.; Xie, S. H.; Stucky, G. D.; Zhao, D. Y. *Nature Mater.* **2003**, *2*, 159-163.
60. Tian, B. Z.; Yang, H. F.; Liu, X. Y.; Xie, S. H.; Yu, C. Z.; Fan, J.; Tu, B.; Zhao, D. Y. *Chem. Commun.* **2002**, 1824-1825.
61. Corriu, R. J. P.; Leclercq, D. *Angew. Chem. Int. Ed.* **1996**, *35*, 1420-1436.
62. Braun, P. V.; Osenar, P.; Tohver, V.; Kennedy, S. B.; Stupp, S. I. *J. Am. Chem. Soc.* **1999**, *121*, 7302-7309.
63. Feng, P. Y.; Bu, X. H.; Gier, T. E.; Stucky, G. D. *Micropor. Mesopor. Mater.* **1998**, *23*, 221-229.
64. Tiemann, M.; Froba, M. *Chem. Mater.* **2001**, *13*, 3211-3217.
65. Fujishima, A.; Honda, K. *Nature* **1972**, *238*, 37-38.
66. Oregan, B.; Gratzel, M. *Nature* **1991**, *353*, 737-740.
67. Bach, U.; Lupo, D.; Comte, P.; Moser, J. E.; Weissortel, F.; Salbeck, J.; Spreitzer, H.; Gratzel, M. *Nature* **1998**, *395*, 583-585.
68. Antonelli, D. M.; Ying, J. Y. *Angew. Chem. Int. Ed.* **1995**, *34*, 2014-2017.
69. Sanchez, C.; Soler-Illia, G.; Ribot, F.; Lalot, T.; Mayer, C. R.; Cabuil, V. *Chem. Mater.* **2001**, *13*, 3061-3083.
70. Kimura, T. *Chem. Mater.* **2005**, *17*, 5521-5528.
71. Hogarth, W. H. J.; da Costa, J. C. D.; Drennan, J.; Lu, G. Q. *J. Mater. Chem.* **2005**, *15*, 754-758.
72. Attard, G. S.; Goltner, C. G.; Corker, J. M.; Henke, S.; Templer, R. H. *Angew. Chem. Int. Ed.* **1997**, *36*, 1315-1317.
73. Dillon, A. C.; Jones, K. M.; Bekkedahl, T. A.; Kiang, C. H.; Bethune, D. S.; Heben, M. J. *Nature* **1997**, *386*, 377-379.

74. Yu, J. S.; Kang, S.; Yoon, S. B.; Chai, G. *J. Am. Chem. Soc.* **2002**, *124*, 9382-9383.
75. Ryoo, R.; Joo, S. H.; Jun, S. *J. Phy. Chem. B* **1999**, *103*, 7743-7746.
76. Al-Muhtaseb, S. A.; Ritter, J. A. *Adv. Mater.* **2003**, *15*, 101-114.
77. Jenekhe, S. A.; Chen, X. L. *Science* **1999**, *283*, 372-375.
78. Pindzola, B. A.; Hoag, B. P.; Gin, D. L. *J. Am. Chem. Soc.* **2001**, *123*, 4617-4618.
79. Jiang, P.; Bertone, J. F.; Colvin, V. L. *Science* **2001**, *291*, 453-457.
80. Zalusky, A. S.; Olayo-Valles, R.; Wolf, J. H.; Hillmyer, M. A. *J. Am. Chem. Soc.* **2002**, *124*, 12761-12773.
81. Jang, J.; Bae, J. *Chem. Commun.* **2005**, (*9*), 1200-1202.
82. Lee, K. T.; Oh, S. M. *Chem. Commun.* **2002**, 2722-2723.
83. Liang, C. D.; Hong, K. L.; Guiochon, G. A.; Mays, J. W.; Dai, S. *Angew. Chem. Int. Ed.* **2004**, *43*, 5785-5789.
84. Tanaka, S.; Nishiyama, N.; Egashira, Y.; Ueyama, K. *Chem. Commun.* **2005**, 2125-2127.
85. Meng, Y.; Gu, D.; Zhang, F. Q.; Shi, Y. F.; Yang, H. F.; Li, Z.; Yu, C. Z.; Tu, B.; Zhao, D. Y. *Angew. Chem. Int. Ed.* **2005**, *44*, 7053-7059.
86. Zhang, F. Q.; Meng, Y.; Gu, D.; Yan, Y.; Yu, C. Z.; Tu, B.; Zhao, D. Y. *J. Am. Chem. Soc.* **2005**, *127*, 13508-13509.
87. Fan, H. Y.; Yang, K.; Boye, D. M.; Sigmon, T.; Malloy, K. J.; Xu, H. F.; Lopez, G. P.; Brinker, C. J. *Science* **2004**, *304*, 567-571.
88. Martin, C. R. *Science* **1994**, *266*, 1961-1966.
89. Yang, H. F.; Zhao, D. Y. *J. Mater. Chem.* **2005**, *15*, 1217-1231.
90. Tian, B. Z.; Che, S. N.; Liu, Z.; Liu, X. Y.; Fan, W. B.; Tatsumi, T.; Terasaki, O.; Zhao, D. Y. *Chem. Commun.* **2003**, 2726-2727.
91. Tian, B. Z.; Liu, X. Y.; Yang, H. F.; Xie, S. H.; Yu, C. Z.; Tu, B.; Zhao, D. Y. *Adv. Mater.* **2003**, *15*, 1370-1374.
92. Tian, B. Z.; Liu, X. Y.; Solovyov, L. A.; Liu, Z.; Yang, H. F.; Zhang, Z. D.; Xie, S. H.; Zhang, F. Q.; Tu, B.; Yu, C. Z.; Terasaki, O.; Zhao, D. Y. *J. Am. Chem. Soc.* **2004**, *126*, 865-875.
93. Liu, X. Y.; Tian, B. Z.; Yu, C. Z.; Tu, B.; Liu, Z.; Terasaki, O.; Zhao, D. Y. *Chem. Lett.* **2003**, *32*, 824-825.
94. Ryoo, R.; Joo, S. H.; Kruk, M.; Jaroniec, M. *Adv. Mater.* **2001**, *13*, 677-681.
95. Yang, H. F.; Shi, Q. H.; Tian, B. Z.; Lu, Q. Y.; Gao, F.; Xie, S. H.; Fan, J.; Yu, C. Z.; Tu, B.; Zhao, D. Y. *J. Am. Chem. Soc.***2003**, *125*, 4724-4725.
96. Yang, H. F.; Yan, Y.; Liu, Y.; Zhang, F. Q.; Zhang, R. Y.; Meng, Y.; Li, M.; Xie, S. H.; Tu, B.; Zhao, D. Y. *J. Phy. Chem. B* **2004**, *108*, 17320-17328.
97. Gao, F.; Lu, Q. Y.; Zhao, D. Y. *Adv. Mater.* **2003**, *15*, 739-742.

98. Shi, Y. F.; Meng, Y.; Chen, D. H.; Cheng, S. J.; Yang, H. F.; Wan, Y.; Zhao, D. Y. *Adv. Funct. Mater.* **2006**, *16*, 561-567.
99. Vix-Guterl, C.; Saadallah, S.; Vidal, L.; Reda, M.; Parmentier, J.; Patarin, J. *J. Mater. Chem.* **2003**, *13*, 2535-2539.
100. Li, Z.; Jaroniec, M. *J. Phys. Chem. B* **2004**, *108*, 824-826.
101. Lacelle, S.; Stevens, W. C.; Kurtz, D. M.; Richardson, J. W.; Jacobson, R. A. *Inorg. Chem.* **1984**, *23*, 930-935.
102. Kamperman, M.; Garcia, C. B. W.; Du, P.; Ow, H. S.; Wiesner, U. *J. Am. Chem. Soc.* **2004**, *126*, 14708-14709.
103. Hartmann, M. *Chem. Mater.* **2005**, *17*, 4577-4593.
104. Fan, J.; Lei, J.; Wang, L. M.; Yu, C. Z.; Tu, B.; Zhao, D. Y. *Chem. Commun.* **2003**, 2140-2141.
105. Yan, X. X.; Yu, C. Z.; Zhou, X. F.; Tang, J. W.; Zhao, D. Y. *Angew. Chem. Int. Ed.* **2004**, *43*, 5980-5984.
106. Xu, X.; Tian, B. Z.; Kong, J. L.; Zhang, S.; Liu, B. H.; Zhao, D. Y. *Adv. Mater.* **2003**, *15*, 1932-1936.
107. Hench, L. L.; Polak, J. M. *Science* **2002**, *295*, 1014-1017.
108. He, X.; Trudeau, M.; Antonelli, D. M. *Adv. Mater.* **2000**, *12*, 1036-1040.
109. Yuan, M. J.; Shan, Z.; Tian, B. Z.; Tu, B.; Yang, P. Y.; Zhao, D. Y. *Micropor. Mesopor. Mater.* **2005**, *78*, 37-41.
110. Lu, A. H.; Schmidt, W.; Matoussevitch, N.; Bonnemann, H.; Spliethoff, B.; Tesche, B.; Bill, E.; Kiefer, W.; Schuth, F. *Angew. Chem. Int. Ed.* **2004**, *43*, 4303-4306.
111. Fan, J.; Wang, T.; Yu, C. Z.; Tu, B.; Jiang, Z. Y.; Zhao, D. Y. *Adv. Mater.* **2004**, *16*, 1432-1436.
112. Yuan, M. J.; Lu, Y. D.; Tian, B. Z.; Yang, H. F.; Tu, B.; Kong, J.; Zhao, D. Y. *Chem. Lett.* **2004**, *33*, 1396-1397.
113. Wan, Y.; Yang, H. F.; Zhao, D. Y. *Acc. Chem. Res.* **2006**, *39*, 423-432.

Chapter 2

Proteins in Mesoporous Silicates

Sarah P. Hudson[1], Simon White[1], Dimple Goradia[1], Hayder Essa[1], Baohong Liu[2], Liang Qiao[2], Yun Liu[2], Jakki C. Cooney[1], B. Kieran Hodnett[1], and Edmond Magner[1,*]

[1]Materials and Surface Science Institute and Department of Chemical and Environmental Sciences, University of Limerick, Limerick, Ireland
[2]Department of Chemistry, Fudan University, Shanghai 200433, People's Republic of China

Mesoporous silicates (MPS) provide a means of immobilizing proteins and enzymes in a stable environment while retaining physiological function. A systematic methodology of assessing the potential of immobilising a given protein on to MPS has been developed from extensive characterization of the surface properties of the silicates and proteins. This approach can be potentially utilised in the design of a solid support for any protein.

Enzymes catalyze reactions with high specificity. Unfortunately, they are not always suited to industrial or medical applications. They are often unstable and show low activity in organic solvents or at high temperatures. Denaturation of the enzyme, which destroys its catalytic activity, can often be induced by pH, mechanical or thermal treatment. In industrial processes, the enzyme is often denatured during its removal from the reaction mixture. This is uneconomical as active enzyme is lost. These disadvantages involving enzymes can be overcome by immobilization of the enzyme onto a solid structure (*1,2*). With increased mechanical stability and potential for recycling, the advantages of enzymes in catalyzing specific reactions under mild conditions can be harnessed via immobilisation.

In the early 1990's Beck et al. demonstrated a significant practical method of synthesising zeolite type materials with larger pore diameters (*3*). A wide range of mesoporous silicates (MPS) has been described since this report (*4-9*). MPS are formed using surfactants such as cetyl-trimethyl ammonium bromide in solution, which act as structure directing agents. Addition of a silane results in silica polymerisation around the surfactant micelle structures, forming a gel. This gel is set thermally, condensing the silane to form a stable material. Upon removal of the surfactant, a mesoporous structure remains. MPS possess large surface areas (up to 1000 $m^2\ g^{-1}$), highly ordered pore structures and very narrow pore size distributions (PSD); properties which have made these materials attractive candidates for a wide range of applications in catalysis, sensor, and separation technologies (*4-8*). The size and connectivity of the pores is determined by the synthesis and surfactant used, allowing channel diameters to be tailored between 3-20 nm, a pore size range similar to that of proteins, making MPS ideal for encapsulation applications. MPS are transparent in the visible region of the spectrum, and possess good mechanical and thermal properties. They possess a number of additional attributes, which make them attractive candidates for the immobilisation of proteins. It is possible to chemically modify their surfaces with various functional groups, enabling electrostatic attraction or repulsion between an MPS and the biological molecule of interest to be optimized (*9,10*). As a result of their silicate inorganic framework, MPS are chemically and mechanically stable and are resistant to microbial attack.

Materials such as sol-gels display similar stability to MPS and have been used to encapsulate proteins for use as biosensors (*11,12*). However, sol-gels suffer from the disadvantage of possessing a highly variable pore size distribution. More importantly, their preparation can involve the use of harsh conditions or reagents, which are detrimental to, and can cause denaturation of some proteins. Using MPS, protein encapsulation occurs after synthesis of the support, avoiding this difficulty. Controlled pore glass (CPG) (*13*) has been used extensively as a means of immobilising enzymes and a range of CPG materials are available. However, adsorption of enzymes to CPG generally entails the use of covalent links between the protein and CPG, which can result in a significant loss in enzyme activity.

In 1996 Balkus and Diaz (*10*) first described the immobilisation of proteins onto mesoporous MCM-41. We, and others, have shown that mesoporous silicates (*14-22*) provide a means of immobilizing proteins and enzymes in a stable environment while retaining physiological function. It is important to note that the process of immobilising enzymes is simple; the protein and MPS are incubated under the appropriate conditions for a period of minutes to hours, after which the MPS adsorbed protein is removed by centrifugation and washing steps. Many factors (*11-17*) have a strong influence on the enzyme loading and on the activity of the resultant biocatalyst, including the relative sizes of the mesopores and the enzyme, surface area, pore size distribution, mesopore volume, particle size, ionic strength, isoelectric point and surface characteristics of both the support and the enzyme. We have recently established a methodical means of evaluating the effect of each variable on the adsorption of a particular enzyme into a particular mesoporous support (*17,18*). Here we describe the parameters that affect the adsorption and activity of a number of proteins and enzymes onto MPS.

Experimental

Pluronic P123 ($EO_{20}PO_{70}EO_{20}$) was obtained from BASF. Tetraethoxysilane (TEOS, 98%), 1,2-bis(trimethoxysilyl)ethane (BTMSE, 96%), 2-cyanoethyltriethoxysilane (CEOS), sodium hydroxide, horse heart cytochrome c (90%), cetyltrimethylammonium bromide (CTAB), 2,2'-azino bis (3-ethylbenzthiazoline-6-sulfonic acid) (ABTS), trizma base (99.9%), acetic acid, sodium acetate, potassium hydrogen phosphate, potassium dihydrogen phosphate, sodium hydrogen carbonate and sodium carbonate were all obtained from Sigma-Aldrich.

SBA-15 (*23*), MSE (*24*), PMO-ICS-50 (*25*), CNS (*17*), CNS-cal (*17*)[17], MCM-41/28 (*26*), MCF (*27*) and AlPO (*28*) were all synthesised following previously published protocols. PMO-PA-50 was synthesised following the protocol for PMO-ICS-50 but using bis[3-(trimethoxysilyl)propyl]amine instead of tris[3-(trimethoxysilyl)propyl] isocyanurate. MCM-41 was synthesised according to a published protocol (*3*) but with the addition of trimethylbenzene as a swelling agent, in a ratio of SiO_2 : TMB = 1:0.8. The surfactants were either removed by Soxhlet extraction with ethanol (SBA-15, MSE, CNS, KIT-6, MCF, PMO-ICS-50 and PMO-PA-50) or calcination (CNS-cal, MCM-41/28, AlPO and MCM-41).

Nitrogen gas adsorption/desorption isotherms were measured at 77 K using a Micromeritics ASAP 2010 system.. The surface area was measured using the Brunauer-Emmett-Teller (BET) method (*29*). The pore size data were analysed by the thermodynamic based Barrett-Joyner-Halenda (BJH) method (*30*) on the adsorption and desorption branches of the N_2 isotherm. Transmission electron microscopy was conducted using a JEOL JEM-2011 electron microscope operated at an accelerating voltage of 200 kV. The powder was placed directly on a Formbar-backed carbon-coated copper grid. Isoelectric points were measured using a Malvern Zetasizer 3000HSA.

The CNS-trypsin nanoreactors were generated by mixing trypsin with a suspension of CNS (final concentration of 1 mg/ml, in 25 mM potassium phosphate, pH 6.5) at room temperature for 16 h. The amount of trypsin adsorbed was 200 ng/μg CNS. All tryptic digests were analyzed on an Applied

Biosystem 4700 proteomics Analyzer. The PMF spectrum of each spot was obtained by accumulation of 2500 laser shots and the laser intensity was 5500.

Results and Discussion

The pore diameter and structure of MPS are illustrated in the TEM images obtained for SBA-15 in Figure 1. MPS with pore diameters of over 20 nm can be synthesized, a size sufficiently large to accommodate large proteins and enzymes.

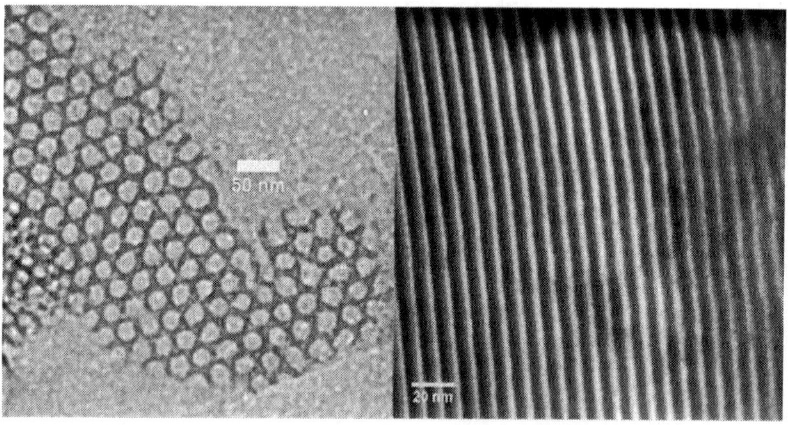

Figure 1. TEM images showing (A) the hexagonal pore structure and (B) the channels in SBA type materials.

In order to ascertain that a particular protein will adsorb onto MPS, it is necessary to first compare the relative sizes of the pores and the protein. Direct proof of inclusion of a protein into the pores is not simple. Comparison of the loading obtained on MPS with a pore diameter larger than that of the protein to that of MPS with a smaller pore diameter is the most facile, though indirect method. Zhi et al. (*31*) have used TEM to show that ferritin, which contains a large core of iron (6 nm diameter), is located in the interior of boron nitride nanotubes. Such an approach is not possible with the majority of enzymes which can not be visualized by TEM. From adsorption studies using cytochrome c and a range of MPS, we have shown that provided the pore diameter is sufficiently large to accommodate the protein, more than 90% of the protein loading is accommodated within the pores. Elliposmetric analysis of the amount of cytochrome c adsorbed onto an MPS film with pore diameters smaller than that of cytochrome c showed that between 5 and 10 layers of the protein could be adsorbed on to the exterior surface of the MPS (*32*).

The charge of the protein and of the MPS can play crucial roles in the adsorption process. With proteins, it is not just the magnitude of the charge, but also the charge distribution. Figure 2 shows the charge distribution of three proteins, cytochrome c, trypsin and choroperoxidase. While these proteins range from positively charged (cytochrome c) to approximately neutral (trypsin) and negatively charged (chloroperoxidase) at pH 7.0, MPS are negatively charged, though the magnitude of their charge varies. Adsorption of protein into the pores can, under optimal conditions occur rapidly, as illustrated in figure 3, where SBA-15 becomes saturated with cytochrome c in a matter of minutes (18). The loading achieved is of the order of 300 mg/g, illustrating the adsorptive capacity of MPS. The highest loading was achieved at pH 10 where cytochrome c bears an overall charge close to neutral. Adsorption is achieved by incubation of the protein with MPS; the extent of adsorption is then monitored by following the absorbance at the appropriate wavelength. It is important to ascertain that the protein itself is stable under the adsorption conditions, something that has not always been monitored in these types of adsorption studies. For instance, a protein such as cytochrome c shows a stable absorbance in solution for periods of 24 hours or more, while this is not the case for unstable enzymes such as xylananse, as has been previously documented (18). Even with the relatively stable protein, myoglobin, the adsorption process can not last longer than 24 hours as after this time period the protein will start to aggregate and precipitate from solution.

Screening of the parameters that affect adsorption can be rapidly performed using a 96 well plate format (*33*). Figure 4 shows how the effect of ionic strength can be ascertained for a range of conditions and demonstrates that electrostatic effects are dominant in the adsorption process (with the exception of MSE and PMO materials), while varying in degree for different MPS.

Introduction of alkyl and modified groups has the potential of assisting in the adsorption of more hydrophobic proteins and increasing the activity of the adsorbed enzyme. Trypsin adsorbs on to cyano-modified MPS and remains highly active. The adsorbed protein retains and somewhat surprisingly is slightly more active than the enzyme in solution. In nonaqueous solvents the adsorbed enzyme is significantly more active and more thermally stable than the corresponding lyophilized powder form (*34-35*). Similar results have been observed with cytochrome c (*36*).

Tryspin-CNS has been used to digest myoglobin (*37-38*), with analysis of the proteolytic products by mass spectrometry (figure 5). After a 10 minute digestion period using trypsin-CNS, comparable results to in-solution digestion (12 hours) were obtained, demonstrating that the rate of reaction was significantly accelerated.

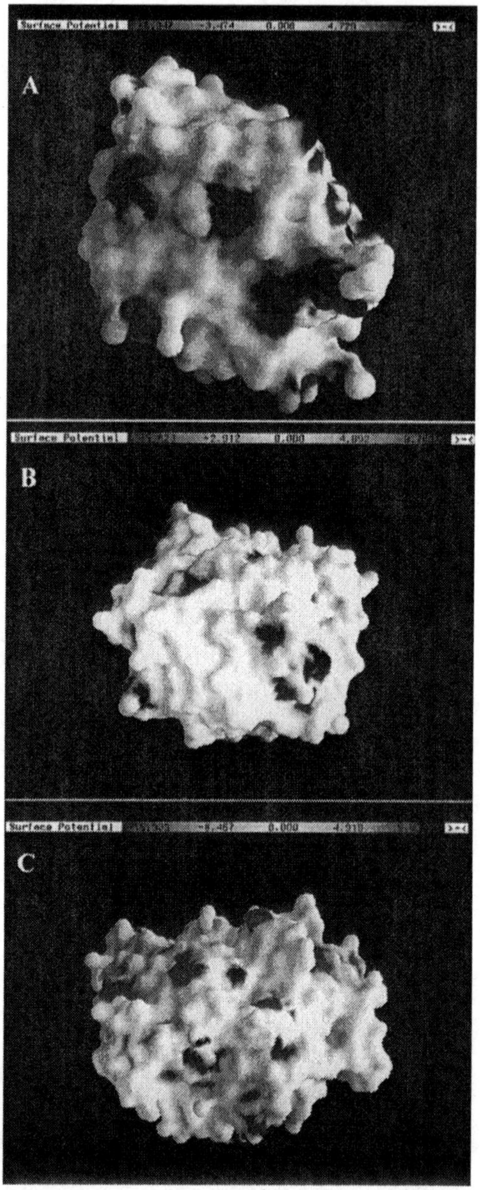

Figure 2. Poisson-Boltzmann electrostatic potentials of (A) cytochrome c, (B) trypsin and (C) chloroperoxidase calculated by GRASP (39).

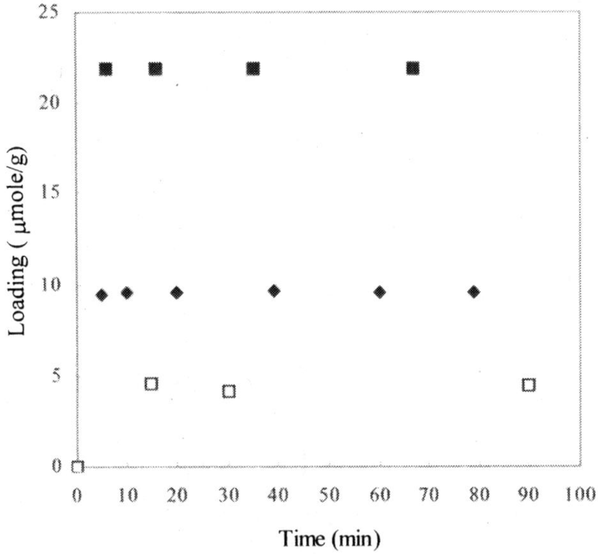

Figure 3. Plot of amount of cytochrome c loaded onto SBA-15 at pH 4 (□), 7 (♦) and 10 (■).

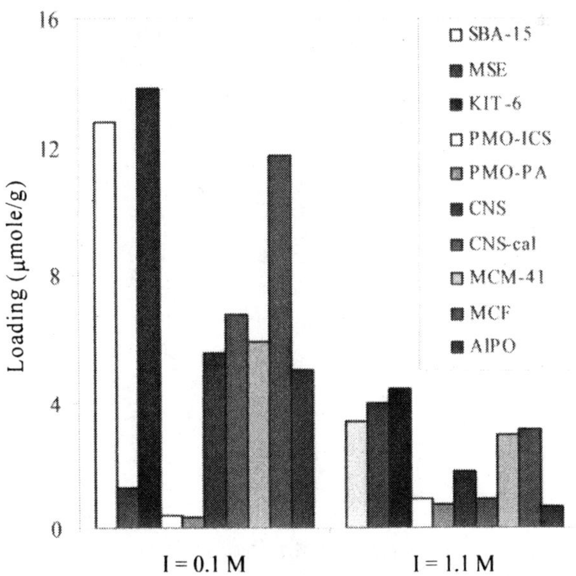

Figure 4. Effect of ionic strength (I) on the amount of cytochrome c adsorbed on to MPS, as determined using a 96 well plate format.

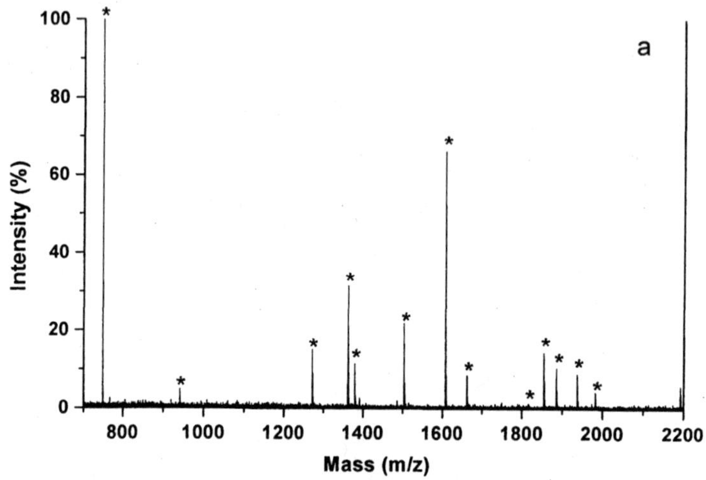

Figure 5. Peptide mass fingerprint spectra of myoglobin proteolysis products after protein digestion by CNS-trypsin (10 min).

MPS encapsulated enzymes display significant potential as bioreactors with MPS providing significant increases in enzyme stability and catalytic efficiency. We have recently outlined a detailed protocol (Figure 6) that describes the steps required to successfully immobilise an enzyme; in principle this protocol may be applied to any combination of enzymes and porous supports. Current work is focussing on using this protocol to develop stable biocatalysts for use in enantio- and region-selective transformations.

Acknowledgement

Financial support of this work by Enterprise Ireland, ITCSET and the Royal Irish academy is gratefully acknowledged. Prof. S. Nakahara and Dr. D. Tanner are thanked for the TEM images.

References

1. *Immobilised Enzymes*; Chibata, I.; Hastel Press: New York, 1978.
2. *Biocatalysis, Fundamentals and Applications*: Bommarius, A.S., Riebel, B.R.: Wiley-VCH, Weinheim, 2004.

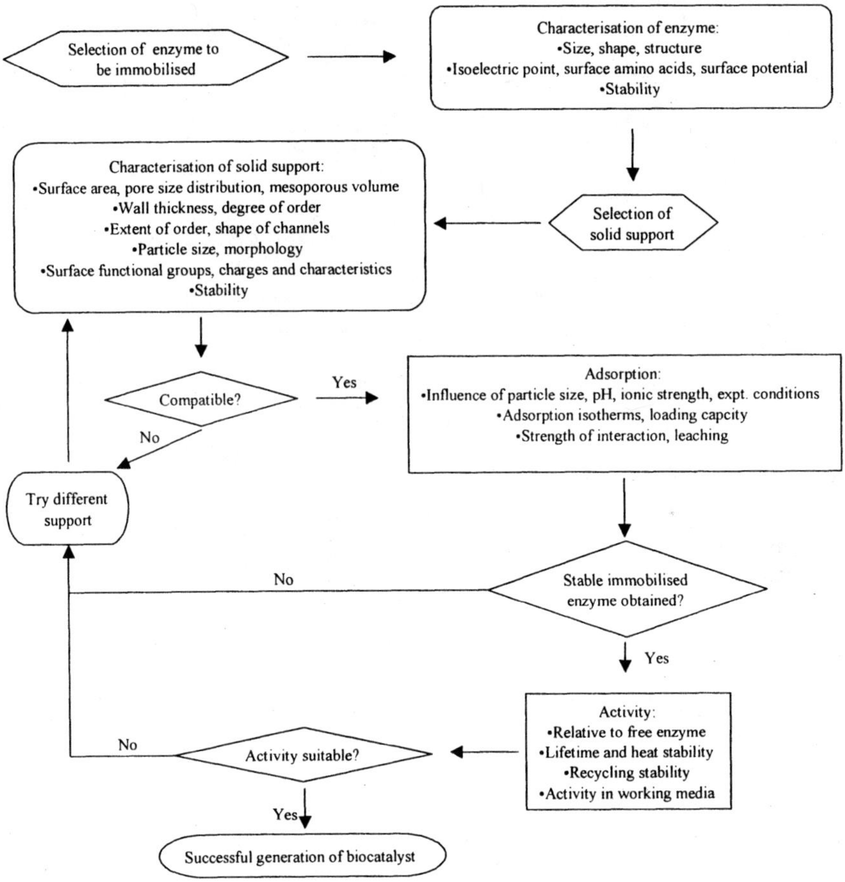

Figure 6. Protocol for the generation of novel immobilised enzyme biocatalysts (18).

3. Beck J.S.; Roth W..J.; Leonowicz, M.E.; Kresge, C.T.; Scmitt, K.D.; Chu, C.T-W.; Olson, D.H.; Sheppard, E.W.; McCullen, S.B.; Higgins, J.B.; Schlenker, J.L A new family of mesoporous molecular sieves prepared with liquid crystal templates. *J. Am. Chem. Soc.* **1992**, *114*, 10834.
4. Grosso, D.; Balkenende A.R.; Albouy; P.A; Ayral, A.; Amenitsch, H.; Babonneau, F. Two-dimensional hexagonal mesoporous silica thin films prepared from block copolymers: Detailed characterization and formation mechanism. *Chem. Mater.* **2001**, *13*, 1848.
5. Soler-Illia, G.J.; Louis, A.; Sanchez, C. Synthesis and characterisation of mesostructured titania-based materials through evaporation-induced self-assembly. *Chem. Mater.* **2002**, *14*, 750.
6. Tian, B. et al. Self-adjusted Synthesis of ordered stable mesoporous minerals by acid–base pairs. *Nature Materials* **2003**, *2*, 159-163.
7. Schulz-Ekloff, G.; Wöhrle, D.; van Duffel, B.; Schoonheydt, R. A. Chromophores in porous silicas and minerals: preparation and optical properties. *Microporous Mesoporous Mater.* **2002**, *51*, 91.
8. Scott, B. J.; Wirnsberger, G.; Stucky, G. D. Mesoporous and mesostructured materials for optical applications. *Chem. Mater.* **2001**, *13*, 3140.
9. Lim, M. H.; Blanford, C. F.; Stein, A. Synthesis and characterisation of a reactive vinyl-functionalised MCM-41: Probing the internal pore structure by a bromination reaction. *J. Am. Chem. Soc.* **1997**, *119*, 4090.
10. Huh, S.; Wiench, J. W.; Yoo, J.-C.; Pruski, M.; Lin, V. S.-Y. Organic functionalization and morphology control of mesoporous silicas via a co-condensation synthesis method. *Chem. Mater.* **2003**, *15*, 4247.
11. Gill, I.;, Ballesteros, A. Encapsulation of biologicals within silicate, siloxane, and hybrid sol-gel polymers: An efficient and generic approach. *J. Am. Chem. Soc.*, **1998**, *120*, 8587-8598.
12. Gill, I.; Ballesteros, A. Bioencapsulation within synthetic polymers; sol-gel encapsulated biologicals. *Trends Biotech.* **2000**, *18*, 282.
13. Weetall, H.H. Preparation of immobilised protein covalently coupled through silane coupling agents to inorganic supports. *Appld. Biochem. Biotech.*, **1993**, *41*, 157
14. Diaz, J. F.; Balkus, J. K. Enzyme immobilization in MCM-41 molecular sieve *J. Mol. Catal. B: Enz.* **1996**, *2*, 115
15. Fiorilli, S.; Onida, B.; Bonelli, B.; Garrone, E. In situ infrared study of SBA-15 functionalized with carboxylic groups incorporated by a co-condensation route. *J. Phys. Chem. B* **2005**, *109*, 16725.
16. Inagaki, S.; Guan, S.; Fukushima, Y.; Ohsuna, T.; Terasaki, O. Novel mesoporous materials with a uniform distribution of organic groups and inorganic oxide in their frameworks. *J. Am. Chem. Soc.* **1999**, *121*, 9611.
17. Deere, J.; Magner, E.; Wall, J. G.; Hodnett, B. K. Mechanistic and structural features of protein adsorption onto mesoporous silicates. *J. Phys. Chem. B* **2002**, *106*, 7340.

18. Hudson, S.; Magner, E.; Cooney, J.; Hodnett, B. K. Methodology for the immobilization of enzymes onto mesoporous materials. *J. Phys. Chem. B* **2005**, *109*, 19496-19506.
19. Deere, J.; Magner, E.; Wall, J. G.; Hodnett, B. K. Adsorption and activity of cytochrome c on mesoporous silicates. *Chem. Commun.* **2001**, 465.
20. Vinu, A.; Murugesan, V.; Tangermann, O.; Hartmann, M. Adsorption of cytochrome c on Mesoporous Molecular Sieves: Influence of pH, Pore Diameter, and Aluminum Incorporation. *Chem. Mater.* **2004**, *16*, 3056.
21. Yiu, H. H. P.; Wright, P. A.; Botting, N. P. Enzyme immobilisation using SBA-15 mesoporous molecular sieves with functionalised surfaces. *J. Mol. Catal. B: Enz.* **2001**, *15*, 81.
22. Fan, J.; Lei, J.; Wang, L.; Yu, C.; Tu, B.; Zhao, D. Rapid and high-capacity immobilization of enzymes based on mesoporous silicas with controlled morphologies. *Chem. Commun.* **2003**, 214
23. Zhao, D.; Huo, Q.; Feng, J.; Chmelka, B. F.; Stucky, G. D. Nonionic triblock and star diblock copolymer and oligomeric surfactant syntheses of highly order, hydrothermally stable, mesoporous silica structures. *J. Amer. Chem. Soc.* **1998**, *120*, 6024.
24. Bao, X. Y.; Zhao, X. S.; Li, X.; Chia, P. A.; Li, J. A novel route towards the synthesis of high-quality large-pore periodic mesoporous organosilicas. *J. Phys. Chem. B* **2004**, *108*, 4684.
25. Olkhovyk O.; Jaroniec M., Periodic mesoporous organosilica with large heterocyclic bridging groups, *J. Am. Chem. Soc.* **2004**, *127*, 60.
26. Kleitz F.; Choi S. H.; Ryoo R.; Cubic *Ia3d* large mesoporous silica: synthesis and replication to platinum nanowires, carbon nanorods and carbon nanotubes. *Chem. Commun.* **2003**, 2136.
27. Katiyar A.; Ji L.; Smirniotis P.; Pinto N.G. Protein adsorption on the mesoporous molecular sieve silicate SBA-15: effects of pH and pore size, *J Chrom. A* **2005**, *1069*, 119.
28. Wang L.; Tian B.; Fan J.; Liu X.; Yang H.; Yu C.; Tu B.; Zhao D. Block copolymer templating syntheses of ordered large-pore stable mesoporous aluminophosphates and Fe-aluminophosphate based on an "acid–base pair" route. Micro. Meso. Mater. **2004**, *67*, 123.
29. Brunauer, S.; Emmett, P.H.; Teller, E. Adsorption of gases in multimolecular layers *J. Amer. Chem. Soc.* **1938**, *60*, 309.
30. Barrett, E.P.; Joyner, L.G.; Halenda, P.P. The determination of pore volume and area distributions in porous substances. I. Computations from nitrogen isotherms *J. Amer. Chem. Soc.* **1951**, *73*, 373.
31. Zhi, C.; Bando, Y.; Tang, C.; Golberg, D. Immobilization of proteins on boron nitride nanotubes, *J. Am. Chem. Soc.*, **205**, *127*, 17144.
32. Deere, J.; Serantoni, M.; Edler, K.J.; Wall, J.G.; Hodnett, B.K.; Magner, E. Measurement of the adsorption of cytochrome c onto the external surface of a thin film mesoporous silicate by ellipsometry. *Langmuir*, **2004**, *20*, 532.

33. Hudson, S.; Cooney, J.; Hodnett, B.K.; Magner, E. Manuscript in preparation.
34. Deere, J., Magner, E., Wall, J.G., Hodnett, B.K. Oxidation of ABTS by silicate immobilized cytochrome c in nonaqueous solutions, *Biotech. Prog.*, **2003**, *19*, 1238-1243.
35. Goradia, D., Cooney, J., Hodnett, B.K., Magner, E. Characteristics of a mesoporous silicate immobilized trypsin bioreactor in organic media. Submitted to *Bioproc. Technol.* 2005
36. Gomez, J. M. Deere, J., Magner, E., Wall, J.G., Hodnett, B.K. Transesterification catalyzed by trypsin supported on MCM-41. *Catal. Lett.* **2003**, *88*, 183.
37. Liu, Y., Lu, H., Zhong, W., Song, P., Kong, J., Yang, P., Girault, H. H., Liu, B. Multi layer-assembled microchip for enzyme immobilization as reactor towards low-level protein identification. *Anal Chem.* **2006**, *78*, 801.
38. Wu, H., Tian, Y., Liu, B., Lu, H., Wang, X., Zhai, J., Jin, H., Yang, P., Xu, Y., Wang, H. Titania and alumina sol-gel-derived microfluidics enzymatic-reactors for peptide mapping: Design, characterization, and performance. *J. Proteome Res.* **2004**, *3*, 1201.
39. Nicholls, A.; Sharp, K.A.; Honig, B. Protein folding and association - insights from the interfacial and thermodynamic properties of hydrocarbons *Proteins* **1991**, *11*, 281.

Chapter 3

Enzyme Stabilization Involving Molecular Evolution and Immobilization in Mesoporous Materials

Haruo Takahashi and Chie Miyazaki-Imamura

Biotechnology Laboratory, Toyota Central R&D Laboratories, Inc., 41–1, Yokomichi, Nagakute, Aichi 480–1192, Japan

Enzymes (proteins) were successfully stabilized in mesoporous materials. When the pore diameter of mesoporous material was controlled to nearly the same as that of the enzyme, the immobilized enzyme sometimes showed high stability. The surface amino acid residues would form hydrogen bond with silanol residues of mesoporous materials during the immobilization. The enzymes could be well immobilized and stabilized at the low pH region below pI (isoelectric point) of the enzyme. A mutant manganese peroxidase (MnP) library involving three randomized amino acid residues in the entry site of the H_2O_2-binding pocket of MnP was produced on a multi-well plate using SIMPLEX (single-molecule PCR-linked in vitro expression). The screening of more than 10^4 samples independently expressed for improved H_2O_2 stability gave four positive mutants, of which the H_2O_2 stability was up to nine times higher than that of the wildtype. An immobilized MnP mutant in mesoporous material showed high H_2O_2 stability, i.e. more than 50-fold that of the wild-type MnP. But the stability of the immobilized wild-type MnP was not improved as much as that of the immobilized mutant MnP. MnP immobilized in mesoporous materials could be applied to a ligneous bleaching system.

Introduction

In recent years, periodic mesoporous materials with uniform pore diameters of 2~30 nm have been synthesized (1-10). Because the pore diameters of these materials approximate those of enzymes molecules, their application as enzyme supports has been suggested.

It has been shown that the surface characteristics of mesoporous materials, and matching of the sizes of protein molecules and the pore diameters of mesoporous materials are essential for protein stabilization (11-14). We have also reported that MnP was successfully stabilized in a mesoporous material (FSM) when the mesopore size of FSM was nearly the same as the diameter of the enzyme (15). MnP catalyzes the oxidation of Mn^{2+} to Mn^{3+} utilizing H_2O_2 and forms a complex with an organic acid. A Mn^{3+}-chelate complex is a highly reactive non-specific oxidant capable of oxidizing a variety of environmental pollutants (16-18). However, MnP is very sensitive to inactivation by H_2O_2 or thermal treatment. An attempt to increase the thermo-stability of MnP has been reported [19], but one to increase its resistance to H_2O_2 had not been reported.

On the other hand, cell-free protein synthesis systems allow the rapid production of proteins directly from plasmid or polymerase chain reaction (PCR)-amplified DNA templates. Another striking advantage of the system is the easy modification of the composition of the reaction mixture according to the requirements for the synthesis of a protein. However, functional expression of heme-containing proteins has not been successful. We have succeeded in synthesizing a functional MnP using a modified *E. coli in vitro* protein synthesis system with disulfide-forming catalysts involving protein disulfide isomerase (PDI) and various chaperones (22).

Recently, Nakano et al. developed single-step single-molecule PCR (SM-PCR) with a homo-primer for amplification from a single-molecule DNA template and proposed a novel protein library construction system named SIMPLEX (single-molecule-PCR-linked *in vitro* expression)(20). The correspondence of a genotype (SM-PCR product) with a phenotype (*in vitro* synthesized protein) has made it possible to screen a large range of mutant enzymes and to increase the enzymatic function efficiently. We have increased the H_2O_2 stability by means of SIMPLEX technology exclusively on 384-well plates. The whole process takes only ~8 h from the beginning of the SM-PCR to the end of the screening, and only a few days were required to screen 10^4 independent clones. The high-throughput construction of mutant MnPs and screening in the presence of H_2O_2 resulted in the finding of mutants with improved stability against H_2O_2 (21-22).

In this paper, we present an excellent enzyme (peroxidase) stabilizing strategy involving molecular evolution and immobilization in mesoporous materials.

Nano-size materials and bio-molecules

Figure 1 shows nano-size materials and bio-molecules. The size of a hydrogen atom is 0.1nm. The small axis of a DNA molecule is about 2nm, and the sizes of protein range from about 2 to 30nm. The diameter of a virus is about 100 nm, and that of a bacterium (*E. coli)* is about 2μm. In materials, the size of fullerene is 0.7nm and that of a carbon nano-tube is about 3nm. On the other hand, the pore diameters of mesoporous materials range from about 2 to 30nm. The pore diameters of mesoporous materials are nearly the same as the sizes of protein molecules.

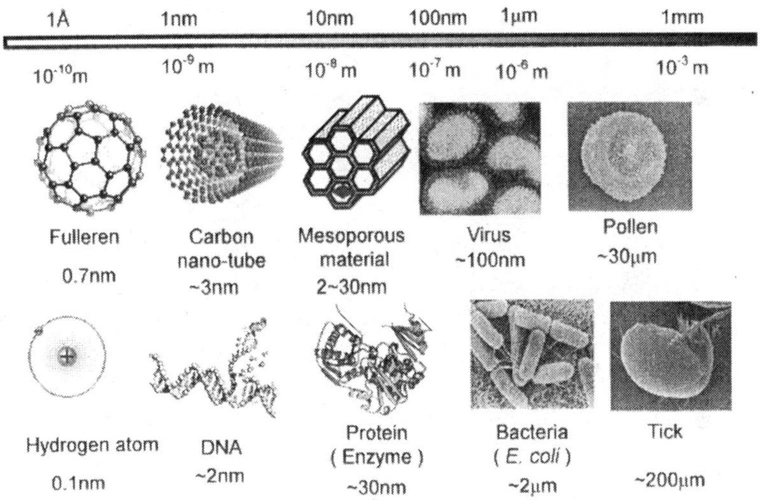

Figure 1. Nano-size materials and bio-molecules

Enzyme immobilization in mesoporous materials

The synthesis (A) and crystal structures (B) of mesoporous materials are shown in Figure 2. Mesoporous materials (FSM) were prepared using hexadecyltrimethylammonium chloride and 1, 3, 5-triisopropylbenzene (TIPB) as described by Inagaki et al. (4). On the surface of micelles, silica molecules became self assembled and the mesoporous materials could be obtained after removal of the surfactant. In recent years, Inagaki et al. synthesized organic-inorganic hybrid type mesoporous materials. Nanosize controlled uniform pores (2- or 3-dimensional hexagonal structure) were formed. When the pore diameter is larger than that of an enzyme molecule, the enzyme molecule would be immobilized in the pore.

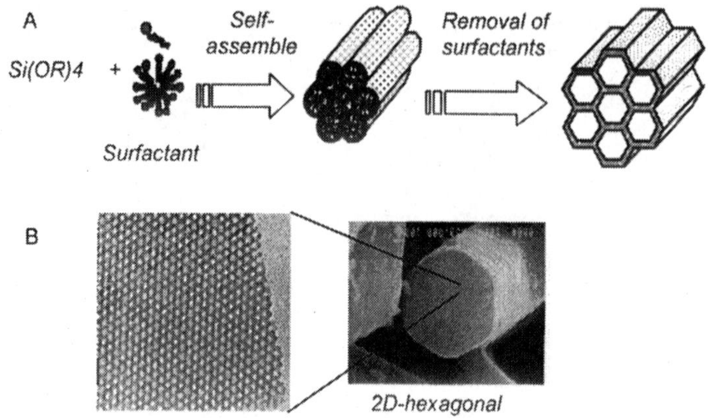

Figure 2. Typical synthesis procedure and structure of mesoporous materials. Silica molecules are self assembled on the surface of micelles and mesoporous materials are obtained after removal of the surfactant.

To determine the effect of pH on enzyme immobization, the pH profile of the adsorbed amount of HRP for mesosorous materials was obtained. The iso-electric point of HRP is 6.2, so a HRP molecule has a positive net charge in a lower pH region, and the charge density gradually decrease with increasing pH. The adsorbed amount of HRP decreases with increasing pH, but in a high pH region, the adsorbed amount does not change so much.

In other experiments, it was concluded that ionic characteristics and ionic strength were important factors for enzyme immobilization. A Mesoporous material (MCM or FSM etc.) was generally prepared with an cationic surfactant, so the cationic substance might be specifically adsorbed in the mesoporous material. Therefore, the large immobilized amount of enzyme (up to 200mg enzyme/ g support) in the pH region below iso-electric point (pI) decreased with increasing pH.

A general procedure for enzyme immobilization in mesoporous materials is shown in Figure 3. First, the axis of an enzyme molecule and its iso-electric point (pI) was measured. The pore diameter of a mesoporous material was controlled at the larger size of an enzyme molecule. By gently mixing in pH region under pI, the enzyme molecule was immobilized in the pores of the material. Hydrogen bonds would be formed between the surface amino residues of the enzyme and silanol residues of the mesoporous material during immobilization, and thus the enzyme was excellently stabilized. When the pore diameter of the mesoporous material was larger than the size of the enzyme, the enzyme would be immobilized in the pores.

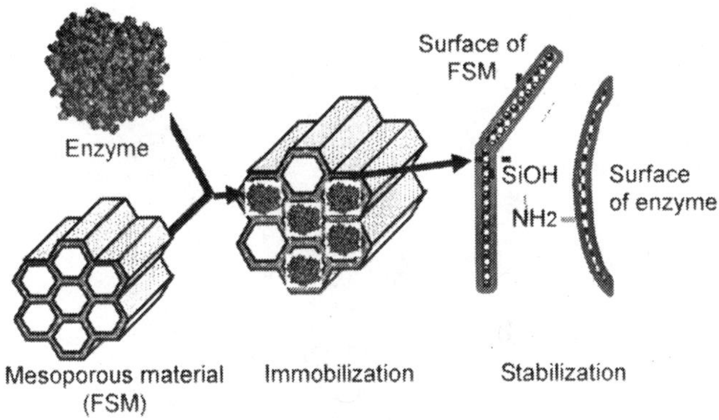

Figure 3. Typical procedure of enzyme stabilization in mesoporous materials.

Image models of immobilized enzymes in mesoporous materials with various pore sizes were constructed using a computer schematic model (Figure 4). When the pore diameter is much smaller than the enzyme molecular size, the enzyme can not enter the mesopore space of material and thus is immobilized on the outside of the particles. When the pore size greater than the enzyme molecular size, the enzymes would be adsorbed on one side of the mesopores. When the pore size is nearly the same as the size of the enzyme molecule, the enzyme can be adsorbed in the mesopores and is sometimes the most stabilized.

Stabilization effect of immobilized bio-molecules in mesoporous materials

Enzyme stabilization

To analyze the effect of the pore size on enzyme stability, HRPs in mesoporous materials with various pore sizes were investigated (Figure 4). The residual activities after thermal treatment with the native HRP, immobilized HRPs and silica gel were compared. Free HRP (i.e. without immobilization) was completely inactivated within 60 minutes at 70°C. When the pore diameter is much smaller than the enzyme molecular size, a HRP immobilized in a mesoporous material is inactivated by more than 50% within 60min, as in the case of silica gel. When the pore diameter is much greater than the size of an enzyme, the immobilized HRP activity was decreased by about 40% after 120min thermal treatment. When the average mesopore size just matched the molecular diameters of enzymes (6nm size), the enzyme molecules showed the best

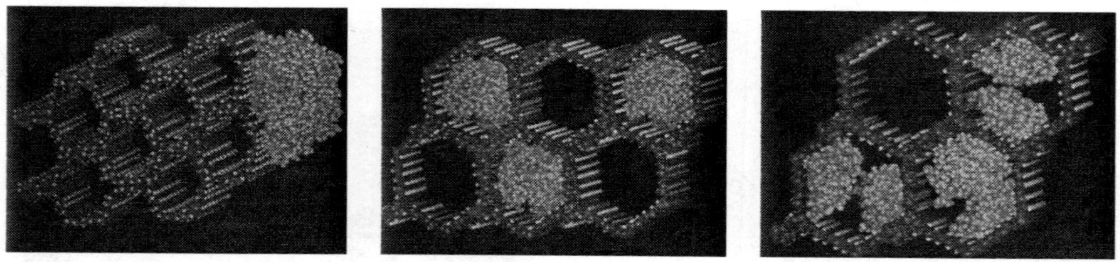

Figure 4. Image models of immobilized enzymes in mesoporous materials with various pore sizes. A: the pore diameter is much smaller than enzyme size, B; the pore diameter is nearly the same size of enzyme molecule C; the pore diameter is much larger than enzyme size

stability, and the immobilized HRP activity was decreased by only 20% even after 120min thermal treatment.

Similar results were obtained in the case of MnP (manganese peroxidase). The H_2O_2 stability of MnP in mesoporous materials with various pore sizes was also investigated. MnP immobilized in mesoporous materials was suspended in the MnP assay buffer (pH 4.5) containing H_2O_2. The suspensions were mixed vigorously at 37°C for 5 minutes. After centrifugation, the supernatants were immediately subjected to MnP activity measurement. With respect to the H_2O_2 stability of free MnP, the optimal H_2O_2 concentration was 0.02 mM, and its activity was completely lost with 1 mM hydrogen peroxide. For MnP immobilized in silica gel, the optimal H_2O_2 concentration was 0.1 mM, and its activity decreased steeply with concentrations of more than 0.5 mM H_2O_2, and it was completely inactivated with over 1 mM. MnP immobilized in the pore size controlled mesoporous materials exhibited high enzymatic activity with 0.1 mM to 3 mM H_2O_2.

Chlorophyll stabilization

Ito et. al also immobilized the chlorophyll molecules extracted from natural *Spirulina* in mesoporous materials (FSM). The chlorophyll-FSM conjugates in water were much more photo-stable in comparison with free chlorophyll. The photo-stabilization values of the chlorophyll conjugates with porediameters 1.6, 2.3, 2.7 and 4.0 nm were investigated. The photo-stability of chlorophyll in FSM pores increased with the pore size (31).

When pore diameter was greater than 2.7nm, two chlorophyll molecules yielded a chlorophyll dimer in the mesopore. The dimer structure was similar to the native form, so the photo-stability would drastically increased between 2.4 and 2.7 nm pore diameter.

Enzyme evolution

E. coli cell-free protein synthesis system

Figure 5 shows the strategy of Molecular evolution using single-molecule PCR-linked *in vitro* expression for short SIMPLEX. The pool of mutated genes is diluted to a single molecule per well, followed by individual amplification by SM-PCR. A mutated gene library is constructed on the plate. A portion of the PCR product is used as a template for cell-free expression, and a mutated protein library is thereby constructed. Subsequently, the protein library is screened under desirable condition. If we can get positive wells, the correspondent DNA sequences can be determined. With this system, the genotype and phenotype can be linked according to the well number.

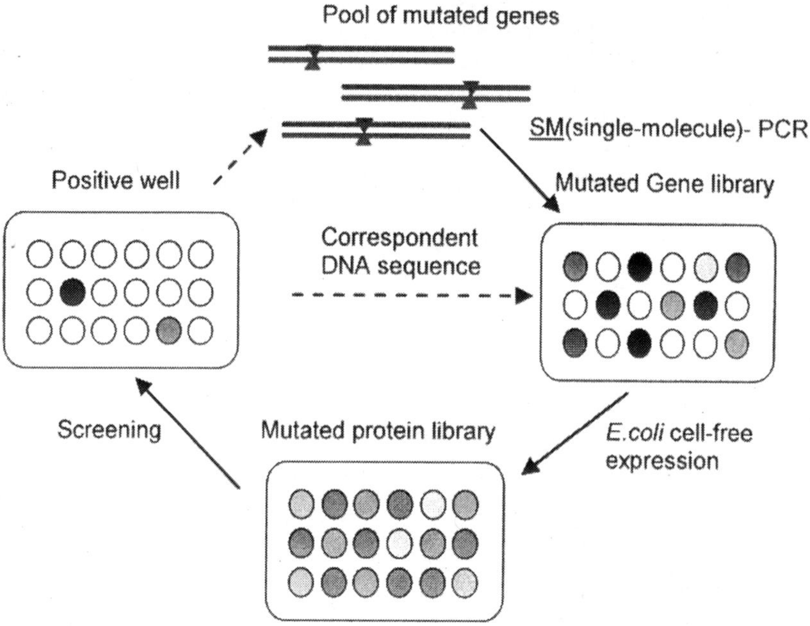

Figure 5. Schematic representation of molecular evolution involving SIMPLEX. The pool of mutated genes is diluted to a single molecule per well followed by individual amplifications by SM-PCR. A.mutated gene library is constructed on the plate. A portion of the PCR product is used as a template for cell-free expression, and a mutated protein library is thereby constructed. Subsequently, the protein library is screened under desirable conditions.

Compared with conventional colony-based screening, SIMPLEX has various benefits because of the use of a cell-free system. Transformation and the cultivation step are not necessary, and selection by means of enzyme assaying is possible, so the total time required for screening is very short, i.e. only 8hrs. Genotype, which is a SM-PCR plate can be frozen for long-term storage in a stable state. Miss folding is a serious problem when heterologous proteins are expressed, for example disulfide bridge or heme containing proteins. To address this issue, we attempted to improve the cell-free expression system, i.e. cell-free expression using chaperone containing extract was examined. GroES/EL operone and DnaK/J and GrpE operone were highly expressed in *E.coli*. When beta-glucosidase (BGL) from *streptmyces* was used as a template, BGL activity in chaperone containing extract increased 16 times higher than that of untreated one.

MnP synthesis with an E. coli cell-free protein synthesis system

MnP has five S-S bonds, Mn^{2+}, Ca^{2+} and heme as the catalytic center. Due to this complicated structure, insoluble inclusion bodies are formed on *E. coli in vivo* expression. For preparation of the *E. coli* S30 extract for MnP synthesis, a reducing agent, dithiothreitol (DTT), was omitted to maintain oxidizing conditions for the formation of the disulfide bridges of the protein, because functional MnP is not formed in the presence of a strong reducing agent (22). To enhance the yield of functional MnP, reduced and oxidized glutathione (GSH and GSSG), PDIs and some chaperones were added to the *in vitro* expression mixture. GSH/GSSG was added to activate the active site of PDI. The addition of only chaperones (DnaK/J, GroES/EL and GrpE) hardly affected the MnP activity. When fungal PDI was present in the reaction mixture, the MnP activity increased up to 4-fold compared with in the case of a control mixture without PDI or chaperones, but interestingly bovine PDI had no such effect. Although MnP activity was high when both chaperones and fungal PDI were added, it was suggested that this was mainly an effect of the fungal PDI. Although an *E.coli* disulfide-forming catalyst, DsbA might be included in the S30 extract, it dose not seem effective in producing active MnP. It is considered that fungal PDI, which has both isomerase and chaperone activities (26), and facilitates the folding of denatured and reduced SH-proteins, played an important role in the folding of MnP into the native form with the correct disulfide formation in this cell-free system.

Design of a mutant MnP library

The three-dimensional structure of MnP isozyme 1 (PDB:1MNP) has been well characterized by X-ray crystallography (27). We have constructed a model of MnP isozyme 2 by means of the homology modeling method using 1MNP as the backbone for the starting structure (21). A structural model of the H_2O_2-binding pocket of MnP is shown in Figure 6 (A). Amino acid residues located near His46 and Arg42, which are conserved in H_2O_2-binding residues (27), were picked as candidates for substitution for the library. Conserved amino acid residues in various peroxidases were excluded as candidates for substitution. To bring about a minute conformational change around the entry site of the H_2O_2-binding pocket, three amino acid residues, i.e., A79, N81 and I83, were selected for replacement with all kinds of amino acid residues (A79X,N81X and I83X). The theoretical number of mutated MnPs with different sequences was 8,000.

Screening of the MnP library

We have thus constructed a mutant MnP library containing randomized A79, N81 and I83 by SIMPLEX and screened more than 10^4 wells by measuring MnP activity in the presence of 1.0mM H_2O_2, with which the wild type exhibited no detectable activity. Fifteen clones showed improved H_2O_2 stability and their whole sequences were confirmed. The clones were classified into four types of sequences according to the amino acid substitutions. The amino acid substitutions and the H_2O_2 stability of the four types of clones are shown in Table I. In the presence of 0.1 mM H_2O_2, the half-life of a type 1 clone was nine times higher than that of the wild type. Also, other clones were five times more stable than the wild type. Furthermore, even in the presence of higher concentrations of H_2O_2 of 0.5 and 1.0mM, all the four types of clones showed significantly higher stability than the wild-type. These four clones have the common substitution of I83L. In addition, A79 was changed to either E or S and N81 for S or L. Since these three amino acid residues were converged from a large variety in the library as a result of selection, this high-throughput screening system is likely to work well. Fine tuning of the local arrangement in the H_2O_2-binding pocket seems to be effective for improving the H_2O_2 stability of MnP.

Table 1. Amino acid substitutions in each mutant

Clone	Amino acid			H_2O_2 stability*		
	79	81	83	0.1mM	0.5mM	1.0mM
Wild type	A	N	I	1.0	1.0	1.0
Clone 1	E	S	L	9.0	6.6	3.9
Clone 4	S	L	L	7.2	6.5	4.4
Clone 6	S	S	L	5.3	4.6	3.2
Clone 8	E	L	L	7.0	3.5	1.8

The H_2O_2 concentrations which gave 50% relative activity are shown. Mutant/wild-type ratio of H_2O_2 concentrations for % residual activity after treatment for 60min at 37°C in the presence of 1mM H_2O_2.

Synergetic effect of H_2O_2 stability

A structural model of H_2O_2 -binding pocket of a MnP molecule (A,B) , in which H_2O_2 was directly supported by His46 and Arg42, and an image model of the immobilized mutant MnP in a mesoporous material (C) are shown in Figure 6. The improved oxidative stabilities of the immobilized native MnP(A) and the immobilized H_2O_2 resistant MnP mutant (B) are shown in Figure 7. The mutant MnP, as to the entry site of the H_2O_2 binding pocket to a stable amino acid residue, showed improved stability against H_2O_2 compared with native MnP (21,22). The immobilized H_2O_2 resistant mutant MnP in a mesoporous material showed excellent H_2O_2 stability (more than 50-fold that of wild type MnP). But the stability of the immobilized wild-type MnP was improved up to 5-fold that of wild type MnP. MnP mutant was stabilized by protein evolution. The whole MnP molecule was immobilized in mesoporous material. Unstable amino acids existing around the surface region would be protected on immobilization in a suitable mesoporous material. But the unstable amino acids facing the H_2O_2 -binding pocket were not protected by mesoporous materials, so protein evolution around the H_2O_2 binding pocket would be effective.

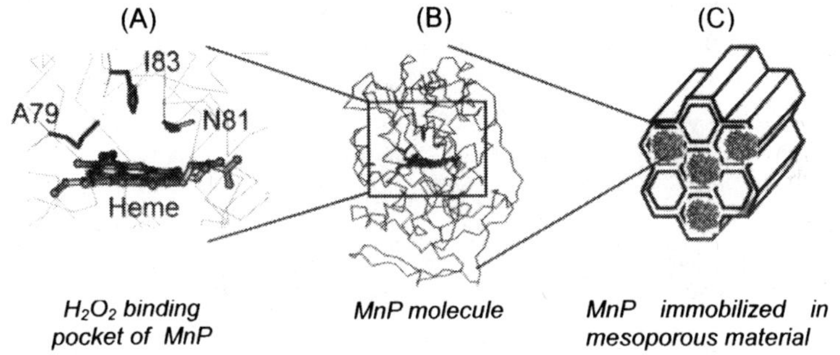

Figure 6. Structural model of the H_2O_2-binding pocket (left), a MnP molecule (middle), and an image model of the immobilized MnP in a mesoporous material (right)

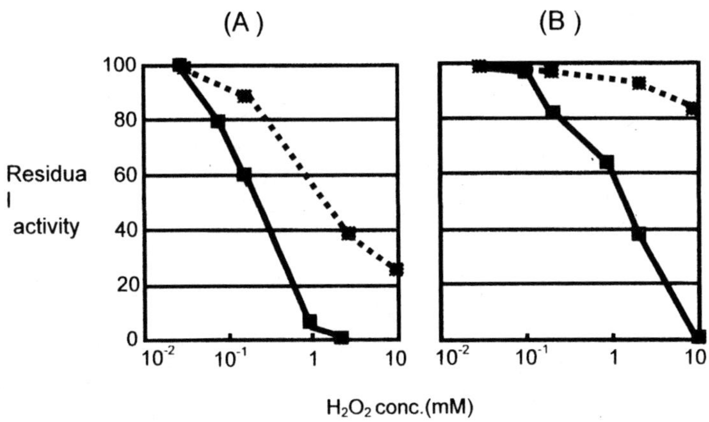

Figure 7. Synergetic effect on H_2O_2 stability.
The oxidative stability of the wild type enzymes (A) and mutant enzymes (B) either non treated (straight line) or immobilized in a mesoporous material (dotted line) is shown. Enzyme mixtures were incubated in the presence of various concentrations of H_2O_2 and then residual activity was measured.

Application to a ligneous bleaching system

We have newly designed a thermally discontinuous immobilized MnP bleaching system as shown in Figure 8 (15), which the Mn^{3+} generation step on an enzyme-mesoporous material column and the bleaching step are separate. In the first stage, the substrate solution comprising Mn^{2+}, hydrogen peroxide, and an organic acid (chelating agent) is introduced into an enzyme column packed with immobilized-MnP to generate a Mn^{3+}-chelate complex. In the next stage, the Mn^{3+}-chelate complex generated with the immobilized MnP is transferred to a bleaching vessel containing unbleached kraft pulp. The reactivity of the Mn^{3+}-malonate complex depends on the temperature. When the MnP reaction and pulp bleaching reaction are performed at 39°C and 70 °C, respectively, MnP activity is maintained throughout the reaction and the brightness of the pulp after 9 hours is increased by over 8 points. The MnP reaction and pulp bleaching with Mn^{3+} are most effective when they are performed under the optimum conditions independently.

Treatment with MnP by means of this two-stage reactor system was repeated in combination with alkaline extraction. The pulp brightness increased to about 85% when the enzyme treatment was repeated three times. The mesoporous material which immobilized MnP molecules was packed into a column. The reaction buffer containing 0.1 mM H_2O_2 was continuously loaded onto the column at 40 °C. Immobilized MnP activity was measured by monitoring the generation of the Mn^{3+}-malonate complex at 270 nm. Immobilized enzyme retained more than 80% of its initial activity even after 30 days of continuous reaction.

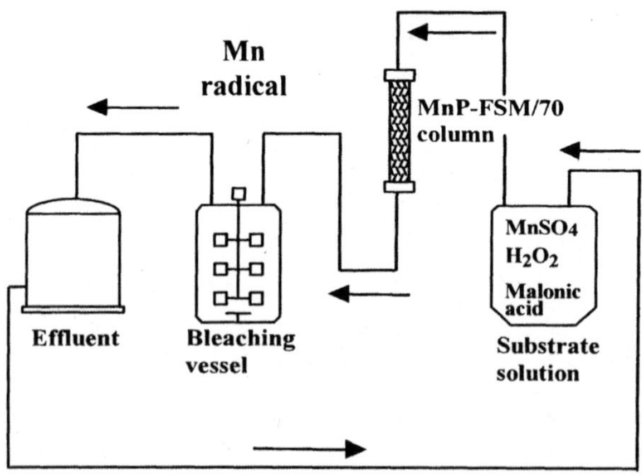

Figure 8. Schematic diagram of the thermally discontinuous immobilized MnP bleaching system.

We compared the hand-sheet properties of the pulp bleached by means of multiple treatments. The tensile properties, tearing resistance, and bursting strength of the immobilized MnP-treated pulp were almost the same as those of the non-treated pulp, thus deterioration as a result of Immobilized MnP treatment was not observed. Kenaf fibers were also bleached well using the immobilized MnP system.

Conclusions

The surface characteristics and the pore diameter of mesoporous materials were suggested to be a critical factor for immobilization and stabilization of the enzyme molecule. When the pore diameter of the mesoporous material was larger than the size of the enzyme, the enzyme would be immobilized in the pores of mesoporous materials. When the average mesopore size of mesoporous materials was suitable to the molecular diameters of the enzyme, the immobilized enzyme sometimes exhibits the good stability. By gently mixing in pH region under pI, the enzyme molecule would be well immobilized in the pores of mesoporous materials. Hydrogen bonds would be formed between the surface amino residues of the enzyme and silanol residues of the mesoporous material during immobilization, and thus the enzyme was excellently stabilized.

To evolve the enzymes, SIMPLEX, which comprises single-molecule PCR followed by cell-free protein synthesis, was used. Compared with conventional *in vivo* colony-based methods (28), this method is extremely rapid and efficient. For example, neither a transformation nor a cultivation step is necessary, and therefore replica plates, which are essential for cell-based screening for linking a genotype with a phenotype, also are not necessary. Instead, after using a portion of the single molecule-PCR product for protein synthesis, the remainder on a 384-well PCR plate can be frozen for long-term storage in a stable state. More importantly, the total time required for protein expression and screening is short, ~8 h, 4 h for SMPCR, 3 h for cell-free protein synthesis, and 1 h for screening of the protein library.

Our cell-free system involving fungal PDI will be applicable to the screening of other heme-containing proteins. Furthermore, the system can be widely applied for various kinds of screening of proteins with desired properties. The inactivation pathway of MnP with excess H_2O_2 has already been studied, suggesting that compound III formed during heme bleaching causes irreversible MnP inactivation (29, 30). However, our results suggest that the MnP instability with small amounts of H_2O_2 is closely related to the susceptibility to a conformational change around the active site, which eventually causes subsequent inactivation or denaturation.

We suggested that prediction of critical restricted mutations in the functional region, using a structural model generated with a computer-driven model-building system or an X-ray crystal structure study, is an effective method for changing the function of the protein efficiently. In this study, we constructed models of MnP and predicted oxidizable and/or solvent-accessible and/or conformationally unstable amino acid residues around the H_2O_2-binding pocket. This suggested that amino acid residues around the pocket critically affected the H_2O_2 resistance. Moreover, fine-tuning for accommodation in the pocket of MnP would require a random mutation process such as the directed evolution method.

We also found the effectiveness of the combination of molecular evolution of critical residues around the functional region and the immobilization in a suitable mesoporous material.

The immobilized evolved MnP in mesoporous material showed excellent H_2O_2 stability (more than 50-fold that of wild type MnP). The evolved MnP would show a synergistic effect on immobilization, because the unstable amino acids in the H_2O_2 binding pocket were changed to stable ones. This new strategy involving enzyme evolution and immobilization in mesoporous materials would be applicable for excellent stabilization of other enzymes which have unstable binding pockets.

Acknowledgement

We thank Dr. Shinji Inagaki and Yoshiaki Fukushima for valuable discussion. This work was supported in part by a Grant-in-Aid for Scientific Research from the Ministry of Education, Culture, Sports, Science and Technology of Japan (No. 16360418).

References

1. Klibanov, A. M. *Science* **1983**, *219*, 722-727.
2. Beck, J. S.; Vartuli, J, C.; Roth, W, J.; Leonowicz, M. E.; Kresge, C. T.; Schmitt, K. D.; Chu, C. T-W.; Dlsou, D. H.; Sheppard, E. W.; McCullen, S. B.; Schlenker, J. L. *J. Am. Chem. Soc.* **1992**, *114*, 10834-10843.
3. Corma, A.; Kan, Q-B.; Navarro, M. T.; Perez-Pariente, J.; Rey, F. *Chem. Mater.* **1997**, *9*, 2499-2506.
4. Yanagisawa, T.; Shimizu, T.; Kuroda, K.; Kato, C. *Bull. Chem. Soc. Jpn.* **1990**, *63*, 1535-1537.
5. Inagaki, S.; Fukushima, Y.; Kuroda, K. *J. Chem. Soc., Chem. Commun.* **1993**, 680-682.
6. Inagaki, S.; Koiwai, A.; Suzuki, N.; Fukushima, Y.; Kuroda, K. *Bull. Chem. Soc. Jpn.* **1996**, *69*, 1449-1457.
7. Zhao, D-Y.; Feng, J-T.; Huo, Q-S.; Melosh, N.; Fredrickson, G. H.; Chmelka, B. F.; Stucky, G. D. *J. Am. Chem. Soc.* **1998**, *120*, 6024-6036.

8. Yang, P-D.; Zhao, D-Y.; Margolese, D. I.; Chmelka, B. F.; Stucky, G. D. *Nature* **1998**, *396*, 152-155.
9. Inagaki, S.; Guan, S.; Fukushima, Y.; Ohsuna, T.; Terasaki, O. *J. Am. Chem. Soc.*, **1999**, *121*, 9611-9614
10. Inagaki, S.; Guan, S.; Fukushima, Y.; Ohsuna, T.; Terasaki, O. *Nature*, **2002**, *416*, 304-307
11. Kato, K.; Irimescu, R.; Saito, T.; Yokogawa,Y.; Takahashi, H. *Biosci, Biotechnol. Biochem.*, **2003**, *67*, 203-206
12. Takahashi, H.; Li, B.; Sasaki, T.; Miyazaki, C.; Kajino, T.; Inagaki, S. *Chem. Mater,* **2000**, *12*, 3301-3305
13. Takahashi, H.; Li, B.; Sasaki, T.; Miyazaki, C.; Kajino, T.; Inagaki, S. *Microporous Mesoporous Mater*.**2001**, *44-45*, 755-762
14. Felipe Diaz, J.; Balkus Jr, K. J. *J. Mol. Catal. B: Enzymatic* **1996**, *2*, 115-125.
15. Sasaki, T.; Kajino, T.; Li, B.; Sugiyama, H.; Takahashi, H.; *Appl. Environ. Microbiol.*, **2001**, *67*, 2208-2212
16. Glenn, J. K.; Akileswaran, L.; Gold, M. H., *Arch. Biochem. Biophys.*, **1986**, *251*, 688-696
17. Hammel, K. E.; Tardone, P. J.; Moen,M. A.; Price, L. A. *Arch. Biochem. Biophys.*, **1989**, *270*, 404-409
18. Popp, J. L.; Kirk, T. K. *Biochem. Biophys.*, **1991**, *288*, 145-148
19. Reading. N. S.; Aust, S.D. *Biotechnol.Prog.*, **2000**, *16*, 326-329
20. Nakano, H.; Kobayashi, K.; Ohuchi, S.; Sekiguchi, S; Yamane, T. *J. Biosci. Bioeng.*, **2000**, *90*, 456-458
21. Miyazaki, C.;Takahashi, H. *FEBS Lett.*, **2001**, *509*, 111-114
22. Miyazaki-Immura, C.; Oohira, K.; Kitagawa, R.; Nakano, H.; Yamane, T.; Takahashi, H. *Protein Engng.* **2003**, *16*, 423-428
23. Jiang, X.; Ookubo, Y.; Fujii, I.; Nakano, H.; Yamane, T., *FEBS Lett.*, **2002**, *514*, 290-294
24. Kajino,T.; Sarai,K.; Imaeda,T.; Idekoba, C.; Asami, O.; Yamada, Y.; Hirai, M.; Udaka, S. *Biosci. Biotechnol. Biochem.*, **1994**, *58*, 1424-1429
25 Majcherczyk, A.; Johannes, C.; Huttermann, A. *Appl. Microbiol. Biotechnol.*, **1999**, *51*, 267-230
26. Wang,C.C. *Methods Enzymol.*, **2002**, *348*, 66-75
27. Sundaramoorthy, M.; Kishi, K.; Gold, M. H.; Poulos, T. L., *J. Mol.Biol.*, **1994**, *238*, 845-848
28. Lin, Z.; Thorsen, T.; Arnold,F. H. *Biotechnol. Prog.*, **1999**, *15*, 467-471
29. Wariishi, H.; Gold,M. H. *J. Biol. Chem.*, **1990**, *265*, 2070-2077
30. Timofeevski, L.; Reading, R. S.; Aust, S. D. *Arch. Biochem. Biophys.*, **1998**, *356*, 287-295
31. Itoh, T., Yano, K., Inada, Y., Fukushima, T. *J. Am. Chem. Soc.*, **2002**, *124*, 13437-13441

Chapter 4

Microperoxidase-11 Immobilized in a Metal Organic Framework

Kenneth J. Balkus, Jr., Thomas J. Pisklak, and Rita Huang

Department of Chemistry and the NanoTech Institute, University of Texas at Dallas, Richardson, TX 75083–0688

Ultra large pore molecular sieves have proven to be viable hosts for biomolecule adsorption and separation. Microperoxidase-11 has been immobilized for the first time in a nano-crystalline metal organic framework (MOF). Microperoxidase-11 was physically absorbed from solution into the 3-dimensional $[Cu(OOC\text{-}C_6H_4\text{-}C_6H_4\text{-}COO)\cdot\frac{1}{2}\,C_6H_{12}N_2]_n$ MOF. The activity of the MOF immobilized peroxidase for the oxidation of methylene blue and α-methylstyrene in organic solvents was determined.

Enzymes are generally more active and show higher selectivities than synthetic, biomimetic catalysts (*1*). While biocatalysts are generally more active, enzymes have some inherent drawbacks as well. These problems include low thermal stability, low stability in organic solvents, and difficulty in recycling (*1-4*). Biocatalyst lifetime can be significantly improved by supporting the biomolecules on polymers and metal oxides. However, the non-uniform porosity of these support materials may limit access to the enzymes or allow leaching. Attempts to overcome these obstacles have resulted in the immobilization of enzymes and proteins in mesoporous materials (*1-47*). Mesoporous molecular sieves, comprised of one- to three-dimensional channel systems, have well defined pores ranging from 2-50 nm. The uniform pores of molecular sieve materials make them excellent materials for the size selective absorption of various biocatalysts (*1-4*). Mesoporous materials can also be compositionally tailored to provide an environment which is favorable for the adsorption of biocatalysts, such as changing the hydrophobicity of the material or adding surface binding sites (*1-4*). Mesoporous materials are often chemically and thermally stable; these attributes provide a protective environment for the biocatalyst after immobilization.

We first demonstrated that mesoporous molecular sieves such as MCM-41, MCM-48, SBA-15, and DAM-1 are capable of immobilizing enzymes (*37*). Subsequently, many different mesoporous hosts and enzymes have been studied (*1-47*). The pore size of mesoporous silicates must be chosen which closely match the size of the enzyme or protein which is to be immobilized (*23*). If the pores are too small, the biocatalyst will not be absorbed and if the pores are too large, the biocatalyst can leach out. To reduce leaching, the hydrophilic interiors of mesoporous silicates have been functionalized with organic groups to enhance hydrophobicity and help retain the biocatalyst in the pores (*42-44*).

In an effort to provide more hydrophobic environments for inclusion of biocatalysts in mesoporous materials, we and others have immobilized biocatalysts in periodic mesoporous organosilicas (PMO) (*11, 45-48*), such as mesoporous benzene silica (MBS) and ethane bridged organosilica (MSE). Even though the hydrophobicity is greater than mesoporous silicas, leaching in PMO's is still highly dependant upon pore size (*45*). Recently, Hudson and coworkers have shown that electrostatic interactions also play an important role in biocatalyst retention (*11*). A material which exhibits all of these properties, such as: hydrophobicity, tunable pore size, and electrostatic interaction with the biocatalyst would provide an ideal environment for the biocatalyst and prevent leaching. Inorganic/organic hybrid framework materials may satisfy these requirements. Metal organic frameworks (MOF) are a family of materials which exhibit these properties and should be excellent hosts for biocatalysts

Metal organic frameworks (MOF) comprise a relatively new class of porous materials (*49-54*). The novelty of these materials arises from the ability to selectively coordinate variable organic linkers to metals or inorganic clusters to produce porous inorganic-organic hybrid networks. When metals are used to connect the frameworks they contain one (*55-60*) or more vacant or labile sites to which the organic linkers can attach. Typically, the organic linkers contain functionalities such as carboxylate, amine, or pyridine groups through which they bind to the open sites on the metal centers (*49, 54, 61-63*). The metals or inorganic clusters act as anchoring points to which the organic linkers are bound. The organic ligands are subsequently connected in such a fashion that voids are formed. The pores are defined by the organic linkers which can impart properties to these materials which are not easily attained with porous metal oxides, such as zeolites. In fact, by tailoring the length of the linkers, MOFs have been synthesized with pore sizes and pore volumes greater than zeolites (*49, 64-67*). Catalytic properties (*61, 68-69*) as well as luminescence (*70-75*), chirality (*64, 69*), adsorption (*64-65, 67, 69, 74-87*) and framework polarity (*49, 54, 77-78, 82*) can also be manipulated by varying the type of organic linker used to form the framework. This nanoporous family of metal organic frameworks exhibits the largest surface areas of any known materials (*67, 88*). With pore sizes ranging as high 28.8 Å (*65*) MOFs show a great potential for adsorption of guest molecules (*64-65, 67, 74-87*) and inclusion chemistry (*69*). Recently, Chael et al. reported the inclusion of polycyclic molecules such as C_{60}, Astrazon orange R, Nile red, and Reichardt's dye in a metal organic framework (*67*). With pore sizes in the mesoporous range MOF's should also be capable of adsorbing small proteins and enzymes. Additionally, MOF pores are generally hydrophobic which could result in an affinity for and promote immobilization of proteins and enzymes. Also redox active MOF's could be synthesized that might operate in concert with proteins and enzymes in catalyzed reactions.

In order to test the immobilization of a biomolecule in a MOF the proteolytic degradation product of cytochrome c, Microperoxidase-11 (MP-11) was selected (Figure 1) (*38*). The shortened peptide chain (11 amino acids) prevents denaturing of MP-11 by organic solvents (*89*). Figure 1 (random coil configuration) shows that MP-11 is composed of a heme group attached to an 11 amino acid peptide chain through thioether linkages of two Cysteines (Cys 14 and Cys 17) as well as to the imidazole group from histadine which is coordinated to the iron center as an axial ligand. The second axial position is coordinated with a molecule of water. Although there are no crystal structures of MP-11 available, the figure is drawn in the lowest energy configuration similar previously performed molecular dynamics calculations for MP-11 in methanol solution (*90*). MP-11's usefulness arises from its ability to reduce hydrogen peroxide to water while oxidizing a substrate. However, free MP-11 has the tendency to aggregate in solution due to both intermolecular attractions and ligation through the metal center (*38*). When MP-11 oligimerizes through coordination to the metal center, the heme becomes less accessible and the activity of the enzyme is adversely affected. Immobilization in a suitable host material prevents aggregation and renders the heme more accessible to substrate molecules. The small size of MP-11 relative to complete enzymes, as well as the peroxidase activity of MP-11 makes it a suitable biocatalyst for immobilization in materials with pore sizes around 20 Å. MP-11 has been shown to degrade organic soluble azo dyes such as Solvent Yellow, Solvent Blue 11, Solvent Green 3, and Solvent Orange 7 (*89, 91*) while entrapped in reversed micelles. Therefore, by immobilizing MP-11 in a hydrophobic matrix such as the MOF $[Cu(OOC\text{-}C_6H_4\text{-}C_6H_4\text{-}COO)\cdot\frac{1}{2}C_6H_{12}N_2]_n$, (Figure 2) the microperoxidase can be dispersed in organic media and the hydrophobic interior of the MOF will allow the substrate access to the active site of the microperoxidase. Synthetic dyes are a large part of the waste stream from industrial manufacture of textiles and various consumer products. The removal of the dyes from the environment is a pressing concern. Although, peroxidases have been shown to degrade dyes via oxidation they have been limited to water soluble dyes. Many of the dyes used are water insoluble and a method of degrading these dyes is needed.

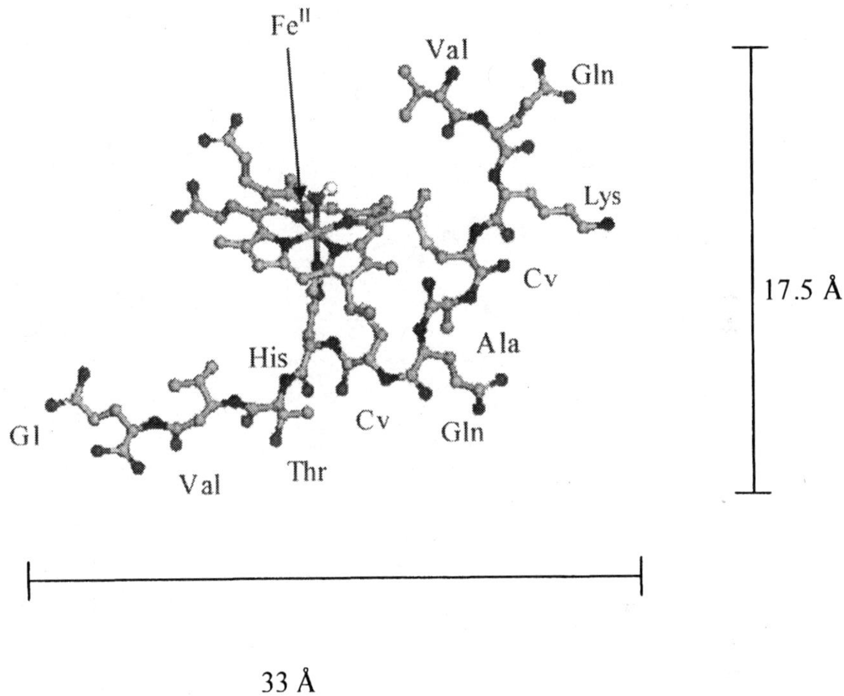

Figure 1. Structure of MP-11 calculated using Materials Studio

Recently, Kadnikova et al. immobilized microperoxidase-11 (MP-11) in sol-gel silica glass (*38*) and reported that the α-helical structure is promoted by a hydrophobic environment. MP-11 has also been immobilized on surfaces such as gold, in hybrid polymers, and mesoporous materials (*98-101*). Immobilization on surfaces and in polymers may not sufficiently sequester the microperoxidase which would allow it to dimerize and lose activity. Also, because MP-11 is relatively small, when immobilized in mesoporous materials there is a high propensity for leaching. The metal organic framework $[Cu(OOC\text{-}C_6H_4\text{-}C_6H_4\text{-}COO)\cdot\frac{1}{2}C_6H_{12}N_2]_n$ (Figure 2) is composed of 2-dimensional sheets of copper dimers connected via biphenyl dicarboxylates. These 2-dimensional sheets are connected together by 1,4–diazabicyclo[2.2.2]octane to form a 3-

dimensional framework. Consequently, the interior of this MOF is composed of phenyl rings which form a hydrophobic environment in the pores. Immobilization of the microperoxidase MP-11 in the pores of this MOF would provide a hydrophobic environment around the MP-11 and may promote the α-helical structure, which would facilitate reactions in organics solvents.

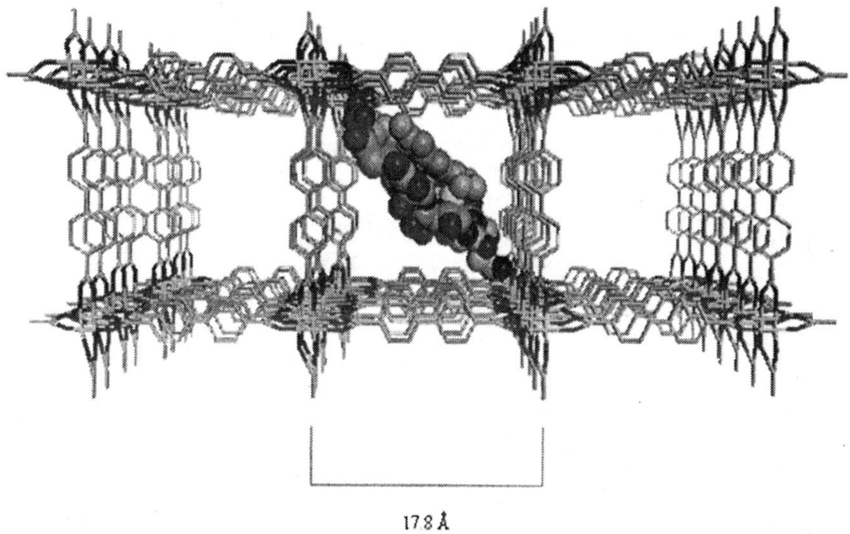

Figure 2. Microperoxidase-11 immobilized in the MOF calculated using Materials Studio

MP-11 is only 17.5 x 30 Å in diameter which should allow the fragment to be physically adsorbed by porous materials with pore sizes ~18 Å and larger. The metal organic framework $[Cu(OOC\text{-}C_6H_4\text{-}C_6H_4\text{-}COO)\ ½\ C_6H_{12}N_2]_n$, shown in Figure 2, has a theoretical pore size of 17.8 Å and is suitable for the immobilization of MP-11. Consequently, for the first time we have successfully immobilized MP-11 in a nano-crystalline metal organic framework.

The reactivity of MOF immobilized MP-11 versus free MP-11 (dissolved) in organic solvents have been studied. MP-11 shows no activity in non-polar organic solvents, consequently the polar organic solvents N,N'-dimethylformamide (DMF) and acetonitrile were used in this study (*89*). The oxidation of methylene blue in DMF under various conditions as well as the oxidation of α-methylstyrene in acetonitrile was used to evaluate the reactivity of supported MP-11. In all cases, the immobilized MP-11 has been shown to be more active than free MP-11 which is unusual for supported enzymes.

Experimental

MOF synthesis

The nanocrystalline MOF $[Cu(OOC\text{-}C_6H_4\text{-}C_6H_4\text{-}COO)\ \frac{1}{2}C_6H_{12}N_2]_n$ was synthesized by modifying the procedure reported by K. Seki et al (*87*). To form the MOF, a layered 2-D copper(II) dicarboxylate was first synthesized and then pillared with 1,4 – diazabicyclo[2.2.2]octane to form a 3-D structure. To synthesize the layered phase, 0.59 g of copper(II) acetate (Fisher Scientific) was dissolved in 25 mL of absolute ethanol (Aaper Alcohol and Chemical Company) and 0.25 g of 4, 4'-biphenyldicarboxylic acid (Aldrich) was dissolved in 70 mL N,N-dimethylformamide (EM Science) and 0.30 mL formic acid (Aldrich). The solutions were mixed together with stirring at room temperature and transferred to a 250 mL Teflon® reaction vessel. The synthesis mixture was heated at 50°C for ~72 hours. The reaction was removed from the oven and allowed to cool to room temperature. The blue product was then filtered and the blue solids were washed with absolute ethanol. The layered precursor was dried in a vacuum oven at 90°C overnight. This material was then pillared by placing 100 mg of the blue powder in a 23 mL Teflon® lined autoclave along with 0.023 g of 1,4 – Diazabicyclo[2.2.2]octane (DABCO) (Aldrich) dissolved in 5.0 mL of toluene (Mallinckrodt). The autoclave was heated at 110°C while rotating at 30 rpm for 6 hours. The green product was then filtered and the green solids were washed with absolute ethanol and dried in a vacuum oven at 90°C overnight. The solids were then washed with DMF to remove any soluble impurities and the green product was filtered and washed again with absolute ethanol and dried in a vacuum oven at 90°C overnight.

Immobilization of MP-11

A 45 μM solution of MP-11 was prepared by dissolving 3.5 mg of MP-11 (Aldrich, 90% purity) in 100 mL of N,N-dimethylformamide (EM Science, 99.99%) or 100 mL absolute ethanol (Aaper Chemical). To immobilize the MP-11 10 mL of this solution was added to 10 mg of the MOF which had been dried in a vacuum oven at 90°C overnight. The mixture was stirred at RT for 3 hours. To isolate the MOF immobilized MP-11 the mixture was centrifuged and the supernatant was decanted. The supernatant was then filtered with a 0.2 μm syringe filter. The uptake of MP-11 was monitored by the disappearance of the Soret band (416 nm) in the supernatant and used to calculate loading.

Oxidation of methylene blue

A 4.5 μM solution of methylene blue (Allied Chemical, 98%) was prepared by dissolving 9.8 mg of methylene blue in 250 mL of N,N-dimethylformamide (DMF), this solution was successively diluted until the desired concentration was obtained. For the oxidation reactions, 5 mL of 4.5 μM methylene blue was added to 10 mg MOF/MP-11 and 135 μL of 0.01 M hydrogen peroxide at room temperature. This mixture was stirred for 24 hours under various conditions such as in air and light, air and dark, N_2 and light, N_2 and dark, and N_2 and dark with heating (40°C). Also, a control reaction in which 5.0 mL 20 μM DMF solution of free MP-11 was reacted with 135 μL of 0.01 M hydrogen peroxide and 2.4 μM methylene blue under the same conditions as above, was performed. Also, a control reaction was performed with 10 mg as-synthesized MOF, 135 μL of 0.01 M hydrogen peroxide, and 5 mL of 2.4 μM methylene blue in DMF. The progress of the reactions were monitored by the disappearance of the visible band at 664 nm associated with methylene blue. Turnovers are based on moles of methylene blue converted per mole of MP-11.

Oxidation of α-methylstyrene

In a typical oxidation reaction, 3.3 mL of Clorox bleach (0.8 M NaOCl) was mixed with 5 mL 0.05 M Na_2HPO_4 (pH = 11.3) and cooled to 0°C in an ice bath. In a separate reaction vessel, 6 mL of acetonitrile (Aldrich, 99%) was mixed with 1 mL α-methylstyrene (Aldrich). To this solution 10 mg of 10 mg MOF/MP-11 was added with stirring and the mixture was cooled to 0°C in an ice bath. Then the bleach solution was added to the α-methylstyrene/MOF/MP-11 mixture and the reaction was stirred under N_2. After four hours, 10 mL of methylene chloride (Aldrich) was added and the reaction mixture was washed twice with deionized water. The organic layer was then separated and dried over magnesium sulfate (Aldrich). The products were analyzed by capillary GC using a CyclodexB column (J&W Scientific) and decane as an internal standard. Turnovers are based on moles of methylestyrene converted per mole of MP-11.

Instrumentation

The concentrations of the methylene blue oxidations were monitored using a Shimadzu UV-1601PC UV-visible spectrophotometer. X-ray diffractograms were obtained using Scintag XDS 2000 x-ray diffractometer with CuKα radiation. FTIR spectra were obtained using an Avatar 360 spectrometer from KBr pellets. Surface area and pore size was measured using Nitrogen adsorption at 77 K on a Quantachrome Autosorb 1. All samples were out gassed at 100°C and 10^{-6} torr prior to measurements. A seven point BET equation was used to calculate surface areas. Pore size distribution was calculated from the desorption data using the DA method. The oxidation of α-methyl styrene to 1,2-epoxy-2-phenylpropane was monitored using Hewlett Packard 5890 Series II gas chromatograph with a CyclodexB column. UV-Vis spectra of solid samples were acquired with a Perkin Elmer Lambda 900 UV/VIS/NIR Spectrometer equipped with a Labsphere PELA-1000 integrating sphere.

Results and Discussion

Immobilization of MP-11

Figure 3a shows the powder XRD pattern of the blue, layered material which had a surface area of 462 m^2/g (Figure 3b). After the intercalation of DABCO to produce the 3-dimensional $[Cu(OOC\text{-}C_6H_4\text{-}C_6H_4\text{-}COO)\ ½\ C_6H_{12}N_2]_n$ MOF the XRD pattern (Figure 3c) of the powder was in accordance with that reported in the literature (*87*). The MOF had a Langmuir surface area of 1260 m^2/g (Figure 3d) and a pore size of 17.8 Å. The particle size of the MOF was determined to be 30 nm by the Scherrer equation. The MOF is insoluble in DMF, but due to its' small particle size is very difficult to remove from solution. This is reflected in the UV-Vis spectra where absorption peaks are present for aromatics (~250 nm) even after centrifugation and filtration.

The size of MP-11 (Figure 1) was calculated to be ~17.5 x 33 Å using Materials Studio and Discover Minimizer. It can be seen from Figure 2 that the MOF's theoretical pore size is large enough to accommodate MP-11. The biocatalyst was adsorbed into the MOF by stirring the MOF in a solution of MP-11. The amount adsorbed was monitored by UV-Vis spectroscopy. Figure 4a shows the UV-Vis spectrum of the 45μM in DMF and Figure 4b is the UV-Vis spectrum of the same solution after absorption of the MP-11 by the metal organic framework. As can be seen in Figure 4 the MOF adsorbed ~98% of the MP-11 from the solution. MP-11 uptake by the MOF occurred rapidly with the maximum loading being reached in less than 45 minutes. The loading of MP-11 in the MOF was calculated to be 5.9 μmol/g. The XRD pattern (not shown) of the MOF with adsorbed MP-11 is unchanged from the as-synthesized MOF. This is expected because absorption of the microperoxidase should not affect the crystallinity of the MOF, but it also shows that the MOF was unaltered during the absorption procedure. The Langmuir surface area after adsorption of MP-11 was 300 m^2/g, representing a 76% decrease (as-synthesized = 1260 m^2/g) in surface area.

By comparing the infrared spectrum of the MOF immobilized MP-11 (Figure 5a) to the as-synthesized MOF (Figure 5b) it can be seen that the MP-11 is present. The IR bands characteristic of MP-11 are located at 1652 cm^{-1}, 1592

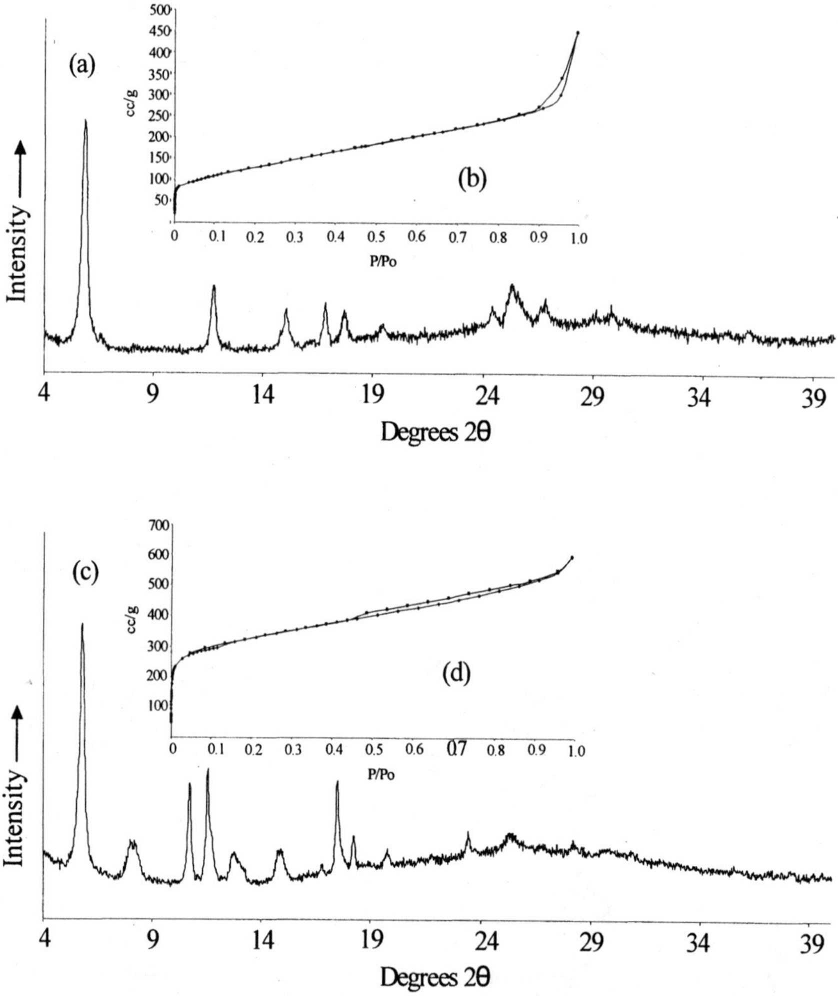

Figure 3 X-ray diffraction patterns and N_2 adsorption isotherms for (a. – b.) as-synthesized 2-D metal organic framework and (c. – d.) the 3-D MOF.

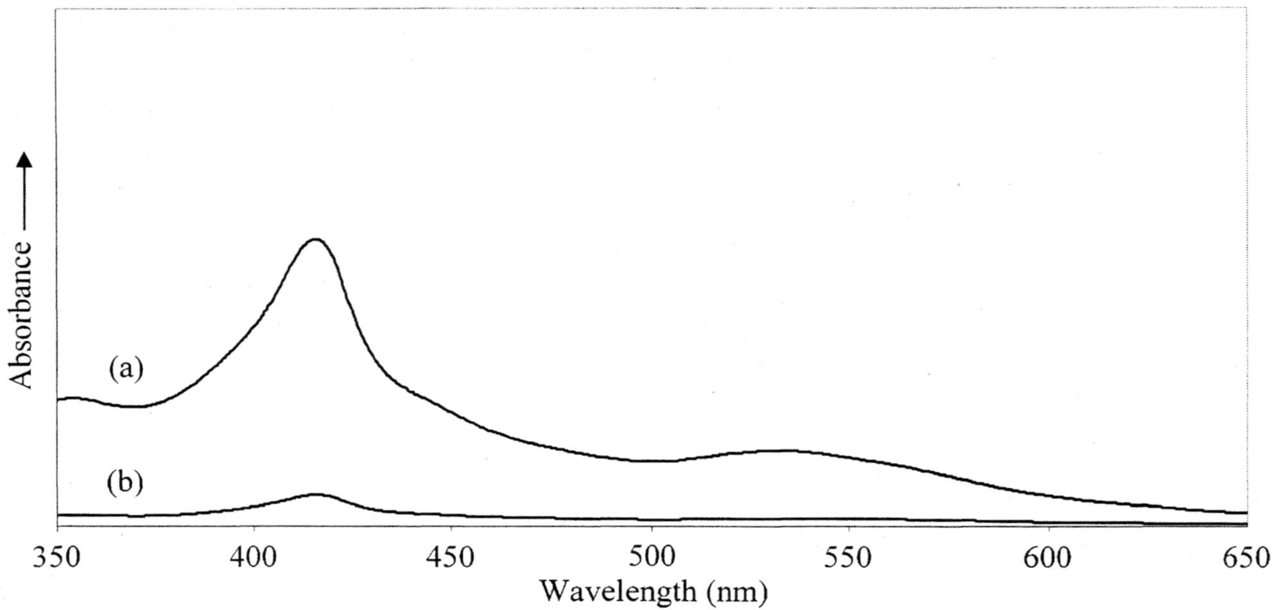

Figure 4. UV-Vis spectra of MP-11 solution a) before and b) after immobilization in MOF

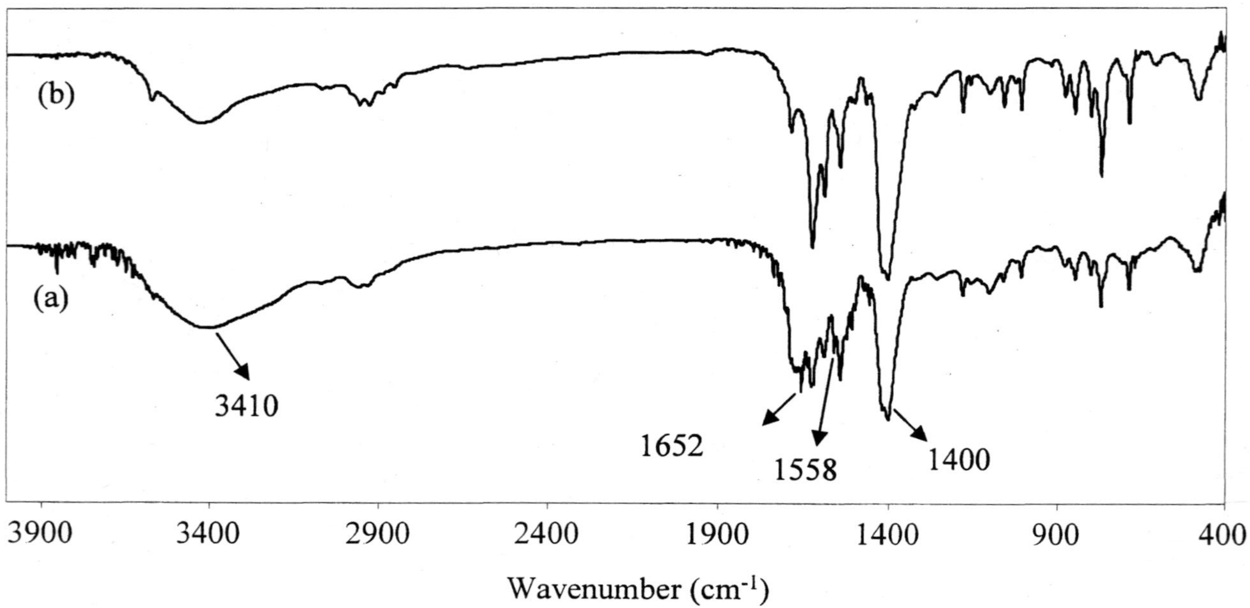

Figure 5. FTIR spectra of a) the MOF immobilized MP-11 and b) the as-synthesized MOF

cm^{-1}, 1567 cm^{-1}, and 1450 cm^{-1} (*92*). The band located at 1652 cm^{-1} is attributed to the amide I' stretch (peptide carbonyl). This band is present in Figure 6a while it is not in Figure 5b, compared to the literature value of from 1642 cm^{-1} (*92*) this band is shifted 10 wavenumbers. MP-11 also has bands located at 1592 and 1558 cm^{-1} which are associated with the asymmetric stretching mode of the carboxylates (*92*). While the band at 1558 cm^{-1} appears in Figure 5a, also shifted relative to free MP-11 (1567 cm^{-1}), the band located at 1592 cm^{-1} is obscured by a stretch in the same location attributed to the MOF. The stretch at 1450 cm^{-1}, due to the amide II' stretch (*92*), has also been blocked by a large band located at 1400 cm^{-1} (asymmetric vibrations of 4,4'-biphenyldicarboxylate) from the MOF. Also, while Figure 5b has a peak located at 3410 cm^{-1}, Figure 5a shows a much broader peak with a shoulder at the same location. This is due N-H stretches from the MP-11. The shifts in the 1652 and 1567 cm^{-1} bands imply that there is interaction between the MOF and MP-11.

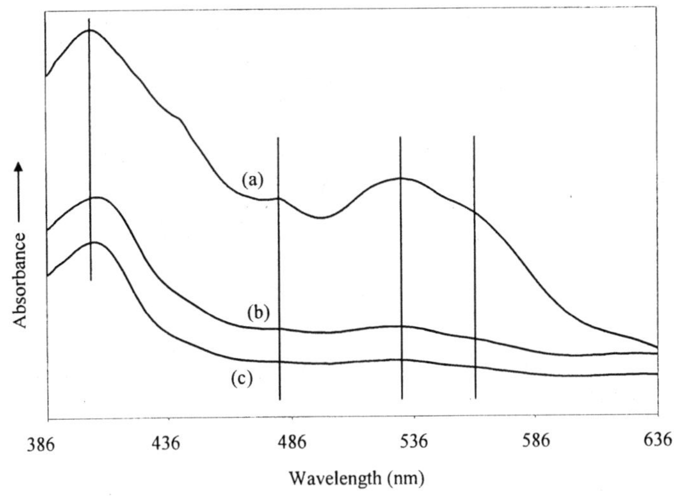

Figure 6. UV-Vis spectra of solid a) MP-11, b) MOF immobilized MP-11, and (c) MOF immobilized MP-11 after oxidation of methylene blue

Figure 6 shows the solid state UV-Vis spectra of MP-11 (Figure 6a), MP-11 immobilized in MOF (Figure 6b), and MP-11 immobilized in MOF after the oxidation of methylene blue (Figure 6c). The Soret band for solid MP-11 is located at 402 nm and the β band at 530 nm. The Soret band of the immobilized MP-11 was red shifted to 406 nm. It has been shown that the Soret bands of porphyrins red shift when confined in zeolites due to distortions of the porphyrin ring (*102*). The shifts in the Soret and β bands may arise from the MP-11 confinement in the pores of the MOF. The immobilized MP-11 after use in the oxidation of methylene blue shows no change from the fresh catalyst.

To determine if the microperoxidase would leach out of the MOF after immobilization, the MOF immobilized MP-11 was stirred at room temperature for 24, 48, and 72 hours in fresh DMF (99.99%). The UV-Vis spectra of the supernatants showed no MP-11 in solution. This indicates that the MP-11 is tightly bound in the pores of the metal organic framework. When the same leaching experiment was performed in a potassium phosphate buffer system the MOF dissolved and the MP-11 was quantitatively released into solution after ~30 minutes. The MOF is not soluble in DI water, but has low stability and reverts to the layered form with time. When the leaching experiment was performed in DI water after 4 hrs ~90% of the immobilized MP-11 was released into solution.

Oxidation of methylene blue

The activity of MOF immobilized MP-11 in a polar organic solvent was evaluated by monitoring the oxidation of methylene blue (Figure 7) to the corresponding sulfoxide (*93*) in N,N-dimethylformamide. The progress of the reaction was monitored by UV-Vis spectroscopy using the absorbance band of methylene blue located at 664 nm. Methylene blue is a photosensitizer which when exposed to light can produce singlet oxygen (*94-95*). Singlet oxygen is a

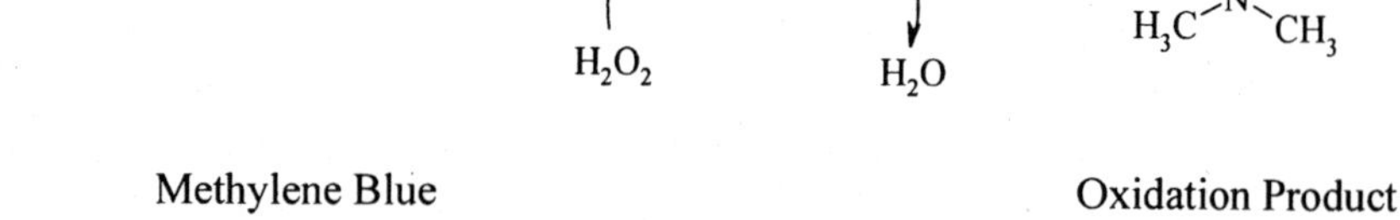

Figure 7. Reaction scheme for the oxidation of methylene blue

powerful oxidant which selectively oxidizes functional groups, such as electron rich double bonds (*95*), which are present in methylene blue. Due to methylene blue's ability to self oxidize, the activity determinations of MOF immobilized MP-11 were carried out in the presence and absence of air and light and

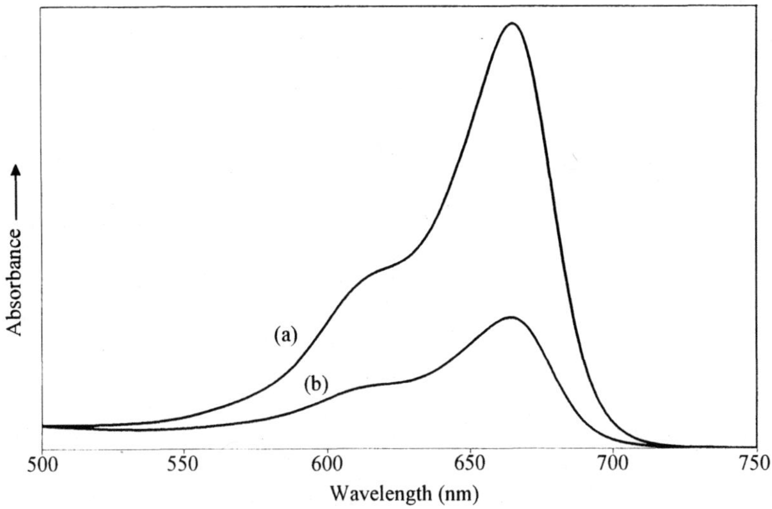

Figure 8. UV-Vis spectra of methylene blue solution a) before and b) after 3 hours of reaction catalyzed by the MOF immobilized MP-11

variations thereof, to ensure the conversions attributed to the microperoxidase were correct. Figure 8 shows a representative UV-Vis spectrum associated with the oxidation reaction. Figure 8a is the UV-Vis spectrum of the methylene blue in DMF solution before the reaction. This solution was added to the MOF containing MP-11 along with hydrogen peroxide and allowed to react at room temperature under ambient lighting. After 3 hours the supernatant was isolated by centrifugation and analyzed by UV-Vis spectroscopy (Figure 8b). Approximately, 63% of the methylene blue was converted to the colorless sulfoxide, corresponding to 35 turnovers per day with a peroxide efficiency of 87%. For the control reaction, in which free MP-11 was reacted with methylene blue under the same conditions (in the presence of air and light) as the encapsulated microperoxidase showed only a 6% percent conversion of the methylene blue which is only 5 turnovers per day. Thus there was an 86% increase in activity after immobilization of the MP-11. Figures 9a and 9b show the turnovers versus time for the immobilized and free microperoxidase, respectively. By monitoring the Soret band of MP-11 during the free MP-11 oxidation of methylene blue it was observed that ~20% of the MP-11 was bleached during the reaction. A reduction in the amount of active microperoxidase present can, in part, contribute to the loss in activity compared with the immobilized microperoxidase. Since the immobilized microperoxidase retains a higher activity this also shows that adsorption in the MOF provides some protection from denaturing of the MP-11. This experiment was repeated three times with the same catalyst, the second and third methylene blue

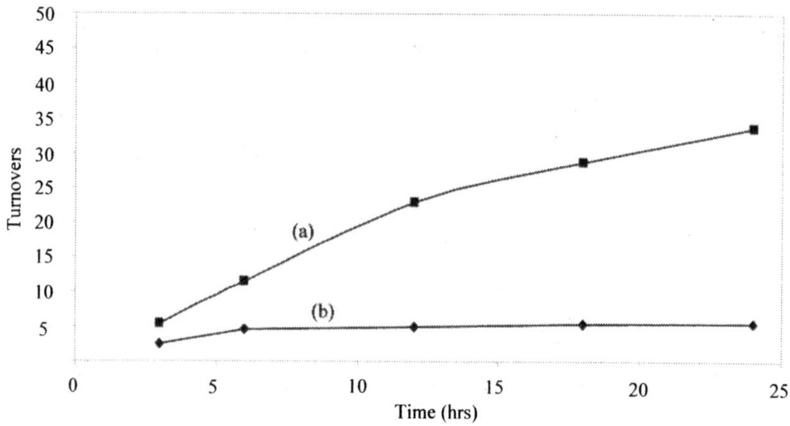

Figure 9. Turnovers versus time for a) the MOF immobilized MP-11 and b) free the free MP-11 dissolved in DMF.

oxidations showed only a slight reduction in activity (33 and 29 turnovers, respectively) . The reaction was also performed with fresh catalyst and the inclusion of a radical trap (hydroquinone), the activity for this reaction was 35 turnovers in a 24 hour period. Since the inclusion of a radical trap does not affect the activity it suggests that the oxidation does not proceed primarily through a radical mechanism. However, the oxygen atom transfer from the heme may occur in a concerted mechanism which is not affected by the radical trap.

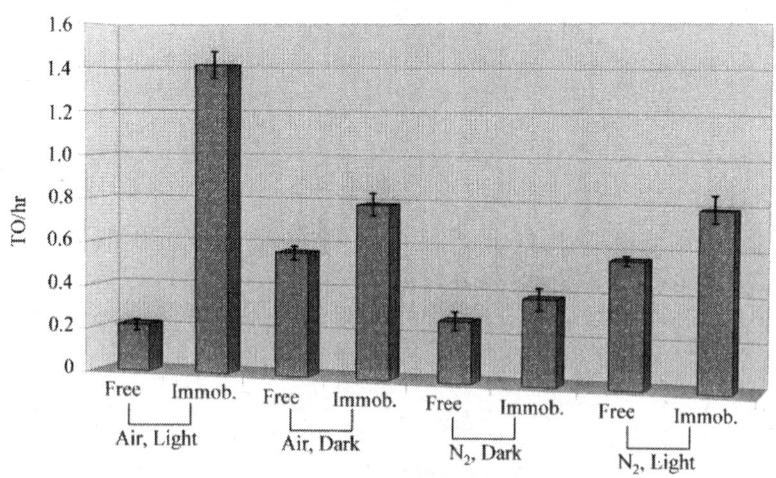

Figure 10. Turnovers per hour for the oxidation of methylene blue catalyzed by free and immobilized MP-11 under air or N_2 and ambient light or dark.

To determine the effect of light on the conversion of methylene blue the reaction was performed in air, under the same conditions as the previous examples except that the reaction vessels were covered to exclude room light. (Figure 10) The free microperoxidase in DMF showed an activity of 14 turnovers per day while that of the immobilized microperoxidase was 20 turnovers per day. The activity of the immobilized microperoxidase was 28% greater than the free microperoxidase. While there was a decrease in the overall activity of the immobilized MP-11 (35 vs. 20 TO/day) the activity of the immobilized microperoxidase was still greater than that of free MP-11. The increased number of turnovers for the free MP-11 is due to microperoxidase concentration, unlike the air-light oxidation where the MP-11 was bleached, the concentration of the free MP-11 did not change during the course of the air-dark oxidation (UV-Vis spectra not shown).

Since light appears to affect the reaction, it may be possible that oxygen also plays a role. Therefore, the methylene blue oxidations catalyzed by free and MOF immobilized MP-11 were conducted in the absence of light and air (under N_2). Again the activities of both catalysts were lower than in the presence of light and oxygen. However, the MOF immobilized microperoxidase still showed a higher activity for the oxidation of methylene blue than the homogeneous catalyst. The concentration of the free MP-11 did not change during the course of the N_2-dark oxidation (UV-Vis spectra not shown). The fact that the turnover values for both the free and immobilized microperoxidase under N_2 and in the dark are lower than in light suggests that ambient oxygen contributes to the oxidation of methylene blue in the dark.

The oxidation of methylene blue was also carried out under N_2 in room light (Figure 10). The turnovers for the free microperoxidase and the immobilized were only 14 and 20 TO/day, respectively. However, the immobilized microperoxidase showed a 29% higher activity than the free MP-11, which is similar to the relative difference between free and immobilized microperoxidase of the air-dark and N_2-dark oxidations. The concentration of the free MP-11 did not change during the course of the N_2-light oxidation (UV-Vis spectra not shown). This evidence suggests that while the presence of ambient oxygen may increase the rate of oxidation, light may also be factor. The combination of oxygen and light clearly results in bleaching of the free MP-11 but not the supported MP-11.

The oxidation of methylene blue in a polar organic solvent by microperoxidase-11 was enhanced by immobilization in the hydrophobic copper MOF. In all instances the immobilized microperoxidase showed a greater activity than the free microperoxidase. Additionally, the supported MP-11 was more stable than the free MP-11. Unfortunately, the activity for methylene blue oxidation at room temperature was marginal even for the supported MP-11. Therefore, the oxidation of methylene blue was performed under the same conditions as the N_2-dark experiment, but with heating at 40°C. In this case, the free MP-11 showed a 45% increase in activity at 40°C versus RT. Similarly, the immobilized MP-11 showed a 38% increase in activity when heated. The greater affect of heat on the homogeneous reaction may reflect MP-11's propensity to oligimerize with itself (*38*) which would be reduced at higher temperatures.

Since, the MOF is a porous material and has the ability to absorb various species, control reactions were performed with the as-synthesized MOF, hydrogen peroxide, and methylene blue in DMF under the same conditions as the air-light oxidation discussed above. A control reaction in which methylene blue was stirred in the presence of hydrogen peroxide, under the same conditions as the air-light experiments, but with no MOF or MP-11 was also performed. Both of the control reactions showed similar low percent conversions of methylene blue (~6%) after 24 hours at room temperature, which were much lower than the typical immobilized MP-11 percent conversions (~68%). This shows that while some methylene blue is oxidized by the photocatalytic self oxidation of methylene blue, the MOF by itself showed no catalytic activity.

Oxidation of α-methylstyrene

While high conversions in the degradation of the dye methylene blue can be achieved with the MOF/MP-11 system, the rate of reaction is relatively slow. Therefore, it was decided to evaluation the MOF supported MP-11 for olefin epoxidation. The model substrate selected was α–methylstyrene in part because the corresponding epoxide has a chiral center as illustrated in Figure 11.

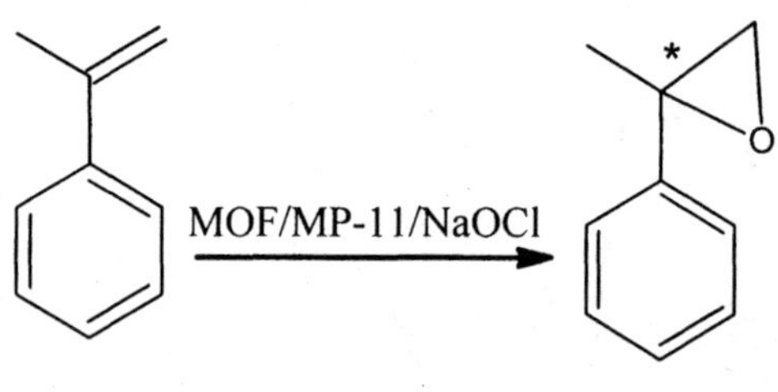

Figure 11. The epoxidation of α-methylstyrene

The reaction conditions were similar to those previously reported for the oxidation of olefins by Jacobsen, et al. (*96*), which utilizes sodium hypochlorite as the oxidant. Leaching experiments were performed in which the MOF immobilized MP-11 was stirred in the organic solvent (acetonitrile) for 24 and 48 hours. No leaching of the microperoxidase was observed. Figure 12 shows the turnovers per hour or the oxidation of α–methylstyrene by the immobilized as well as the free MP-11. The MOF immobilized microperoxidase is 71% more active for the RT oxidation of α-methylstyrene than the free MP-11. This may reflect isolation of the MP-11 in the MOF pores as well as hydrophobic nature of the pores which promote interaction of the organic substrate and MP-11. The primary product of the oxidation of α-methyl styrene by the MOF supported MP-11 is the epoxide (Figure 12) with lesser amounts of 2-phenylpropanol and 2-phenylpropanal. In contrast, the free MP-11 exhibits very low selectivity for the epoxide.

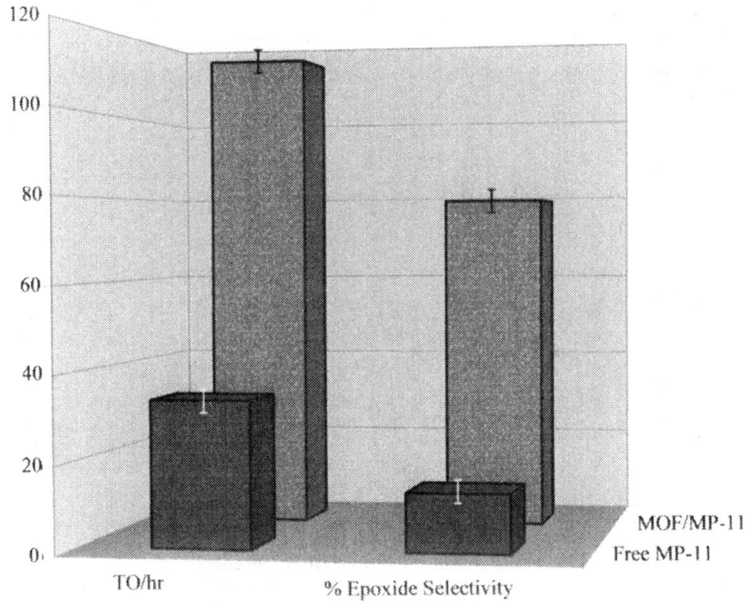

Figure 12. Turnovers per hour and percent epoxide selectivity for the free and immobilized MP-11 during the RT oxidation of α-methylstyrene

The epoxide has a chiral center so the reactions were also evaluated for enantioselectivity. Interestingly, the MOF immobilized microperoxidase shows a larger enantiomeric excess (ee) of 42% ee than the free microperoxidase (36% ee). This increase in % ee over the free MP-11 is most likely due to steric constraints imposed on the microperoxidase and substrate by encapsulation in the small pores of the metal organic framework. While, the enantiomeric excess is not greater than that reported for Jacobsen's catalyst (51% ee) (*97*) the enantioselectivity is higher for the immobilized MP-11 than for the free microperoxidase. The origin of the enantioselectivity must be related to the arrangement of the peptide chain. Kadnikova et al. recently reported the asymmetric oxidation of alkyl sulfides by supported MP-11 (*38*). In this case the observed enantioselectivity was explained by the conformation of the microperoxidase which could be a random coil or α-helix. In both situations, the heme group is equally accessible to substrate molecules leading to no change in activity, but with loss of the α -helix conformation, chiral induction was greatly reduced. They observed a loss in enantioselectivity with repeated cycling of the catalyst. Therefore, MOF immobilized MP-11 catalyst was cycled to determine the stability. The activity and epoxide selectivity was essentially unchanged after 3 cycles, however, the enantioselectivity dropped to 28% ee. These results suggest that after catalysis, separation and washing, the immobilized MP-11 can change conformation in the MOF reducing enantioselectivity. While the pores of the MOF would seem to be a tight fit, it will be interesting to see if other MOF structures with smaller pore dimensions can retain the enantioselectivity by further restricting movement of the peptide chain.

Conclusions

The physical immobilization of microperoxidase-11 in a copper(II) dicarboxylate metal organic framework has been demonstrated for the first time. The activity of the MP-11 for the oxidation of methylene blue in DMF was shown to be higher than that of free MP-11 under various reaction conditions. Also, the metal organic framework immobilized MP-11 was shown to have a greater activity and selectivity for the epoxide for the oxidation of α-methyl styrene. The initial enantioselectivity observed for this reaction was not significantly lower than that observed for Jacobsen's catalyst but this is reduced with repeated cycling of the catalyst. These results suggest that the previously unexplored metal organic framework family of molecular sieves may be promising supports for biomolecules.

Acknowledgements

We thank the Robert A Welch Foundation for support of this research.

References

1. Moelans, D.; Cool, P.; Baeyens, J.; Vansant, E.F., *Catal. Commun.* **2005**, *6*, 307-311.
2. Kim, J.; Grate, J.W.; Wang, P. *Chem. Eng. Sci.* **2006**, *61*, 1017-1026.

3. Hartmann, M. *Chem. Mater.* **2005**, *17*, 4577-4593.
4. Yiu, H.H.P.; Wright, P.A. *J. Mater. Chem* **2005**, *15*, 3690-3700.
5. He, J.; Li, X.; Evans, D.G.; Duan, X.; Li , C. *J. Mol. Catal. B: Enzymatic*, **2000**, *11*, 45-53.
6. Yiu, H.H.P.; Wright, P.A.; Botting, N.P., *Micropor. Mesopor. Mater.* **2001**, *44-45*, 763-768.
7. Deere, J.; Magner, E.; Wall, J.G.; Hodnett, B.K. *J. Phys. Chem. B* **2002**, *106*, 7340-7347.
8. Fadnavis, N.W.; Bhaskar, V.; Kantam, M.L.; Choudary, B.M., *Biotechnol. Prog.* **2003**, *19*, 346-351.
9. Deere, J.; Magner, E.; Wall, J.G.; Hodnett, B.K., *Chem. Comm.* **2001**, *5*, 465-466.
10. Lee, C.-H.; Lang, J.; Yen, C.-W.; Shih, P.-C.; Lin, T.-S.; Mou, C.-Y., *J. Phys. Chem. B* **2005**, *109*, 12277-12286.
11. Hudson, S.; Magner, E.; Cooney, J.; Hodnett, B.K., *J. Phys. Chem. B* **2005**, *109*, 19496-19506.
12. Wang, Y.; Caruso, F., *Chem. Mater.* **2005**, *17*, 953-961.
13. Zhang, B.P.; Janicke, M.T.; Woodruff, W.H.; Bailey, J.A., *J. Phys. Chem. B.* **2005**, *109*, 19547-19549.
14. Dai, Z.; Liu, S.; Ju, H.; Chen, H., *Biosens. Bioelectr.* **2004**, *19*, 861-867.
15. Maria Chong, A.S.; Zhao, X.S., *Catalysis Today* **2004**, *93-95*, 293-299
16. Knops-Gerrits, P.-P. H. J. M.; Goddard III, W.A., *Catal. Today* **2003**, *81*, 263-286.
17. Wei., Y.; Dong, H.; Xu, J.; Feng, Q., *Chem. Phys. Chem.* **2002**, *3*, 802-808.
18. Ravindra, R.; Zhao, S.; Gies, H.; Winter, R., *J. Am. Chem. Soc.* **2004**, *126*, 12224-12225.
19. Wei, Y.; Zu, J.; Feng, Q.; Lin, M.; Dong, H.; Zhang, W.-J.; Wang, C., *J. Nanosci. Nanotech.* **2001**, *1*, 83-93.
20. Itoh, T.; Ouchi, N.; Nishimura, Y.; Hui, H.S.; Katada, N.; Niwa, M.; Onaka, M., *Green Chemistry* **2003**, *5*, 494-496.
21. Coradin, T.; Livage, J., *Comptus Rendus Chimie* **2003**, *6*, 147-152.
22. Macario, A.; Calabro, V.; Curcio,S.; De Paola, M.; Giordano, G.; Lorio, G.; Katovic, A., *Stud. Surf. Sci. Cat.* **2002**, *142B*, 1561-1568.
23. Li, A.; Takahashi, H., *Biotech. Lett.* **2000**, *22*, 1953-1958.
24. Katsuya, K.; Roxana, I.; Takao, S.; Yoshiyuki, Y.; Haruo, T., *Biosci. Biotech. Biochem.* **2003**, *67*, 203-206.
25. Deere, J.; Magner, E.; Wall, J.G.; Hodnett, B.K. *Biotech. Progress* **2003**, *19*, 1238-1243.
26. Wang, Y.; Caruso, F., *Chem. Comm.* **2004**, *13*, 1528-1529.
27. Fan, J.; Lei, J.; Wang, L.; Yu, C.; Tu, B.; Zhao, D. *Chem Comm.* **2003**, *17*, 2140-2141.
28. Maria Chong, A.S.; Zhao, X.S., *J. Phys. Chem. B* **2003**, *107*, 12650-12657.

29. Vinu,A.; Murugesan, V.; Hartmann, M., *J. Phys. Chem B.* **2004**, *108*, 7323-7330.
30. Ma, H.; He, J.; Evans, D.G.; Duan, X., *J. Mol. Catal. B: Enzymatic* **2004**, *30*, 209-217.
31. Blanco, R.M.; Terreros, P.; Fernandez-Perez, M.; Otero, C.; Diaz-Gonzalez, G., *J. Mol. Catal. B: Enzymatic* **2004**, *30*, 83-93.
32. Han, Y.-J.; Watson, J.T.; Stucky, G.D.; Butler, A., *J. Mol. Catal. B: Enzymatic* **2002**, *17*, 1-8.
33. Yiu, H.H.P.; Wright, P.A.; Botting, N.P., *J. Mol. Catal. B: Enzymatic* **2001**, *15*, 81-92.
34. Lei, J.; Fan, J.; Yu, C.; Zhang, L.; Jiang, S.; Tu, B.; Zhao, D., *Micropor. Mesopor. Mater.* **2004**, *73*, 121-128.
35. Pandya, P.H.; Jasra, R.V.; Newalkar, B.L.; Bhatt, P.N., *Micropor. Mesopor. Mater.* **2004**, *77*, 67-77.
36. Takahashi, H.; Li, B.; Sasaki, T.; Miyazaki, C., Kajino, T.; Inagaki, S., *Micropor. Mesopor. Mater.* **2001**, *44-45*, 755-762.
37. Diaz, J.F.; Balkus, Jr., K.J., *J. Mol. Catal. B: Enzymatic* *1***996**, *2*, 115-126.
38. Kadnikova, E.N.; Kostic, N.M., *J. Org. Chem..* **2003**, *68*, 2600.
39. Yan, A.X.; Li, X.W.; Ye, Y.H., *Appl. Biochem. Biotechnol.* **2002**, *101*, 113-129.
40. Droghetti, E.; Smulevich, G., *J. Biol. Inorg. Chem.* **2005**, *10*, 696-703.
41. Tortajada, M.; Ramon, D.; Beltran, D.; Amoros, P., *J. Mater. Chem.* **2005**, *15*, 3859-3868.
42. Vinu, A.; Hossain, K.; Ariga, K., *J. Nanosci. Nanotech.* **2005**, *5*, 347-371.
43. Han, U.; Lee, S.S.; Ying, J.Y., *Chem. Mater.* **2006**, A.S.A.P.
44. Yadav, G.D.; Jadhav, S.R., *Micropor. Mesopor. Mater.* **2005**, *86*, 215-222.
45. Pisklak, T.J.; Macias, M.; Coutinho, D.H.; Huang, R.S.; Balkus, Jr., K. J., *Top. Catal.* **2006**, *38*, 1-10.
46. Qiao, S.Z.; Yu, C.Z.; Xing, W.; Hu, Q.H.; Djojoputro, H.; Lu, G.Q., *Chem. Mater.* **2005**, *17*, 6172-6176.
47. Tortajada, M.; Ramon, D.; Beltran, D.; Amoros, P., *J. Mater. Chem.* **2005**, *15*, 3859-3868.
48. Diaz, J.F.; Bedioui, F.; Kurshev, V.; Kevan, L.; Balkus, Jr., K. J., *Chem. Mater.* **1997**, *9*, 61-67.
49. Rowsell , J.L.C.; Yaghi, O.M., *Micropor. Mesopor. Mater.* **2004**, *73*, 3-14.
50. Hoskins, B.F.; Robson, R., *J. Am. Chem. Soc.* **1990**, *112*, 1546.
51. Abrahams, B.F.; Hoskins, B.F.; Michail, D.M.; Robson, R., *Nature* **1994**, *369*, 727-729.
52. Wu, Z.Y.; Lee, S.; Moore, J.S., *J. Am. Chem. Soc.* **1992**, *114*, 8730-8732.
53. Macgillivray, L.; Subramanian, S.; Zaworotko, M.J., *Chem. Commun.* **1994**, 1325-1326.
54. Rosseinsky, M.J., *Micropor. Mesopor. Mater.* **2004**, *73*, 15-30.

55. Dinolfo, P.H.; Hupp, J.T., *Chem. Mater.* **2001**, *13*, 3113-3125.
56. Stang, P.F.; Olenyuk, B., *Acc. Chem. Res.* **1997**, *30*, 502-518.
57. Czaplewski, K.; Hupp, J.; Snurr, R.Q., *Adv. Mater.* **2001**, *13*, 1895-1897.
58. Be'langer, S.; Hupp, J.T.; Stern, C.L.; Slone, R.V.; Watson, D.F.; Carrell, T.G., *J. Am. Chem. Soc.* **1999**, *121*, 557-563.
59. Be'langer, S.; Hupp, J.T., *Angew. Chem. Int. Ed.* **1999**, *38*, 2222-2224.
60. Slone, R.V.; Hupp, J.T.; Stern, C.L.; Albrecht-Schmitt, T.E., *Inorg. Chem.* **1996**, *35*, 4096-4097.
61. Schlichte, K.; Kratzke, T.; Kaskel, S., *Micropor. Mesopor.* **2004**, *73*, 81-88.
62. Wei, Q.; Nieuwenhuyzen, M.; James, S.L., *Micropor. Mesopor. Mater.* **2004**, *73*, 97-100.
63. James, S.L., *Chem. Soc. Rev.* **2003**, *32*, 276-288.
64. Maeda, K., *Micropor. Mesopor. Mater.* **2004**, *73*, 47-55.
65. Eddaoudi, M.; Kim, J.; Rosi, N.; Vodak, D.; Wachter, J.; ÕKeeffe, M.; Yaghi, O.M., *Science* **2002**, *295*, 469-472.
66. Chen, A.; Eddaoudi, M.; Hyde, S.T.; ÕKeeffe, M.; Yaghi, O.M., *Science* **2001**, *291*, 1021-1023.
67. Chael, H.K.; Siberio-Pe'rez, D.Y.; Kim, J.; Go, Y.; Eddaoudi, M.; Matzger, A.J.; O'Keeffe, M.;. Yaghi, O.M., *Nature* **2004**, *427*, 523-527.
68. Evans, O.R.; Manke, D.R.; Lin, W., *Chem. Mater.* **2002**, *14*, 3866-3874.
69. Mori, W.; Takamizawa, S.; Kato, C.N.; Ohmura, T.; Sato, T., *Micropor. Mesopor. Mater.* **2004**, *73*, 31-46.
70. Jiang, A.; Yu, Z.; Wang, S.; Jiao, C.; Li, J.; Wang, Z.; Cui, Y., *Eur. J. Inorg. Chem.* **2004**, 3662-3667.
71. de Lill, A.T.; Gunning, N.S.; Cahill, C.L., *Inorg. Chem.* **2005**, *44*, 258-266.
72. Chen, W.; Wang, J.-Y.; Chen, C.; Yue, Q.; Yuan, H.-M.; Chen, J.-S.; Wang, S.-N., *Inorg. Chem.* **2003**, *42*, 944-946.
73. Shi, X.; Zhu, G.; Wang, X.; Li, G.; Fang, Q.; Zhao, X.; Wu, G.; Tian, G.; Xue, M.; Wang, R.; Qiu, S., *Crystal Growth Design* **2005**, *5*, 341-346.
74. Eddaoudi, M.; Moler, D.B.; Li, H.; Chen, B.; Reineke, T.M.; O'keeffe, M.; Yaghi, O.M., *Acc. Chem. Res.* **2001**, *34*.
75. Fang, Q.; Zhu, G.; Xue, M.; Sun, J.; Tian, G.; Wu, G.; Qiu, S., *Dalton Trans.* **2004**, 2202-2207.
76. Choi, E.-Y.; Park, K.; Yang, C.-M.; Kim, H.; Son, J.-HLee, .; S.W.; Lee, Y. H.; Min, D.; Kwon, Y.-U,. *Chem. Eur. J.* **2004**, *10*, 5535-5540.
77. Zhao, W.; Fan, J.; Okamura, T.; Sun, W.-Y.; Ueyama, N., *New J. Chem.* **2004**, *28*, 1142-1150.
78. Yang, G.; Raptis, R.G. *Chem. Commun.* **2004**, 2058-2059.
79. Kesanli, B.; Cui, Y.; Smith, M.R.; Bittner, E.W.; Lin, W., *Angew. Chem. Int. Ed.* **2005**, *44*, 72-75.

80. Dybtsev, A.N.; Chun,H.; Yoon, S.H.; Kim, D.; Kim, K., *J. Am. Chem. Soc.* **2004**, *126*, 32-33.
81. Ratcliffe, C.I.; Soldatov, D.V.; Ripmeester, J.A., *Micropor. Mesopor. Mater.* **2004**, *73*, 71-79.
82. Wan, S.-Y.; Huang, Y.-T.; Li, Y.-Z.; Sun, W.-Y., *Micropor. Mesopor. Mater.* **2004**, *73*, 101-108.
83. Barea, A.; Navarro, J.A.R.; Salas, J.M.; Masciocchi,N.; Galli, S.; Sironi, A., *Polyhedron* **2003**, *22*, 3051-3057.
84. Tao, J.; Yin, X.; Huang, R.; Zheng, L.; Ng, S.W., *Inorg. Chem. Commun.* **2002**, *5*, 975-977.
85. Choi, A.J.; Suh, M.P., *J. Am. Chem. Soc.* **2004**, *126*, 15844-15851.
86. Seki, K., *Chem. Commun.* **2001**, 1496-1497.
87. Seki, K.; Mori, W., *J. Phys. Chem. B* **2002**, *106*, 1380-1385.
88. Millward, A.R.; Yaghi, O.M., *J. Am. Chem. Soc.,* **2005**, *127*, 17998-17999.
89. Okazaki, S.; Nagasawa, S.; Goto, M.; Furusaki, S.; Wariishi, H.; Tanaka, H., *Biochem. Eng. J.* **2002**, *12*, 237-241.
90. Melchionna, S.; Barteri, M.; Ciccotti, G., *J. Phys. Chem.* **1996**, *100*, 19241-19250.
91. Wariishi, A; Kabuto, M; Mikuni, J.; Oyadomari, M.; Tanaka, H., *Biotechnol. Prog.* **2002**, *18*, 36-42.
92. Wright, W.W.; Laberge, M.;Vanderkooi, J.M., *Biochemistry* **1997** *36*, 14724-14732.
93. Benitez, F.J.; Beltran-Heredia, J.; Gonzalez, T.; Pascual, A., *I. Chem. Eng. Comm.* **1993**, *119*, 151-165.
94. Anthony, B.R.; Warczak, K.L.; Donohue, T.J., *Proc. Nat. Acad. Sci.* **2005**, *102*, 6502-6507.
95. Pickering, C.D.; Wiesner, M.R., *Environ. Sci. Technol.* **2005**, *39*, 1359-1365.
96. Jacobsen, E.N.; Zhang, W.; Muci, A.R.; Ecker, J.R.; Deng, L., *J. Am.Chem. Soc.* **1991**, *113*, 7063.
97. Gbery, G.; Zsigmond, A.; Balkus, Jr., K.J., *Catal. Letters* **2001**, *74*, 77-80.
98. Liu, Y.; Wang, M.; Zhao, F.; Guo, Z.; Chen, H.; Dong, S., *J. Electroanal. Chem.* **2005**, *581*, 1-10.
99. Xu, Z.; Gao, N.; Chen, H.; Dong, S, *Langmuir* **2005**, *21*, 10808-10813.
100. Liu, Y.; Wang, M.; Zhao, F.; Guo, Z.; Chen, H.; Dong, S., *J. Electroanal. Chem.* **2005**, *581*, 1-10.
101. Katz, E.; Willner, I., *Angew. Chem. Int. Ed.* **2005**, *44*, 4791-4794.
102. Balkus, Jr., K.J.; Welch, A.A.; Gnade, B.E., *J. Incl. Phenom. Mol. Recog. Chem.* **1991**, *10*, 141-151.

I: Nanostructures for Biomolecular Catalysis

I–2: Nanotubes, Nanofibers, and Nanoparticles

Chapter 5

Protein–Carbon Nanotube Conjugates

Sandeep S. Karajanagi, Prashanth Asuri, Edward Sellitto, Bilge Eker, Shyam Sundhar Bale, Ravi S. Kane*, and Jonathan S. Dordick*

Department of Chemical and Biological Engineering, Rensselaer Polytechnic Institute, Troy, NY 12180

Proteins have been conjugated to carbon nanotubes for applications in biosensing, biorecognition, delivery, and functional composites. Despite the growing interest in these carbon nanotube-protein hybrids, very little is known about how carbon nanotubes affect the structure and function of bound proteins. Here we provide an overview of our recent efforts to gain a more fundamental understanding of how proteins interact with carbon nanotubes. We also discuss recent results from our laboratories which suggest several new opportunities for protein-carbon nanotube conjugates to address problems in materials science and biotechnology.

The recent emergence of nanoscale materials with precise sizes, shapes, and surface chemistries has revolutionized the fields of materials, biomedical imaging, diagnostics, drug delivery, and biosensing *(1-4)*. Carbon nanotubes, in particular, have attracted significant attention for their useful and intriguing physicochemical properties *(5)*. Apart from offering numerous opportunities in materials and electronics research, carbon nanotubes hold great potential for applications in biosensing, functional surfaces and coatings, biorecognition, and drug delivery *(6-9)*. A wide range of biomolecules including oligonucleotides, proteins, peptides, and carbohydrates have been interfaced with carbon nanotubes for realizing these applications *(10-16)*. Herein we describe recent results from our labs on the interaction of proteins with carbon nanotubes and the potential opportunities that may emerge from this approach.

Proteins have been attached to carbon nanotubes using both covalent and noncovalent approaches *(14-20)*. In one of the earliest studies in this area, proteins such as streptavidin and HupR, a DNA binding protein derived from *Rhodobacter capsulatus*, appeared to adsorb noncovalently on the external surface of multi-walled carbon nanotubes (MWNTs) to form helical structures *(21)*. Furthermore, numerous proteins have been shown to adsorb noncovalently onto single-walled carbon nanotubes (SWNTs) *(15-16,22,23)*. For example, Erlanger *et al.* developed an antifullerene IgG antibody, which selectively recognized and adsorbed onto SWNTs *(17)*. Recently, cytochrome C was found to adsorb noncovalently onto a semiconducting SWNT device that was used to detect the presence of protein molecules with a sensitivity as high as 20 protein molecules per nanotube *(15)*. Wang *et al.* adsorbed electrochemically active catalase and xanthine oxidase onto SWNTs deposited on gold electrodes *(16)*. Azamian *et al. (21)* and Lin *et al. (24)* reported strong adsorption of ferritin onto SWNTs. In contrast, incubation of chemical vapor deposition-produced SWNTs with a ferritin solution had previously been found to lead to no observable adsorption *(25)*. These results suggest that protein adsorption onto SWNTs is complex and depends on many factors. In particular, protein-nanotube interactions remain poorly understood and their study has served as the goal of much of our work in this arena.

Protein-SWNT Interactions

Despite the growing interest in preparing protein-nanotube conjugates, relatively little is known about how the SWNT surface affects the structure and function of adsorbed proteins. We, therefore, undertook the first in-depth evaluation of both the structure and function of proteins adsorbed onto carbon nanotubes. Using enzymes as highly sensitive probes of protein function, we demonstrated that protein structure and function are strongly influenced by the hydrophobic, nanoscale environment of an SWNT *(26)*. The structure and function of two unrelated enzymes, α-chymotrypsin (CT) and soybean peroxidase (SBP), adsorbed onto SWNTs were examined using enzyme kinetic analysis, FT-IR spectroscopy, and atomic force microscopy (AFM). Both enzymes adsorbed strongly onto the SWNTs, presumably via hydrophobic interactions with the SWNTs. Most of the SBP adsorbed within the first minute of exposure to the SWNTs (Fig. 1A), and followed a pseudo-saturation behavior (Fig. 1B). Similar results were obtained for CT. SBP retained up to 30% of its native activity upon adsorption, while CT retained only 1% of its native activity; hence, considerable difference in the outcome of adsorption of the two proteins was evident. Analysis of the secondary structure of the proteins using FT-IR spectroscopy revealed that structural changes occur upon adsorption for both

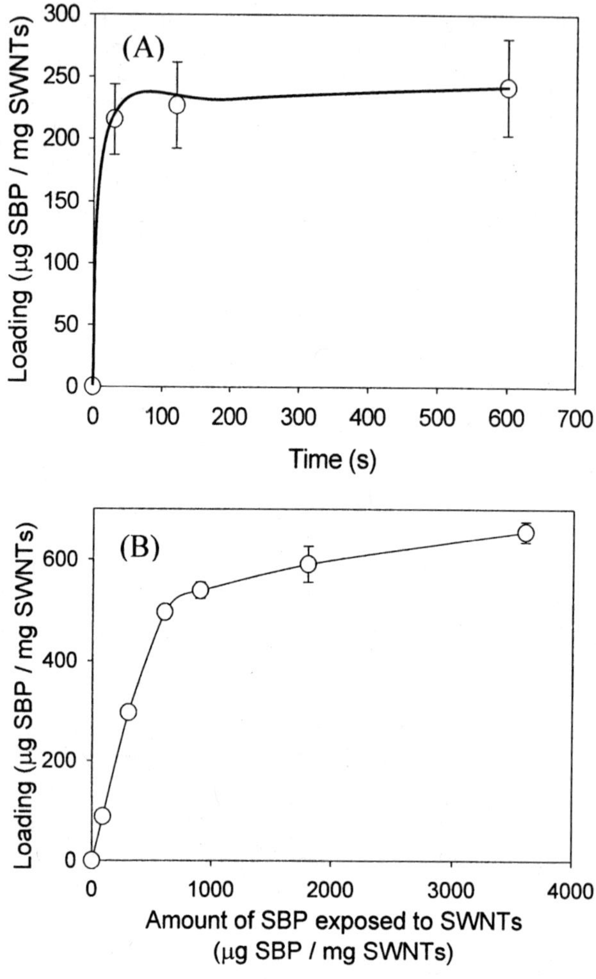

Figure 1. Loading of SBP on SWNTs: (A) Enzyme loading on SWNTs as a function of time using a SBP solution of concentration 0.25 mg/mL . (B) Enzyme loading as a function of amount of SBP exposed to the SWNTs. Error bars indicate the standard deviation of triplicate measurements.

proteins. A change in the secondary structure between the soluble (e.g., native) and adsorbed states of the enzymes, as represented by the simple sum of magnitudes of changes in α-helix and β-sheet contents, were ca. 13% and 44% for SBP and CT, respectively (Table I). Thus, both enzymes undergo a change in secondary structure upon adsorption; however, substantial secondary structural perturbation was only observed with CT.

Table I. Quantitative secondary structure of SBP and CT in the solution phase and adsorbed state

Sample	*% α-helix*	*% β-sheet*
Solution CT	13.6 ± 3.5	50.0 ± 2.4
SWNT-CT	31.5 ± 2.9	23.5 ± 5.1
Solution SBP	36.1 ± 1.2	25.1 ± 2.5
SWNT-SBP	27.9 ± 4.1	20.6 ± 6.9

NOTE: Adapted from Karajanagi *et al.* *(26)*

Atomic force microscopy (AFM) was used to further investigate the structural perturbation of the two adsorbed enzymes. In case of SWNT-SBP, distinct globular SBP molecules were seen to be adsorbed on the SWNTs (Fig. 2A). Line scans revealed that the height of the globular structures ranged from 4 to 6 nm, which closely matches the molecular dimensions of SBP (6.1 x 3.5 x 4.0 nm) *(27)*. The AFM image of SWNT-CT, however, presented a vastly different picture with no distinct enzyme molecules seen on the SWNTs; rather, a layer of CT was observed. Line scans of the image revealed a highly unfolded and aggregated CT layer with the height varying from 2 to 20 nm, values which are far different from the molecular size of CT (5.1 x 4.0 x 4.0 nm) *(28)* (Fig. 2B). Thus, consistent with the activity and FT-IR results, AFM images of the adsorbed enzymes indicated that SBP retains its native three-dimensional shape, while CT appears to unfold on the SWNT surface. The SWNTs may, therefore, significantly alter the structure and activity of a protein and the extent of perturbation from native-like properties depends on the nature of the protein.

Effect of nanotube curvature on protein stability

We have discovered a novel property of SWNTs – their ability to stabilize proteins at elevated temperatures and in organic solvents to a greater extent than conventional flat supports *(29)*. We observed that the half-life, $\tau_{1/2}$, of SBP adsorbed onto SWNTs at 95 °C was ca. 90 min, 10-fold higher than that of the native enzyme (Table II). To evaluate whether the stabilization of SBP on SWNTs was simply a result of immobilization, we adsorbed the enzyme onto flat

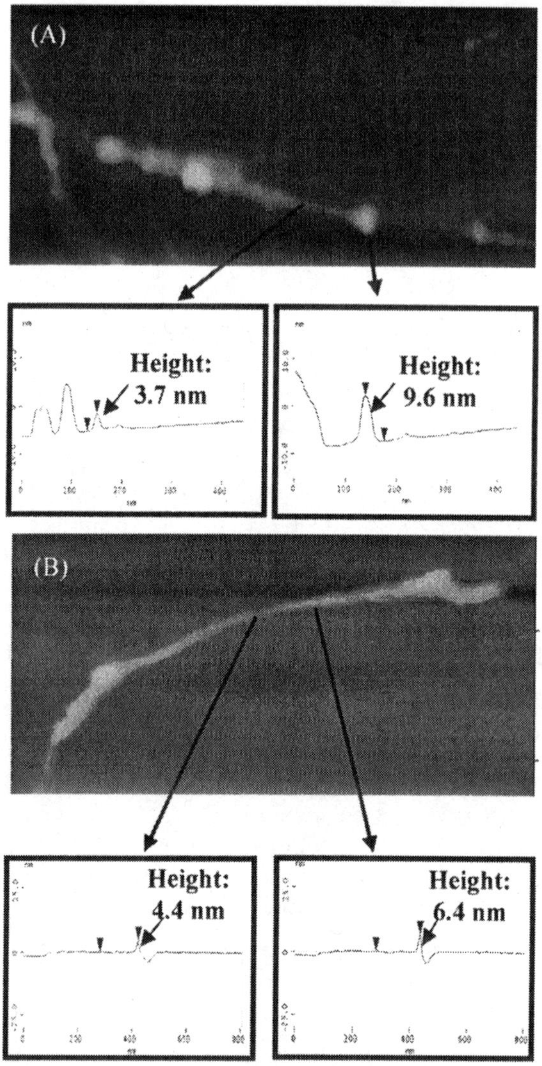

Figure 2. AFM images of enzymes adsorbed onto SWNTs: (A) SBP on SWNTs (B) CT on SWNTs. Arrows point at line sections that reveal the height above the underlying mica surface. This figure is adapted from Karajanagi et al. (26).

graphite flakes, which represent a chemically similar conventional micro/macroscale surface for comparison. The deactivation constant for SBP on SWNTs was ca. half that of SBP on graphite indicating that SBP was ca. 1.9 ±

Table II. Stability of SBP containing samples under harsh conditions

Sample	*Conditions*	*Half-life $\tau_{1/2}$ (min)* [a]	*Deactivation constant $x\ 10^3\ (min^{-1})$* [b]
Native SBP	Aqueous buffer, 95°C	9.0 ± 0.8	77.3 ± 0.7
Graphite-SBP	Aqueous buffer, 95°C	43.9 ± 1.6	15.8 ± 0.6
SWNT-SBP	Aqueous buffer, 95°C	87.9 ± 5.1	7.9 ± 0.5
Native-SBP	Neat Methanol, room temperature	See note	See note
Graphite-SBP	Neat Methanol, room temperature	35.0 ± 2.4	19.8 ± 1.4
SWNT-SBP	Neat Methanol, room temperature	64.5 ± 4.3	10.7 ± 0.7

NOTE: No activity was detected for native SBP in neat methanol. This table is adapted from Asuri *et al. (29)*.

a: Half-life of an enzymatic preparation is based on measurements of initial reaction rates as a function of the incubation time in a given environment.

b: The deactivation constant is defined as the first-order exponential decay constant in the equation $A/A_o = \exp(-kt)$, where A and A_o are the enzymatic activities at time t = t and t = 0, respectively. The negative slope of a straight line fit through a plot of $\ln(A/A_o)$ vs t gives the value of the deactivation constant.

0.2 times more stable at 95 °C on SWNTs than on graphite flakes (Table II). Even in nearly neat methanol, SBP adsorbed onto SWNTs had 2.0 ± 0.2 fold greater $\tau_{1/2}$ than that of the enzyme adsorbed onto flat supports (Table II). These results indicate a statistically significant increase in stability in harsh environments for SBP adsorbed onto SWNTs. Similar enhancement in stabilization on SWNTs was observed for the unrelated protease subtilisin Carlsberg (SC). These results suggest that proteins adsorbed onto this nanoscale support are more stable to deactivation under the harsh conditions of high temperature and organic solvents when compared to those attached onto macroscale supports. A thorough investigation of this interesting result using a combination of experiments and theoretical analysis has revealed that the stabilization results from the intrinsically high surface curvature of SWNTs, which suppresses unfavorable protein-protein lateral interactions *(29)*.

Briefly, at the same separation along the protein-substrate interface, and at the same surface coverage, the curvature of a cylindrical support results in an increase in the center-to-center distance between adjacent adsorbed proteins (Fig. 3A). If unfavorable interactions between adjacent adsorbed proteins contribute to their deactivation in harsh environments, then this increase in separation on nanoscale supports should result in a decrease in the deactivation rate of the proteins, and may result in increased protein stability on SWNTs relative to flat supports. We determined the deactivation rate constants in aqueous buffer at 95 °C and in methanol for SBP adsorbed onto SWNTs and graphite flakes at different fractional surface coverages. Figures 3B and C reveal a strong dependence of the enzymatic deactivation rate on surface coverage, with identical deactivation constants on SWNTs and graphite flakes at very low surface coverage. If unfavorable "lateral" interactions between adsorbed proteins contribute significantly to protein deactivation, then these "lateral" interactions, and hence the rate of enzymatic deactivation, should decrease on all supports with decreasing surface coverage (i.e. with an increase in the average separation between adsorbed proteins). Furthermore, if this hypothesis is valid then the enhancement in stability on SWNTs relative to graphite flakes should disappear at very low surface coverages. Indeed, the strong dependence of the enzymatic deactivation rates on surface coverage and the identical deactivation constants of SBP adsorbed onto SWNTs and graphite flakes at low surface coverage suggest that the increased protein stability on the SWNTs is most likely due to decreased lateral interactions on the nanoscale support relative to macroscale supports. If the enhanced stability on SWNTs relative to graphite flakes were a result of protein-surface interactions being different or disfavored on the curved SWNTs relative to the flat graphite support, then the difference in the enzymatic deactivation rates on SWNTs and graphite flakes should have persisted even at low surface coverages *(29)*.

Applications of carbon nanotube-protein conjugates

In addition to an improved knowledge of protein-nanotube conjugate structure and function, we have demonstrated several applications of such conjugates, which may impact a variety of fields ranging from tools for nanoscale assembly to biocatalyst formulations in aqueous and nonaqueous media.

Solubilization of SWNTs using proteins. While our previous work focused on the ability of SWNTs to increase protein stability, we have shown that proteins can these solutions by a variety of complementary techniques including UV-Vis in turn help solubilize SWNTs in water *(30)*. A variety of proteins differing in

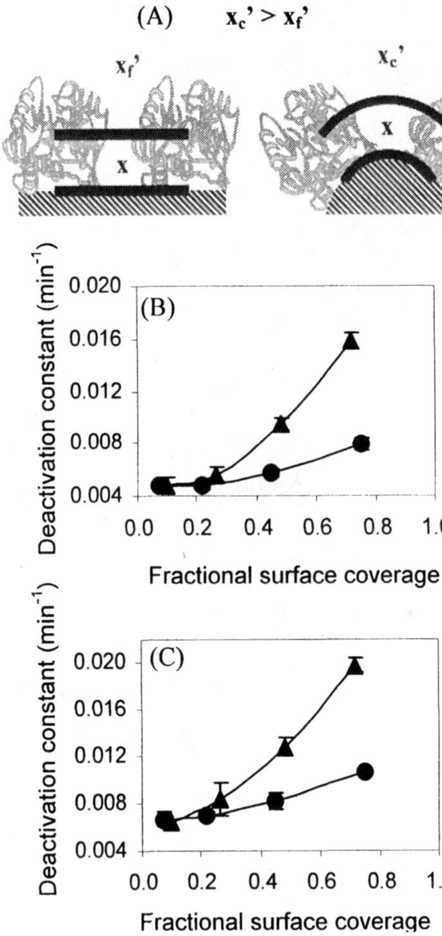

Figure 3. Effect of lateral interactions on the deactivation constants of SBP adsorbed onto different supports. (A) Schematic (drawn approximately to scale) depicting SBP molecules adsorbed onto a "flat" support (left) and a cylindrical support (right). Deactivation constants for SBP adsorbed onto SWNTs (circles) and graphite flakes (triangles) as a function of surface coverage – (B) at 95 °C and (C) in 100% methanol. In (B) and (C), error bars indicate the standard deviation of triplicate measurements. This figure is adapted from Asuri et al. (29).

size and structure were used to generate individual nanotube solutions by this noncovalent functionalization procedure. Ultrasonication of an aqueous dispersion of SWNTs in a protein solution followed by ultracentrifugation yielded a black colored solution of SWNTs (Fig. 4A inset). We characterized spectroscopy, Raman spectroscopy, and AFM, which confirmed dispersion in water at the individual nanotube level. The UV-Vis absorption spectra of dispersions of SWNTs are known to be sensitive to their aggregation state *(31)*. The UV-Vis spectrum for a solution of SWNTs obtained using BSA (SWNT-BSA) exhibited sharp and well-resolved peaks (Fig. 4A), which are characteristic of aqueous solutions containing debundled, individually dispersed SWNTs *(31,32)*. Visualization using atomic force microscopy revealed that most of the SWNTs were indeed individually dispersed in solution (Fig. 4B). The radial breathing mode Raman spectrum for soluble (i.e., individually dispersed) SWNT-BSA obtained using a 785 nm laser excitation showed a prominent peak at 234 cm^{-1} and no peak at 267 cm^{-1} (data not shown); a peak at 267 cm^{-1} is observed if the SWNTs are aggregated *(33)*. Characterization using Raman spectroscopy thus further supported our claim that proteins can be used to individually disperse SWNTs in an aqueous solution. These results may result in a general method to prepare aqueous solutions of SWNTs for use in self-assembled materials and biomedical applications. Furthermore, the wide variety of functional groups present on the adsorbed proteins may be used as orthogonal reactive handles for the functionalization of carbon nanotubes.

Interfacial assembly of SWNT-enzyme conjugates. We have shown that SWNTs can be used as "nanoscale conveyors" to direct enzymes from an aqueous phase to an aqueous-organic interface and enable interfacial biocatalysis *(34)*. SWNTs can be assembled at aqueous-organic interfaces with the aid of anionic, cationic, or neutral surfactants. For example, when an aqueous dispersion of SWNTs is contacted with an equal volume of hexane containing 2 mM Aerosol-OT(AOT), SWNTs from the aqueous phase are transferred to the aqueous-organic interface.

We used this phenomenon to direct enzymes to an aqueous-organic interface (Fig. 5A) and perform interfacial biotransformations. Specifically, we adsorbed SBP onto SWNTs and assembled the SWNT-SBP conjugates at the hexane-water interface using AOT. While aqueous-organic systems are advantageous for biotransformations involving water-insoluble substrates *(35)*, biphasic biotransformations are typically limited by the poor transport of reactants from the organic phase into the aqueous enzyme phase and the loss of enzyme activity, possibly due to denaturation at the interface *(36)*. The assembly of SWNT-enzyme conjugates at the aqueous-organic interface may help overcome these transport limitations. Furthermore, as we discussed earlier, the nanoscale environment of SWNTs can enhance the function and stability of adsorbed proteins relative to microscale or macroscopic supports.

The catalytic activity of the SWNT-SBP adsorbed at the interface was measured using *p*-cresol, a substrate soluble in hexane, and H_2O_2, which is

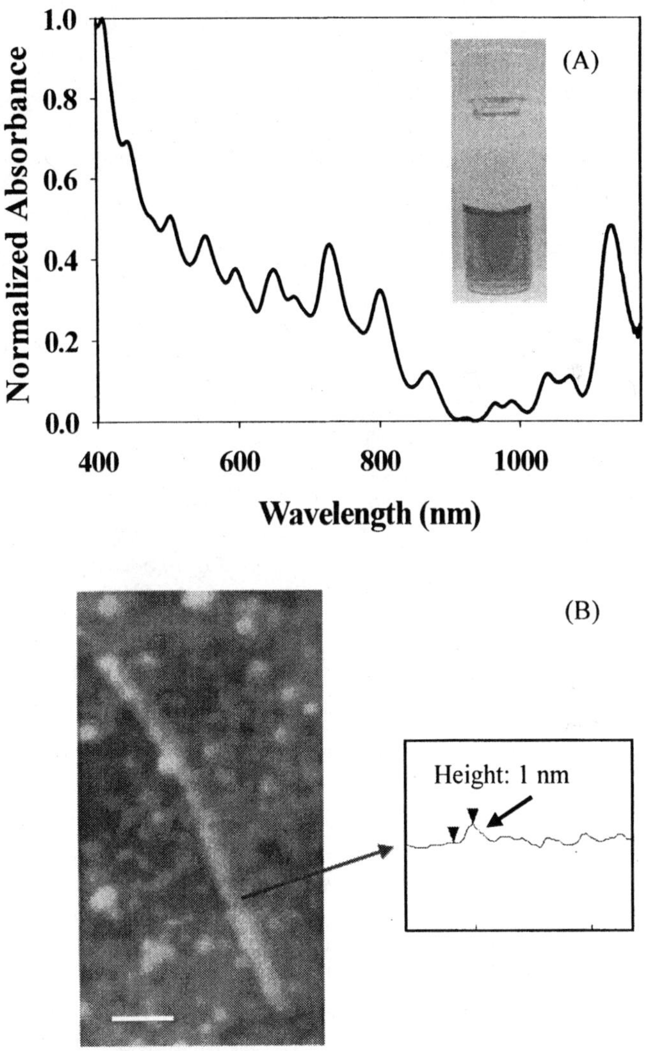

Figure 4. Characterization of solutions of SWNTs obtained using BSA (SWNT-BSA). (A) UV-Vis absorption spectrum of SWNT-BSA. Inset shows a photograph of an SWNT-BSA solution. (B) AFM micrograph of SWNT-BSA. The scale bar represents 100 nm. Line scan shown in the box reveals that the diameter of the SWNT is 1 nm. This figure is adapted from Karajanagi et al. (30).

soluble only in water. SWNT-SBP was highly active at the interface with reaction rates that were at least *three orders of magnitude* greater than those observed with identical enzyme concentrations for either native SBP or SWNT-SBP in the aqueous phase of a biphasic system in the absence of AOT, or for native SBP in the aqueous phase in the presence of AOT. Non-nanoscale supports, e.g., graphite flakes, did not assemble at the interface and those that did (e.g., hydrophobic nanoporous glass beads) showed significantly lower enzyme activity, presumably due to intraparticle diffusion limitations (Fig. 5B). Interestingly, a bell-shaped curve was observed for interfacial activity of SWNT-SBP (measured under quiescent conditions without disturbing the interfacial SWNT-enzyme) plotted against the amount of SWNT-SBP assembled at the interface (Fig. 5B). At sub-monolayer coverages, the reaction rate increased linearly with increasing SWNT-SBP interfacial coverage. For higher amounts of SWNT-SBP at the interface, the absolute reaction rates decreased with increasing amounts of interfacial SWNT-SBP. These results suggest that multiple layers form at the interface for higher amounts of SWNT-SBP, which may limit the rate of diffusion of the reactants across the interface. In addition to enhancing the rate of interfacial biotransformations, the nanotubes also significantly enhanced the thermostability of the interfacial SBP at 95 °C by prolonging the half-life of SBP by 8-fold over that of native SBP under the same conditions. Further extension of this method to direct a broader range of enzymes at the aqueous-organic interface may result in a new form of highly efficient interfacial biocatalysis to carry out biphasic reactions, including natural product derivatization, polymer synthesis and degradation, and bioremediation.

Water-soluble MWNT-enzyme conjugates. We have also attached enzymes covalently to carbon nanotubes to prepare unique *water-soluble* multi-walled carbon nanotube-enzyme conjugates that possess the virtues of both soluble and immobilized enzymes *(37)*. Structurally and functionally different enzymes such as SBP, SC, *Candida antarctica* Lipase B (CALB), proteinase K (PrK), and tyrosinase (Tyr) were covalently attached to the MWNTs using carbodiimide chemistry. These conjugates were soluble in aqueous buffer, followed Michaelis-Menten kinetics, and retained high catalytic efficiencies (values of V_{max}/K_M). Figure 6 shows enzyme kinetic behavior and parameters for SBP covalently attached to the MWNTs (MWNT-SBP). High loadings of enzymes on the MWNTs were obtained and varied from ca. 150 to ca. 200 mg enzyme/ g MWNT. All immobilized enzymes retained activities as high as ca. 40 to 60% when compared to their solution phase activities. In addition to high loading and activity, the immobilized enzymes were highly thermostable – up to 13-fold higher half-life at elevated temperatures of 70°C (95 °C for SBP) over their solution phase counterparts. What is particularly interesting is that the enzymes covalently attached to MWNTs retain their high intrinsic activity along with their increased thermostability while in a *soluble* form. The high surface area of MWNTs led to high enzyme loadings, yet the intrinsic high length of the

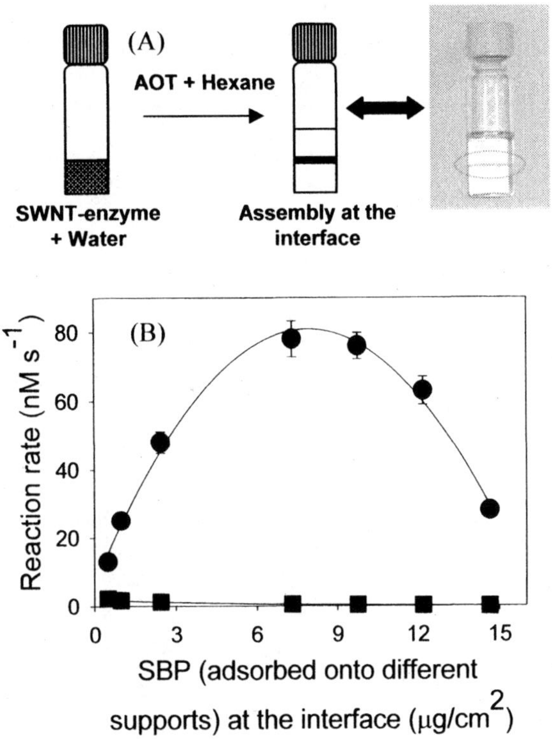

Figure 5. Interfacial assembly of SWNT-enzyme at the aqueous-organic interface. (A) Schematic demonstrating the assembly of SWNT-enzyme at the hexane-water interface. The photograph on the right shows SWNT-SBP assembled at the hexane-water interface. (B) Absolute reaction rates for SBP adsorbed onto SWNTs (circles) and hydrophobic OTS-functionalized nanoporous glass beads (squares) – at the interface. The error bars indicate the standard deviation of triplicate measurements. This figure is adapted from Asuri et al. (34).

MWNTs afforded facile filtration. As a result, these MWNT-conjugates were highly reusable and retained ca. 70% of their initial activity even after being reused more than 100 times. These conjugates thus provide a unique combination of useful attributes, including low mass transfer resistance, high activity and stability, and reusability in biocatalytic applications.

Carbon nanotube-enzyme conjugates in ionic liquids. SWNT-enzyme conjugates are not only active in buffer and organic solvents but also in room temperature ionic liquids (RTILs). RTILs are intriguing solvents, which are recognized as 'green' alternatives to volatile organics *(38)*. However, low enzyme activity in these solvents, particularly at very low water contents, limits

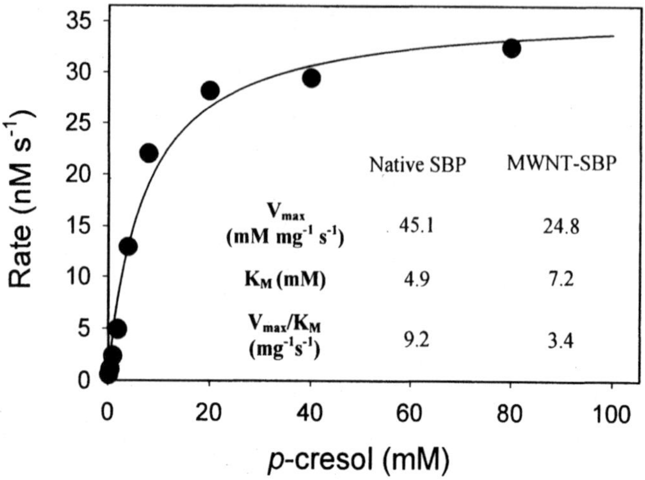

Figure 6. Effect of substrate concentration on the initial rate of oxidation of p-cresol catalyzed by MWNT-SBP. The kinetic parameters were obtained by using a nonlinear Michaelis-Menten fit to the data. The error bars are smaller than the symbols shown. The numbers in the inset represent the kinetic parameters for native SBP and MWNT-SBP. This figure is adapted from Asuri et al. (37).

the potential commercial use of RTILs for biotransformations. It was of interest, therefore, to explore whether SWNT-enzyme conjugates remain active in RTILs. We adsorbed PrK onto SWNTs (0.5 mg PrK/mg SWNTs) using a procedure adapted frm Karajanagi *et al. (26)*. As a proof-of-concept, we then performed the SWNT-PrK catalyzed transesterification of N-acetyl-L-phenylalanine ethyl ester (NAPEE) with 1-propanol in the hydrophobic ionic liquid 1-butyl,3-methylimidazolium hexafluorophosphate (BMIM(PF_6)) to yield N-acetyl-L-phenylalanine propyl ester (NAPPE).

Specifically, PrK (1 mg/mL) in the native and adsorbed state was used to catalyze the transesterification of 40 mM NAPEE in BMIM(PF_6) containing 7% (v/v) 1-propanol. Different concentrations of water (2 and 5%, v/v) in the RTIL were employed to assess the influence of water on the reactivity of the conjugate. Two competing reactions occur in such a system; the hydrolysis of NAPEE to N-acetyl-L-phenylalanine (NAPA) and the transesterification of NAPEE to NAPPE. The reaction was allowed to proceed for 24 h at room temperature and the products formed were detected using HPLC. For the same concentration of enzyme in a reaction mixture containing 2% (v/v) water, ca. four-fold higher conversion of NAPEE to its products was observed with SWNT-PrK as compared to native PrK (Table III). A detailed investigation into this intriguing result is warranted and is currently underway in our labs. Interestingly, when the water content of the system was increased to 5% (v/v), the total conversion of NAPEE achieved using native PrK increased to nearly 50% while that using

Table III. Proteinase K catalyzed hydrolysis and transesterifaction of NAPEE

	% Converison of NAPEE in 2 % water (v/v)			*% Converison of NAPEE in 5 % water (v/v)*		
	NAPEE to NAPA	NAPEE to NAPPE	Total conversion	NAPEE to NAPA	NAPEE to NAPPE	Total conversio n
Native PrK	7.5 ± 2.9	8.0 ± 3.4	15.6 ± 6.1	29.8 ± 0.2	16.9 ± 0.4	46.7± 0.6
SWN T-PrK	39.8 ± 9.6	30.1 ± 4.8	69.9 ± 4.8	58.5 ± 6.0	20.8 ± 3.0	79.2 ± 3.0

SWNT-PrK increased to nearly 80%. This increased conversion may be due to the higher activity of the native and immobilized enzyme, presumably because of the higher water content of the reaction mixture. However, the presence of higher water content in the reaction mixture also led to a higher yield of hydrolysis, as shown in Table III. Nevertheless, these initial results indicate that the SWNT-proteinase K conjugates are active in ionic liquids. Studies are underway to explore the utility of SWNT-enzyme conjugates in ionic liquids in more detail.

Future Outlook

While the extent of structural and functional perturbation of proteins on carbon nanotubes may be protein specific, a detailed understanding of the parameters that control protein structure and function on nanotubes is still lacking. Control over the amount, location, orientation, and function of proteins on carbon nanotubes will be critical in order to design optimal conjugates of proteins and carbon nanotubes for applications in biotechnology, self-assembled materials, and medicine. Opportunities abound in preparing multifunctional surfaces, devices, and materials using carbon nanotube-protein conjugates that also make use of attractive features of the nanoscale material, including high surface areas per unit weight, electrical conductivity, and high mechanical strength. Substantial gains in our knowledge of protein immobilization onto macroscale supports, including novel methods of attachment, will be important to translate such approaches to the nanoscale. We continue to explore ways to understand and control the molecular level interactions that govern protein behavior on carbon nanotubes and exploit the numerous possibilities that will surely arise by interfacing proteins with carbon nanotubes in materials and biotechnology.

Acknowledgments

We acknowledge support by grants to J.S.D. and R.S.K. from NSF (Nanoscale Science and Engineering Center; DMR-0117792) and to J.S.D. (BES-0227783).

References

1. Peng, X.; Manna, L.; Yang, W.; Wickham, J.; Scher, E.; Kadavanich, A.; Alivisatos, A. P. *Nature*, **2000**, *404*, 59.
2. Strano, M. S.; Dyke, C. A.; Usrey, M. L.; Barone, P. W.; Allen, M. J.; Shan, H.; Kittrell, C.; Hauge, R. H.; Tour, J. M.; Smalley, R. E. *Science*, **2003**, *301*, 1519.
3. Bachtold, A.; Hadley, P.; Nakanishi, T.; Dekker, C. *Science*, **2001**, *294*, 1317.
4. Peigney, A. *Nature Mater.*, **2003**, *2*, 15.
5. Ajayan, P. M. *Chem. Rev.*, **1999**, *99*, 1787.
6. Barone, P. W.; Baik, S.; Heller, D. A.; Strano, M. S. *Nature Mater.* **2005**, *4*, 86.
7. Chen, R. J.; Bangsaruntip, S.; Drouvalakis, K. A.; Kam, N. W. S.; Shim, M.; Li, Y.; Kim, W.; Utz, P. J.; Dai, H. *Proc. Natl. Acad. Sci. U.S.A.* **2003**, *100*, 4984.
8. Kam, N. W. S.; Jessop, T. S.; Wender, P. A.; Dai, H. *J. Am. Chem. Soc.* **2004**, *126*, 6850.
9. Graff, R. A.; Swanson, J. P.; Barone, P. W.; Baik, S.; Heller, D. A.; Strano, M. S. *Adv. Mat.* **2005**, *17*, 980.
10. Chen, X.; Lee, G. S.; Zettl, A.; Bertozzi, C. R. *Angew. Chem. Int. Ed.*, **2004**, *43*, 6111.
11. Pompeo, F.; Resasco, D. E. *Nano Lett.*, **2002**, *2*, 369.
12. Zheng, M.; Jagota, A.; Semke, E. D.; Diner, B. A.; Mclean, R. S.; Lustig, S. R.; Richardson, R. E.; Tassi, N. G. *Nature Mater.* **2003**, *2*, 338.
13. Li, J.; Ng, H. T.; Cassell, A.; Fan, W.; Chen, H.; Ye, Q.; Koehne, J.; Han, J.; Meyyappan, M. *Nano Lett.*, **2003**, *3*, 597.
14. Dieckmann, G. R.; Dalton, A. B.; Johnson, P. A.; Razal, J.; Chen, J.; Giordano, G. M.; Munoz, E.; Musselman, I. H.; Baughman, R. H.; Draper, R. K. *J. Am. Chem. Soc.* **2003**, *125*, 1770.
15. Boussaad, S.; Tao, N. J.; Zhang, R.; Hopson, T.; Nagahara, L. A. *Chem. Comm.*, **2003**, *13*, 1502.

16. Wang, L.; Yuan, Z.; *Anal. Sci.*, **2004**, *20*, 635.
17. Erlanger, B.F.; Chen, B. -X.; Zhu, M.; Brus, L.; *Nano Lett.*, **2001**, *1*, 465.
18. Fu, K.F.; Huang, W. J.; Lin, Y.; Zhang, D. H.; Hanks, T. W.; Rao, A. M.; Sun, Y. P. *J. Nanosci. Nanotech.*, **2002**, *2*, 457.
19. Gooding, J.J.; Wibowo, R.; Liu, J. Q.; Yang, W. R.; Losic, D.; Orbons, S.; Mearns, F. J.; Shapter, J. G.; Hibbert, D. B. *J. Am. Chem. Soc.* **2003**, *125*, 9006.
20. Pantarotto, D.; Partidos, C. D.; Graff, R.; Hoebeke, J.; Briand, J. P.; Prato, M.; Bianco, A. *J. Am. Chem. Soc.*, **2003**, *125*, 6160.
21. Balavoine, F.; Schultz, P.; Richard, C.; Mallouh, V.; Ebbesen, T. W.; Mioskowski, C. *Angew. Chem. Int. Ed.*, **1999**, *38*, 1912.
22. Azamian, B.R.; Davis, J. J.; Coleman, K. S.; Bagshaw, C. B.; Green, M. L. H. *J. Am. Chem. Soc.*, **2002**, *124*, 12664.
23. Shim, M.; Kam, N. W. S.; Chen, R. J.; Li, Y.; Dai, H. *Nano Lett.*, **2002**, *2*, 285.
24. Lin, Y.; Taylor, S.; Li, H. P.; Fernando, K. A. S.; Qu, L. W.; Wang, W.; Gu, L. R.; Zhou, B.; Sun, Y. P. *J. Mater. Chem.*, **2004**, *14*, 527.
25. Chen, R.J.; Zhang, Y. W.; Wang, D. W.; Dai, H. J. *J. Am. Chem. Soc.*, **2001**, *123*, 3838.
26. Karajanagi, S. S.; Vertegel, A. A.; Kane, R. S.; Dordick, J. S. *Langmuir* **2004**, *20*, 11594.
27. Henriksen, A.; Mirza, O.; Indiani, C.; Teilum, K.; Smulevich, G.; Welinder, K. G.; Gajhede, M. *Protein Sci.* **2001**, *10*, 108.
28. Zoungrana, T.; Findenegg, G. H.; Norde, W. *J. Colloid Interface Sci.* **1997**, *190*, 437.
29. Asuri, P.; Karajanagi, S. S.;Yang, H.; Yim, T.-J.; Kane, R. S.; Dordick, J. S. *Langmuir*, **2006**, *22*, 5833.
30. Karajanagi, S. S.; Yang, H.; Asuri, P.; Sellitto, E.; Dordick, J. S.; Kane, R. S. *Langmuir*, **2006**, *22*, 1392.
31. O'Connell, M. J.; Bachilo, S. M.; Huffman, C. B.; Moore, V. C.; Strano, M. S.; Haroz, E. H.; Rialon, K. L.; Boul, P. J.; Noon, W. H.; Kittrell, C.; Ma, J.; Hauge, R. H.; Weisman, R. B.; Smalley, R. E. *Science* **2002**, *297*, 593.
32. Islam, M. F.; Rojas, E.; Bergey, D. M.; Johnson, A. T.; Yodh, A. G. *Nano Lett.* **2003**, *3*, 269.
33. Heller, D. A.; Barone, P. W.; Swanson, J. P.; Mayrhofer, R. M.; Strano, M. S. *J. Phys. Chem. B* **2004**, *108*, 6905.
34. Asuri, P.; Karajanagi, S. S.; Kane, R. S.; Dordick, J. S. *J. Am. Chem. Soc.*, **2006**, *128*, 1046.
35. Schmid, A.; Dordick, J. S.; Hauer, B.; Kiener, A.; Wubbolts, M.; Witholt, B. *Nature*, **2001**, *409*, 258.
36. Ghadge, R. S.; Sawant, S. B.; Joshi, J. B. *Chem. Eng. Sci.*, **2003**, *58*, 5125.
37. Asuri, P.; Karajanagi, S. S.; Sellitto, E.; Kim, D.-Y.; Kane, R. S.; Dordick, J. S. *Biotech. Bioeng.*, **2006**, *95*, 804.
38. Welton, T. *Chem. Rev.*, **1999**, *99*, 2071.

Chapter 6

Layer-by-Layer Assembly of Enzymes on Carbon Nanotubes

Jun Wang, Guodong Liu, and Yuehe Lin*

Pacific Northwest National Laboratory, 902 Battelle Boulevard, Richland, WA 99352
***Corresponding author: yuehe.lin@pnl.gov**

The use of layer-by-layer techniques for immobilizing several types of enzymes, e.g., glucose oxidase, horse radish oxidases, and choline oxidase, on carbon nanotubes (CNTs) and the applications of these techniques for biosensing are presented. The enzyme is immobilized on the negatively charged CNT surface by alternatively assembling a cationic polydiallyl-dimethyl-ammonium chloride (PDDA) layer and an enzyme layer. The sandwich-like layer structure (PDDA/enzyme/PDDA/CNT) formed by electrostatic assembly provides a favorable microenvironment to retain the bioactivity of the enzyme and to prevent enzyme molecule leakage. The morphologies and electrocatalytic activity of the resulting enzyme films were characterized using transmission electron microscopy and electrochemical techniques, respectively. It was found that these enzyme-based biosensors are very sensitive and selective for the detection of biomolecules, e.g., glucose, choline.

Enzymes are widely used for biosensor development. The effective immobilization of enzymes on the surface of a transducer is critical for developing a sensitive biosensor. There are already some methods, e.g., covalent binding,(*1*) entrapment,(*2*) and electrostatic adorption,(*3*) that have been developed for immobilizing enzymes on a transducer. However, covalent binding or entrapment methods cannot avoid the loss of enzyme activity caused by the harsh reaction process, and the active sites in the enzyme may be damaged or blocked. Electrostatic adsorption is an effective way to retain the catalytic activity of enzymes, but it is easy for the enzyme to de-associate off the surface of the transducer because of weak electrostatic interaction. The layer-by-layer (LBL) technique is a possible alternative to retain the activity of enzymes and reduce the loss of the enzymes on the transducer.

The LBL assembly of oppositely charged species is a simple and powerful method to construct a composite at nanoscale.*(4)* The LBL technique for ultrathin film assembly via alternate adsorption of oppositely charged polyions and proteins on variety of charged materials has been developed in recent decades. In fact, there are many reports on using the LBL technique to immobilize biomolecules such as enzymes on a substrate.(*5-9)* The LBL technique for immobilizing enzymes shows some advantages over other methods: besides those aforementioned, the amount of immobilized enzymes is controllable and can be increased by increasing the layer numbers.

There are many substrates for immobilizing enzymes using LBL techniques, such as clay nanoparticles, plysterene beads, silica, and pyrolytic graphite electrodes.*(6-9)* However, there are few reports on the use of carbon nanotubes (CNTs) as a substrate for the immobilization of enzymes via the LBL technique. CNTs represent a new class of nanomaterials, composed of graphitic carbon with one or several concentric tubules, and they are receiving considerable interest because of their unique structure and mechanical and electronic properties.(*10-12*) One promising application of CNTs is to construct chemical sensors and biosensors. Researchers have demonstrated that CNTs have a high electrocatalytic effect, a fast electron-transfer rate, and a large specific surface area. For example, Ivnitski and coworkers reported recently that carbon nanotubes can efficiently promote direct electron transfer (DET) between the active site of a glucose oxidase (GO(x)) and electrode. *(13)* the immobilization of enzymes on CNTs shows some advantages: (i) the large surface area of CNTs makes loading a large amount of enzyme possible, (ii) the excellent electronic properties of CNTs facilitate electron transfer between the electroactive center of the enzyme and the electrode, and (iii) because of the catalytic effect of CNTs, the small electroactive molecules generated by the catalytic reaction of enzymes with a substrate in solution are easy to be oxidized or reduced at low potential, which will avoid interference from other substances, e.g., ascorbic acid.

Previously, immobilizing glucose oxidase (GOx) enzymes on CNTs was developed by using a polymer, such as nafion, that entraps the enzyme and covers the CNT in soluble polymer. And the resulting glucose sensor was found to be very sensitive for detecting glucose.(*12*) We have developed the process for immobilizing GOx enzymes on arrayed CNTS by covalent binding.(*14*) The CNTs array was fabricated by Plasmon–enhanced chemical vapor deposition. Recently, we used the LBL technique to immobilize some enzymes on a glassy carbon (GC) electrode for biosensing. (*15*) A glucose biosensor was fabricated based on immobilizing GOx on the negatively charged CNT surface by alternatively assembling a cationic polydiallyldimethylammonium chloride (PDDA) layer and a GOx layer. The excellent electrocatalytic activity toward H_2O_2 of the fabricated PDDA/GOx/PDDA/CNT electrode indicated that the polyelectrolyte-protein multilayer does not retard the electrocatalytic properties of CNTs, enabling sensitive determination of glucose. Flow-injection amperometric detection of glucose is carried out at -100 mV (vs Ag/AgCl) in 0.05 M phosphate buffer solution (pH 7.4) with a wide linear response range of 15 μM to 6 mM and a detection limit of 7 μM.(*16*) The biosensor for organophosphate pesticides and nerve agents was fabricated based on the self-assembly of acetylcholinesterase (AChE) on a CNT electrode using the same LBL approach described above. The electrocatalytic activity of CNT leads to a greatly improved electrochemical detection of the enzymatically generated thiocholine product, including a low oxidation overpotential (+150 mV), higher sensitivity, and stability.(*17*) The biosensor was used to measure as low as 0.4 pM paraoxon with a 6-min inhibition time under optimal conditions. A choline biosensor was also fabricated based on an LBL assembly of a bienzyme of choline oxidase (ChO) and horseradish peroxidase (HRP) onto a CNT/GC electrode.(*18*) The measurement of faradic responses resulting from enzymatic reactions has been realized at low potential with a good sensitivity because of the electrocatalytic effect of CNTs. It has been found that the biosensor based on a bienzyme (ChO/HRP/CNT) electrode is more sensitive than the biosensor based on one enzyme (the ChO/CNT) for choline detection.

Construction of multilayer of enzymes by LBL

The LBL technique can be used to immobilize enzymes onto CNT using a cationic polymer, e.g. PDDA. In the work, the enzyme-modified electrode (PDDA/[enzyme/PDDA]$_n$/MWCNT/GC) was fabricated using the LBL approach. N is an enzyme layer number here. Figure 1 schematically illustrates the procedure for the attachment of enzymes on CNTs through the LBL technique. Briefly, the PDDA was firstly coated on the negative charged MWCNT by soaking the MWCNT/GC electrode in an aqueous solution of 1 mg/ml PDDA containing 0.5 M NaCl for 20 min. Then the resulting PDDA/MWCNT/GC electrode was rinsed with water to remove any weakly adsorbed polycations and dried under a nitrogen stream. The electrode was immersed in a enzyme solution for 20 min to attach the enzyme on the GC electrode surface. The resulting enzyme modified electrode was rinsed with water and quickly dried in a nitrogen stream. More layers of PDDA and enzyme can be prepared by repeating the same approach. The final layer is the PDDA layer which is used to protect the enzyme from desorption from the surface. However, two different enzymes, e.g., HRP and CHO, can also be sequentially immobilized on the CNTs using this technique.*(15, 17-18)*

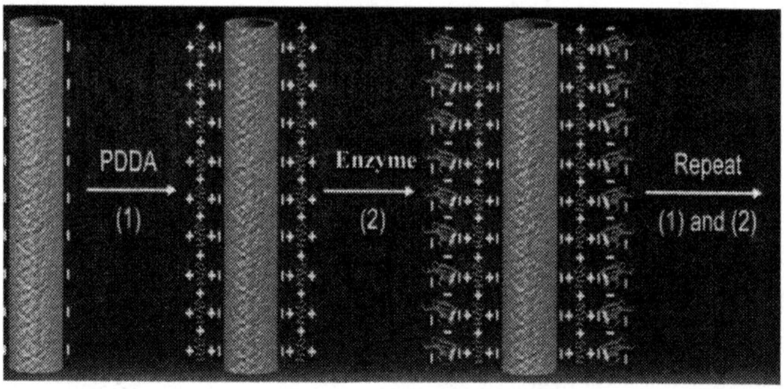

Figure 1. Schematic illustration of the procedure for the attachment of enzymes on a carbon nanotube surface through LBL technique

Flow-injection analysis

A laboratory-built flow-injection system, which consists of a carrier, a syringe pump (Model 1001, BAS), a sample injection valve (Valco Cheminert VIGI C2XL, Houston, TX, USA), and a laboratory-built wall-jet based flow-through electrochemical cell, was used for the amperometric measurement of enzymatic reaction products.. The amperometric measurements were conducted at different potentials. All potentials are referred to the Ag/AgCl reference. The laboratory-built microelectrochemical cell integrates three electrodes and is based on a wall-jet (flow-onto) design (Figure. 2A). A laser-cut Teflon gasket is sandwiched between two acrylic blocks to form a flow-cell. The working electrode is placed into the bottom piece and is sealed with an O-ring; an Ag/AgCl reference electrode and a platinum wire counter electrode are placed into the upper piece inside a groove (Figure. 2B). The solution flows onto the working electrode surface and exits through the groove to the outlet. This cell design allows the working electrode to be quickly installed.

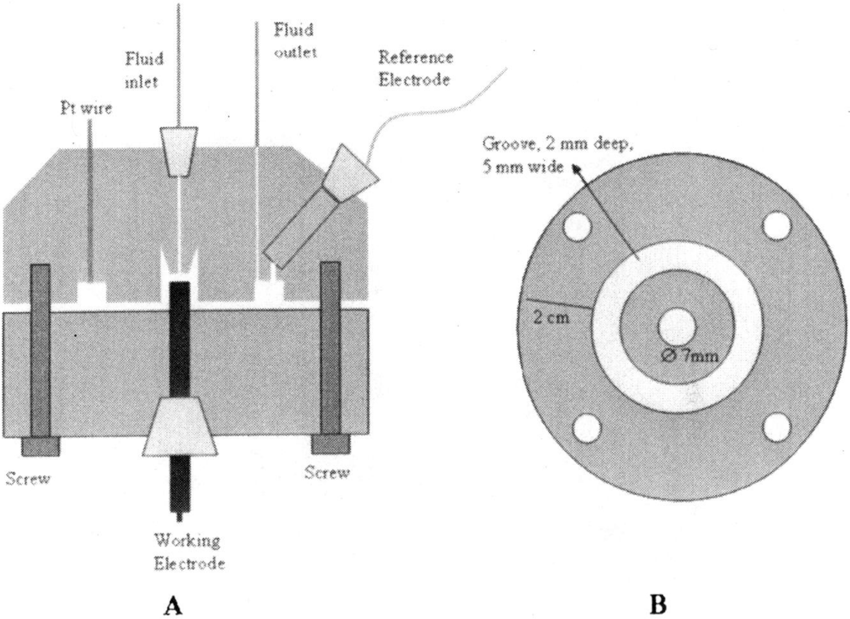

Figure 2. Schematic illustration of (A) a lab-built wall-jet flow cell; (B) bottom scheme of upper piece of wall-jet flow cell.

Results and Discussion

TEM Characteristic of multilayer of enzymes on CNTs

The sample for transmission electron microscope (TEM) experiments was prepared by peeling (GOx-PDDA)n-CNT from the GC electrode and dispersing it on the Cu-C grid. Figure 3 shows the TEM images of an LBL assembly of GOx on CNT with different layers. From these images, we can see that the thickness of CNT increases with the enzyme layers. We can see a series of bamboo-like closed graphite shells along the tube axis because the graphitic layers are not perfectly parallel to the tube axis, and the thickness of the PDDA-GOx bilayer increases with the numbers of layers. Such TEM images provide strong evidence for the formation of an LBL composite. Similar structures have been observed on polymer multilayers on functionalized multiwalled CNTs. There was also relatively high roughness of the protein-polyion nanostructures, which is a general feature observed on polymers adsorbed onto flat surfaces.

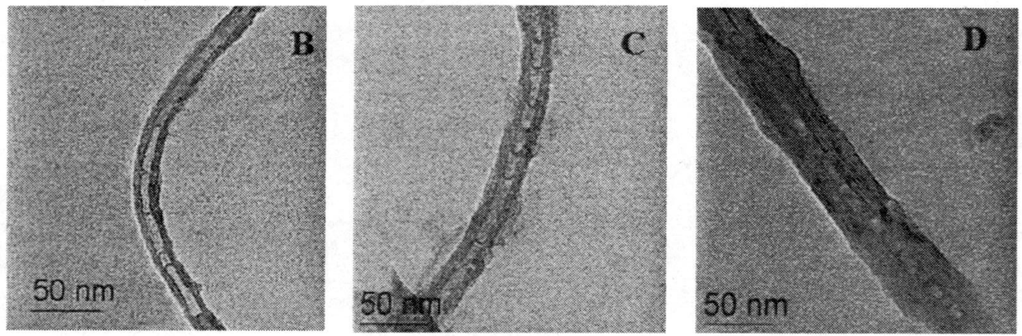

Figure 3. TEM images of (PDDA-GOx) n-CNT at the same magnification, (B) n =1; (c) n = 2; (D) n=6

Electrochemical study of multilayer of enzymes on CNTs

The direct electrochemistry of redox protein on the electrode surface may be used to evaluate the activity of enzymes on a CNT-modified electrode surface. Cyclic voltammetry (CV) was used to study the direct electrochemistry of GOx, and the recorded redox peak currents were used to monitor the assembly process of (PDDA/GOx)n films on the CNT-modified glassy carbon electrode. We used these (GOx-PDDA)n-CNT/GC electrodes with varying numbers of bilayers of adsorbed protein-polyion to study the LBL process and direct electron transfer of immobilized redox protein (GOx) (Figure. 4 [left]). From Figure. 4 (left), no redox peaks were observed with the PDDA-CNT/GC electrode in the potential range of interest, but a pair of well-defined and nearly symmetrical redox peaks was obtained with the (GOx-PDDA)4-CNT/GC electrode. This suggests that the redox peaks are resulted from the electrochemical reaction of GOx immobilized on the surface of the PDDA-CNT/GC electrode. This favorable redox signal (current), which is related to the amount of GOx on the electrode surface, can be used to monitor the LBL process. The electrode was washed with pure water and then transferred to the GOx-free phosphate buffer (pH7) after each double adsorption cycle created a new GOx-PDDA bilayer on the CNT. Figure 4 (right) shows the plot of cathodic current vs. enzyme layer number obtained from cyclic voltammograms of various (GOx-PDDA) bilayer-coated CNT/GC electrodes. The reduction peak currents of GOx grew with the numbers of GOx-PDDA bilayers (n) until $n = 4$. Increasing the redox peak currents benefited from increasing the amount of GOx on the CNT surface. The amplitude from increasing the redox peak currents became smaller with the increasing numbers of GOx-PDDA bilayers. The nonlinear increase in activity corresponds to the increasing of the electron transfer distance between GOx and the graphite sheet of CNT with an increase of the bilayer number. At $n > 4$, no further increase in the peak currents was observed (not shown) which is due to the electron transfer distance becoming too large.

Analytical performance of biosensors based on LBL assembly of enzymes on CNTs

Flow injection amperometric measurements of glucose

Figure 5 shows the calibration curves from typical amperometric responses of the PDDA/GOx/PDDA/CNT/GC electrode for the successive injection of various concentrations of glucose at an applied potential of -100 mV in a flow injection system. The signal was expressed by the area under each peak. It can be seen that the response signal linearly increased with the increase of the concentration of glucose (Figure 5 A) from 15 μM up to 6 mM with a

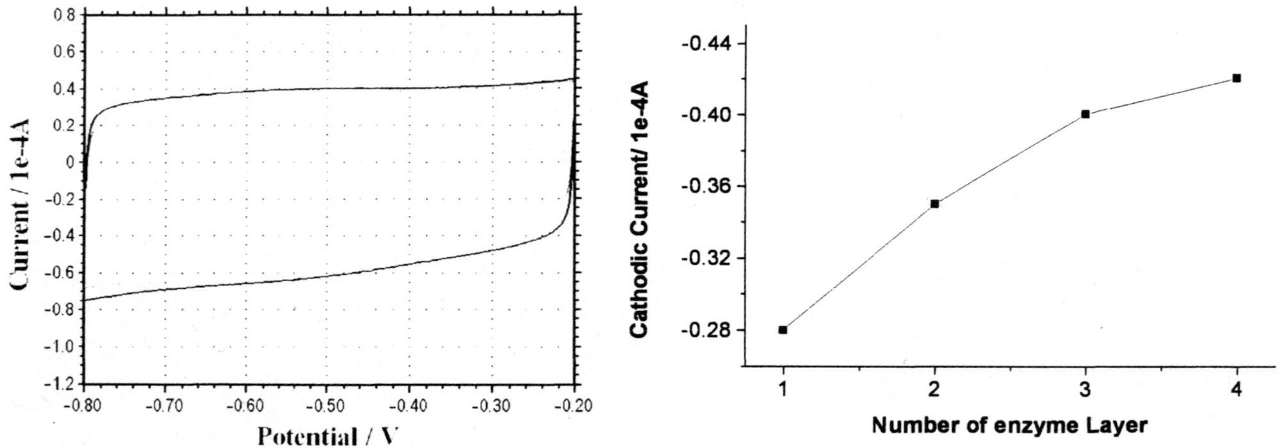

Figure 4. (left) CV of electrode modified with GOx/CNT (light grey curve)and without GOx (dark grey curve); (right) the plot of cathodic current vs. number of enzyme layers.

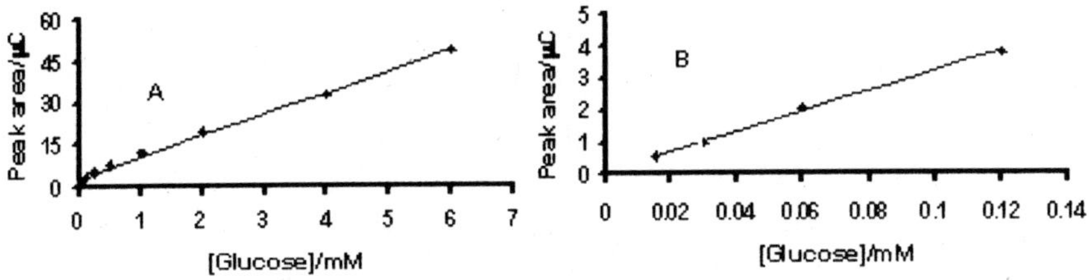

Figure 5. (A) The calibration curve from an electrochemical response of flow-injection of different glucose solutions on GOx/CNTs-modified GC electrodes; concentrations range from 15 μM to 6 mM; (B) the calibration curve at low concentration ranges from 15 μM to 120 μM. (Reproduced with permission from reference 16)

correlation coefficient of 0.992. The high sensitivity is accompanied by a low noise level, allowing convenient detection of micromolar concentrations.

Figure 5 B presents the calibration curve of glucose at low concentrations ranging from 15 μM to 120 μM. We can see that the calibration curve at such a low concentration range also displays a good linear relationship (correlation coefficient, 0.997) between the peak area and glucose concentration. A detection limit of 7.0×10^{-6} M can be estimated based on the signal-to-noise characteristics of these data (S/N=3).

Flow injection amperometric measurements of Choline

Figure 6 compares the Amperometric response of the ChO/CNT-modified electrode and the ChO/HRP/CNT-modified electrode in the same concentration of choline solution. It can be seen that the magnitude of peak current for the ChO/HRP/CNT electrode is higher than that for the ChO/CNT electrode at the same concentration of choline. This is reasonable because HRP catalyzes the reduction of H_2O_2. This suggests that direct electron transfer took place between HRP and CNTs. Here, CNTs can also facilitate electron transfer between HRP and the GC electrode because of their excellent electronic properties. This result is consistent with some literature in which the CNTs were reported to promote the electron transfer of proteins, e.g., cytochrome c, GOx, and catalace on electrode surfaces.*(13)* Figure 7 shows amperometric responses for the ChO/HRP/CNT electrode to injecting a series of 0.1 M Tris buffer solutions (pH 8) containing different concentrations of choline with a flow-injection system. A plug of 40 μl choline solution was injected when the baseline became stable. The measurements were performed at a potential of -0.2 V. It can be seen from Figure 7 that the reduction current increases with the increase of the

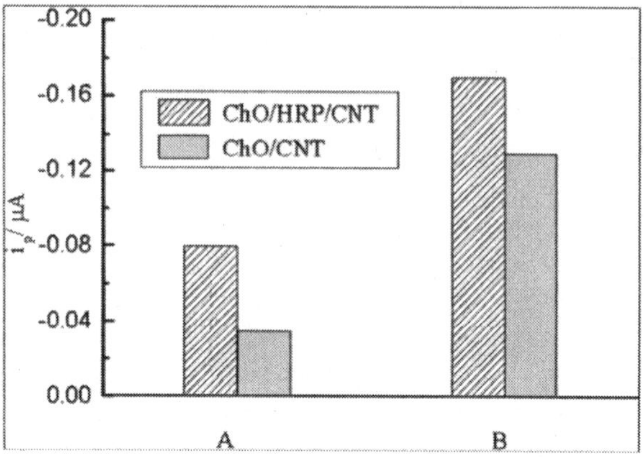

Figure 6. Amperometric responses of two types of enzymes modified GC electrode for choline detection in 0.1 M Tris buffer solution containing 2.0 mM (A) and 9.0 mM (B) choline (pH 8). (Reproduced with permission from reference 18. Copyright 2006.)

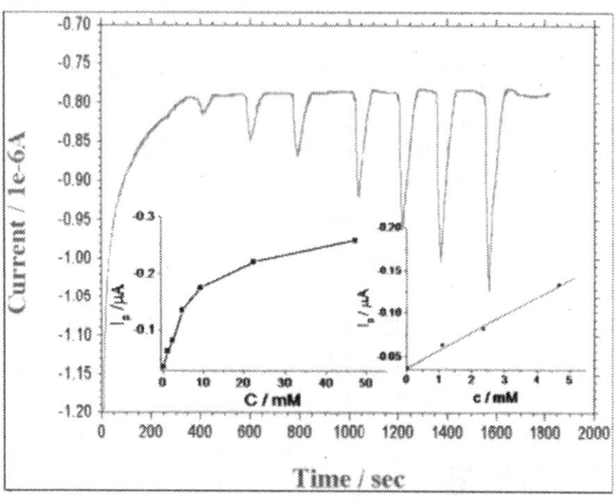

Figure 7. Amperometric response of choline oxidization at a ChO/HRP/CNT/GC electrode by a series of injections of different concentrations of choline solution (from left to right: 0.05 mM, 1.1 mM, 2.4 mM, 5.0 mM, 9.4 mM, 22.4 mM, and 47 mM) in 0.1 M Tris buffer (pH 8). Inset is the plot of peak current vs. choline concentration. (Reproduced with permission from reference 18. Copyright 2006.)

choline concentration, which indicates that the responses resulted from the enzymatic reactions of ChO and HRP on choline. The inset is a plot of peak current vs. choline concentration. The peak current increased linearly with the increase of the choline concentration below 5 mM. The detection limit of this method is estimated to be 5.0×10^{-5}M.

Conclusions

We have demonstrated that the LBL is an effective technique for assembling enzymes on CNTs. Using the CNT template on the electrode surface allows for a 3D nanocomposite to form, which minimizes the inherent limitation in CNT self assembly on flat surfaces. The resulting nanocomposite has been characterized with TEM and electrochemical techniques. The present method can be applied to study direct electron transfer between the redox protein and the electrode surface and to develop enzyme-based biosensors. The concept can be extended to assemble other biological molecules. The flow-injection analysis system has been employed for testing the enzyme/CNT-based biosensors. It has been found that these biosensors are very sensitive and selective. This technique provides a simple, cost-effective way to develop an enzyme-based biosensor.

Acknowledgements

This work was partially supported by CDC/NIOSH grant R01 OH008173-01 and partially by a laboratory directed research and development program at Pacific Northwest National Laboratory (PNNL). The research described in this paper was performed at the Environmental Molecular Sciences Laboratory, a national scientific user facility sponsored by the U.S. Department of Energy's (DOE's) Office of Biological and Environmental Research and located at PNNL. PNNL is operated by Battelle for DOE under Contract DE-AC05-76RL01830.

References

1. Garguilo, M. G.; Huynh, N.; Proctor, A.; Michael, A. C. *Anal. Chem.* **1993**, *65*, 523.
2. Langer, J. J.; Filipiak, M.; Kecinska, J.; Jasnowska, J.; Wlodarczak, J.; Buladowski, B. *Surf. Sci.* **2004**, *573*, 140.
3. Razala, S. S.; Pochet, S.; Grosfils, K.; Kauffmann, J. M. *Biosens. Bioelectron.* **2003**, *18*, 185.
4. Campas, M.; O'Sullivan, C. *Anal. Lett.* **2003**, *36*, 2551.

5. Zhou, Y.; Li, Z.; Hu, N.; Zeng, Y.; Rusling, J. F. *Langmuir* **2002**, *18*, 8573.
6. Lvov, Y.; Lovo, Y., Mohwald, H., Eds.; Marcel Dekker: New York, 2000, pp p. 125.
7. Sun, H.; Hu, N. *Analyst* **2005**, *130*, 76.
8. Ma, H.; Hu, N.; Rusling, J. F. *Langmuir* **2000**, *16*, 4969.
9. Li, Z.; Hu, N. *J. Electroanal. Chem.* **2003**, *558*, 155.
10. Ajayan, P. M. *Chem. Rev.* **1999**, *99*, 1787.
11. Iijima, S. *Nature* **1991**, *354*, 56.
12. Wang, J.; Musameh, M.; Lin, Y. *J. Am. Chem. Soc.* **2003**, *125*, 2408.
13. Ivnitski, D.; Branch, B.; Atanassov, P.; Apblett, C. Electrochem. Commun. **2006**, *8*, 1204-1210.
14. Lin, Y. H.; Lu, F.; Tu, Y.; Ren, Z. F. *Nano. Lett.* **2004**, *4*, 191-195.
15. Liu, G. D.; Lin, Y. H. *J. Nanosci. Nanotech.* **2006**, *6*, 948-953.
16. Liu, G. D.; Lin, Y. H. *Electrochem. Commun.* **2006** *8*, 251-256.
17. Liu, G. D.; Lin, Y. H. *Anal. Chem.* **2006** *78*, 835-843.
18. Wang, J.; Liu, G. D.; Lin, Y. H. *Analyst* **2006** *131*, 477-483.

Chapter 7

Sustainable Biocatalytic Nanofibers: Synthesis, Characterization, and Applications of Enzyme-Coated Polymeric Nanofibers

Seong H. Kim[1], Sujith Nair[1], Byoung Chan Kim[2], Man Bok Gu[3], and Jungbae Kim[4]

[1]Department of Chemical Engineering, The Pennsylvania State University, University Park, PA 16802
[2]Advanced Environmental Monitoring Research Center, Gwangju Institute of Science and Technology, Gwangju, South Korea
[3]College of Life Sciences and Biotechnology, Korea University, Seoul, South Korea
[4]Pacific Northwest National Laboratory, 902 Battelle Boulevard, Mailstop K8–93, Richland, WA 99352

The fabrication and enzymatic stability of two types of sustainable biocatalytic nanofinbers – enzyme-polymer composite nanofibers and enzyme-coated polymer nanofibers – are reviewed in this paper. These biocatalytic nanofibers, produced by coating enzymes on polymeric nanofibers, can make recycling and handling of enzymatic materials easy in bio-catalysis, bio-sensing, bio-remidiation applications. This paper also describes a simple way of dispersing hydrophobic biocatalytic nanofibers in aqueous solution.

Introduction

The substrate-specificity is a great advantage of enzymatic reactions in bio-catalysis, bio-sensor, and bio-remidation applications (*1-4*). Although free enzymes can be utilized in these applications, their storages and handlings are sometimes quite cumbersome. Moreover, these free enzymes are not easily recyclable. From the economic point of view, the catalysts should be reused as long as their activity is preserved. However, there is no simply way of recovering catalyst (enzyme) after use. These difficulties can be circumvented when the enzymes are immobilized or anchored on solid substrates. Then, the catalytically active enzyme species can be easily separated from the liquid-phase reaction media, and handled or reused for multiple times. Figure 1 displays pictures of biocatalytic nanofibers being used in one reaction vial, recovered with tweezers, and transferred to the next reaction vial. The biocatalytic nanofibers can be deposited directly on device surfaces and used as a component that react with analytes and generates sensing signals. Non-woven mats of biocatalytic nanofibers can also be used as a membrane reactor in bio-remediation applications. However, all these potential advantages would not be realized if the immbolilized enzymes significantly lose their activity over time due to denaturation. So, the enzyme stability is a very important issue. In this review, we will discuss the prepration and modification of electrospun nanofibers as well as the enzyme immobilization and stabilization.

Fabrication of Polymer Nanofibers

In heterogeneous enzymatic systems, the solid support material should meet several important requirements. First, it should have a large surface area to maximize the enzyme loading. Second, it should have a distribution of pores or open spaces in the material to achieve the high mass-transfer rate of substrates to the active site of an enzyme. Third, it should be durable and easily recoverable by a simple physical method. Finally, it should have the right functional groups at the surface suitable for covalent attachment of enzymes. *Polymeric nanofibers* can meet all these requirements. In addition, the choice of polymeric materials is quite diverse, so that the material properties as well as surface chemistry can be varied easly.

Polymer nanofibers can easily be fabricated via an electro-spinning process as shown in Figure 2. In the electro-spinning process, a polymer solution is charged in a syringe with a metallic needle or a pulled glass pipette with a fine tip which is located from a certain distance from a grounded collector. Then a high voltage bias is applied to the needle or the solution in the glass pipette. As the electrostatic repulsion force of the eletric charge at the solution surface of the

Figure 1. A series of photographs showing a non-woven nanofiber bundle being used in one reaction vial (a, b), recovered with tweezers (c,d), and transferred to another reaction vial (e, f).

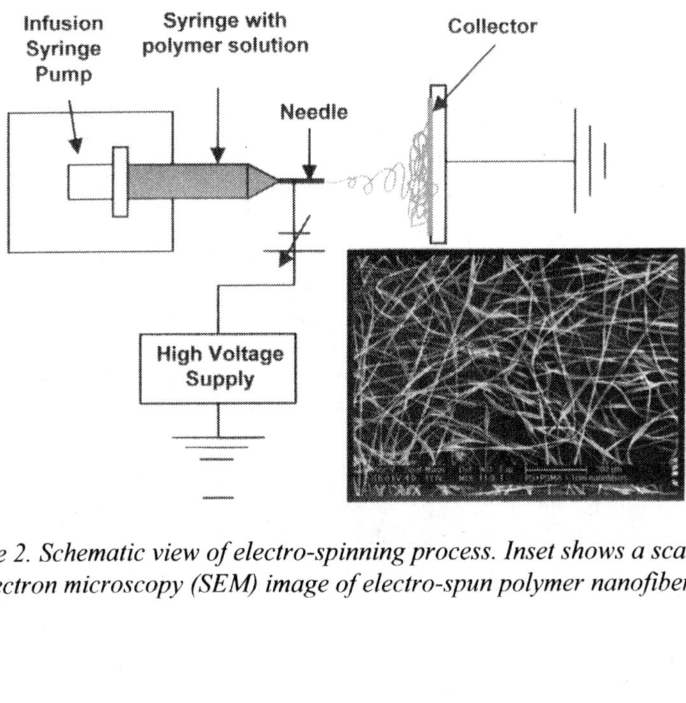

Figure 2. Schematic view of electro-spinning process. Inset shows a scanning electron microscopy (SEM) image of electro-spun polymer nanofibers.

tip becomes larger than the surface tension force of the solution, then a jet of polymer solution is ejected. The solvent in the jet evaporates fast and the dissolved polymers are collapsed into continuous fibers. This electrospinning process has been widely used for fabrication of various polymer nanofibers. The polymer concentration, solvent vapor pressure, bias voltage, collector location, etc. can be controlled to change the diamer and surface texture of nanofibers (*5-7*).

For the enzyme immobilization and bio-catalytic applications, the polymer nanofibers should be insoluble in aqueous solution and mechanically flexible. They should also have the functional groups to which enzymes can be anchored. For these reasons, we chose a mixture of polystyrene and poly(styrene-*co*-maleic anhydride) [PS-PSMA]. Figure 3 shows SEM images of PS-PSMA nanofibers produced by electrospinning from a toluene solution. The surface morphology of the nanofibers can vary depending on the solvent property and the humidity of the environment (*8, 9*).

Figure 4 compares the infrared spectrum of PS-PSMA nanofibers with that of PS nanofibers. The IR spectrum of the PS nanofiber clearly shows the characteristic bands of polystyrene: C-H stretch of the aromatic ring in the 3000 ~ 3100 cm^{-1}, aromatic C-H deformation of the aromatic ring at 1450 and 1490 cm^{-1}, C=C stretch in the aromatic ring at 1605cm^{-1}, and aromatic overtones in the range 1700 ~ 2000 cm^{-1}. The IR spectrum of the PS+PSMA fiber exhibits two additional peaks from the anhydride group: asymmetric C=O stretch at 1860 cm^{-1} and symmetric C=O stretch at 1780 cm^{-1}. The maleic anhydride group can react with the free amine group at the enzyme surface and form a covalent bond.

Enzyme-Polymer Composite Nanofibers

The enzyme-polymer composite nanofibers can be fabricated by electrospinning a enzyme-containing polymer solution (*11*). Since the PS and PSMA dissolve only in organic solvents, the enzyme should be solubilized in organic solvents by using surfactants (Figure 5). In this study, α-chymotrypsin (CT) is used as a test enzyme. Using dioctyl sulfosuccinate (AOT) as a surfactant, CT can be dissolved in toluene into which PS-PSMA is dissolved (*12*). The CT encapsulated in PS are known to be stable (*13-17*).

The activity of the CT embedded in the PS-PSMA nanofiber is measured by monitoring the 410 nm absorbance of *p*-nitroaniline which is the product from the CT digestion of N-succinyl-Ala-Ala-Pro-Phe *p*-nitroanilide (TP). Table 1 summarizes the activity data of the CT-embeded PS-PSMA nanofibers and CT–PS-PSMA composite films prepared by the flash solvent evaporation method. As expected, the activity of the nanofiber is much higher than that of the film. The CT activity of the enzyme-polymer composites can be increased by a simple

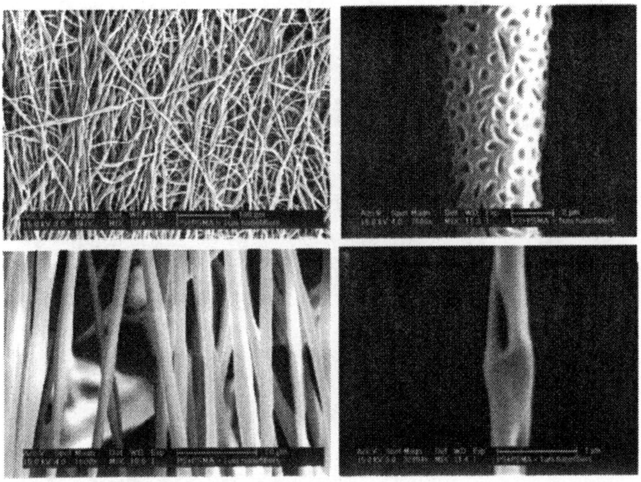

Figure 3. SEM images of PS-PSMA nanofibers produced by electrospinning of a toluene solution of PS-PSMA. (Reproduced with permission from reference 10. Copyright 2005 Instititute of Physics Publishing.)

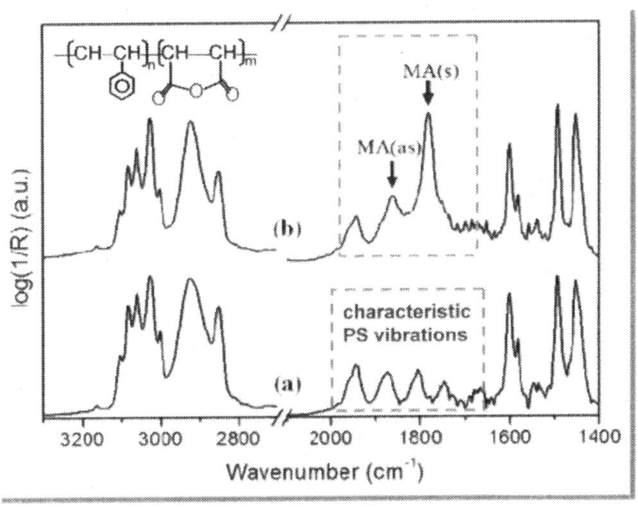

Fgiure 4. Infrared spectra of PS and PS-PSMA nanofibers. The maleic anhydride symmetric and asymmetric vibration peaks are maked with arrows.

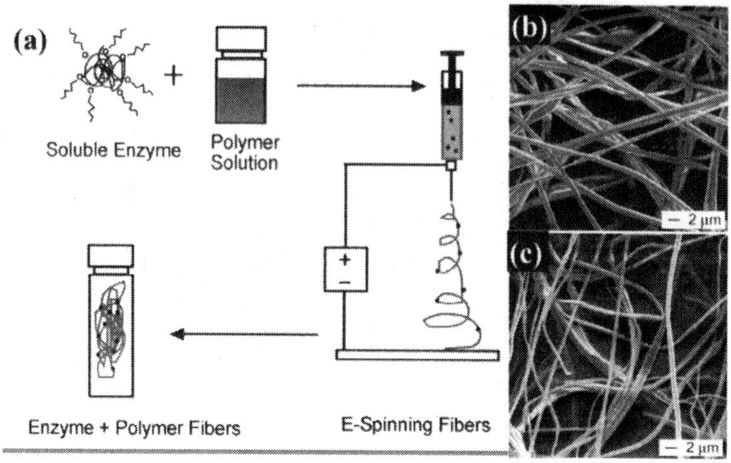

Figure 5. (a) Electrospinning of enzyme-polymer composite nanofibers and (b, c) SEM images of electrospun enzyme-polymer composite nanofibers. (Reproduced with permission from reference 11. Copyright 2005 The Royal Society of Chemistry.)

Table 1. Enzymatic activity of CT-PS-PSMA composite nanofibers and films

Sample Initially loaded with 1.3% CT	**Activity** nM/min per mg of composite in 1 mL
Nanofibers	80
Nanofibes with GA treatment	229
Films	12
Films with GA treatment	49

treatment of the composite with glutaraldehyde (GA) before capping of the unreacted maleic anhydride group and rigorous washing. The GA treatment cross-links enzymes, which enhances the retention of the loaded enzymes. After the GA treatment, the CT activity of the enzyme-polymer composites is incresed by 3 ~ 4 times. These results indicate that a significant portion of enzymes embedded in the PS-PSMA polymer are not chemically attached to the polymer. The GA treatment before rigorous washing can cross-link these enzymes to the strongly bound enzymes.

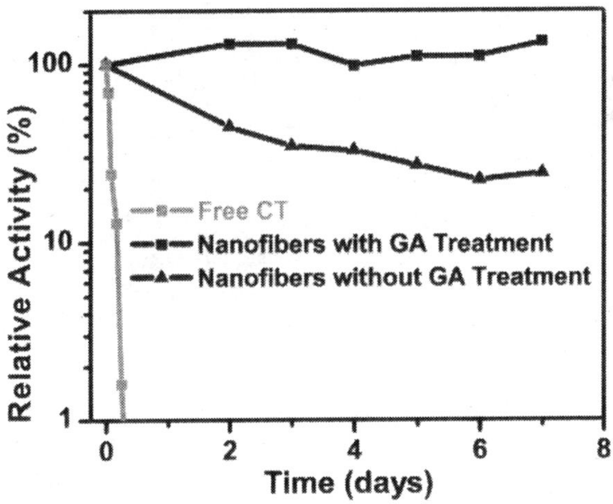

Figure 6. Stability of CT-loaded composite nanofibers treated with and wihtout GA under 250 rpm shaking conditions in phosphate buffer. The deactivation of free CT under the same reaction conditions is also shown for comparison. (Reproduced with permission from reference 11. Copyright 2005 The Royal Society of Chemistry.)

In addition to the activity, the stability of the immobilized enzyme is very important. If the immobilized enzymes lose their activity over time significantly, there is no reason that the enzyme-immobilized solid materials should be recycled or no reliable bio-sensing or bio-remediation performance expected over a long period of time. Figure 6 compares the stability (enzyme activity as a function of time) of CT loaded biocatalytic nanofibers over 7 days as well as free CT. The samples were kept in buffer solution with continuous shaking at 250 rpm. In this condition, the free CT activity decreases fast due to autolysis (half-life of 5 hours). However, the CT immobilized in the composite keeps its activity for much longer period of time. Without the GA tratment, the apparent half-life of the CT loaded biocatalytic nanofiber is about 2 days. This might be due to gradual leaching of the loaded enzyme over time. In the GA-treated baiocatalyst nanofibers, there is no activity decrease observed over 7 days.

Enzymes-Coated Polymer Nanofibers

As in heterogeneous catalytic reactions, only the enzymes exposed at the surface of the support material will be available for bio-catalytic reactions. In the enzyme-polymer composite nanofiber system, the enzyms embeded *in* the nanofiber may not be utilized effecitively in the reaction. This situation can be avoided if the enzymes are coated *on* the nanofiber surface, rather than embedded in the nanofiber. The enzyme coating on the polymr nanofiber can be attained through chemical reactions between the functional groups at the enzyme and the nanofiber surface (*18*).

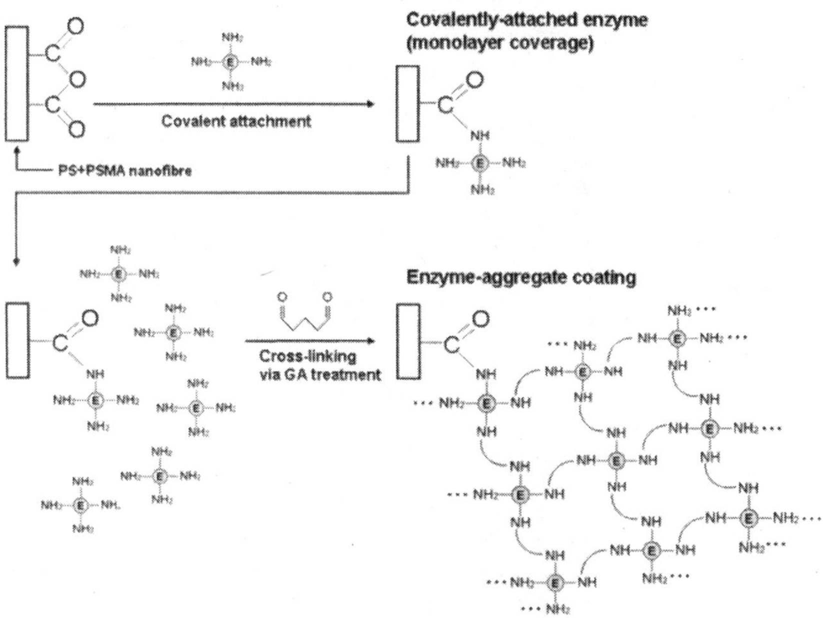

Figure 7. Schematic representation of enzyme-coating on the surface of polymer nanofibers. (Reproduced with permission from reference 10. Copyright 2005 Institute of Physics Publishing.)

In the CT and PS-PSMA system, this reaction is the amide bond formation between the free amine group at the CT enzyme and the maleic anhydride group exposed at the PS-PSMA nanofiber (Figure 7). This reaction can produce a monolayer of CT on the PS-PSMA surface. The surface coverge of CT will be determined by the concentration of the maleic anhydride group available at the PS-PSMA surface. We call this system CA-CT hereafter. As in the case of the enzyme-polymer composite nanofibers, one can utilize the GA tratment to cross-link enzymes from the solution onto the chemically-anchored CT at the polymer surface, forming enzyme multilayers. In this way, the total enzyme loading per unit mass of nanofiber support can be increased significantly. We call this system EC-CT hereafter.

The surface morphology of the CA-CT nanofiber is very close to the bare PS-PSMA nanofiber within the resolution of typical SEM. This is expected since the size of CT is only a few nanometers (Figure 8a). In the case of the EC-CT nanofibers, the surface morphology varies depending on position (Figures 8b,c,d). The desposition of multilayers of CT seems to mask the original texture of the polymer nanofibers. The degree of the surface morphology change is expected to be a function of the thickness of the multilayer coating formed by cross-linking of excess CT molecules on CA-CT.

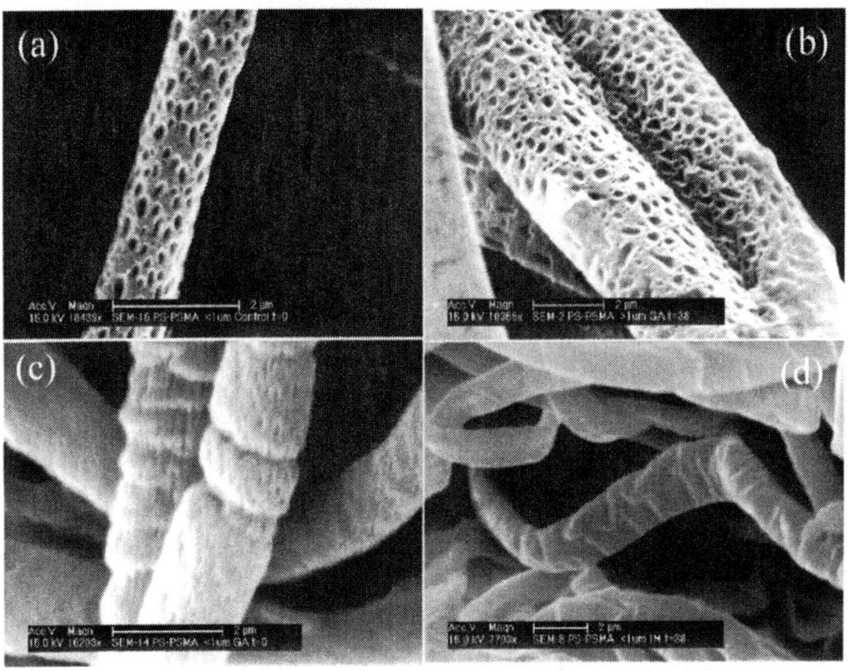

Figure 8. SEM images of (a) CA-CT nanofiber and (b,c,d) EC-CT nanofibers.

The apparent activities of the CA-Ct and EC-CT biocatalytic fibers loaded enzyme are compared in Table 2 for two different thicknesses of PS-PSMA fiber diameter. For the same type of fibers (CA-CT or EC-CT) the nanofibers gives a slighlty higher activity per unit mass of the fiber than the corresponding microfibers due to a higher surface area which provides a higher enzyme loading. The enzymatic activity of EC-CT is increased by 8 ~ 9 times higher than that of CA-CT. These results demonstrate that the significant enhancement of the bio-catalytic nanofibers can be accomplished by multilayer coatings of enzymes through the GA treatment.

Table 2. Activity of CA-CT and EC-CT on PS-PSMA nanofibers (<1 μm) and microfibers (>1 μm).

Sample description	Initial activity mM/min per mg fibers
CA-CT on PS+PSMA nanofibers	0.10
CA-CT on PS+PSMA microfibers	0.08
EC-CT on PS+PSMA nanofibers	0.87
EC-CT on PS+PSMA microfibers	0.63

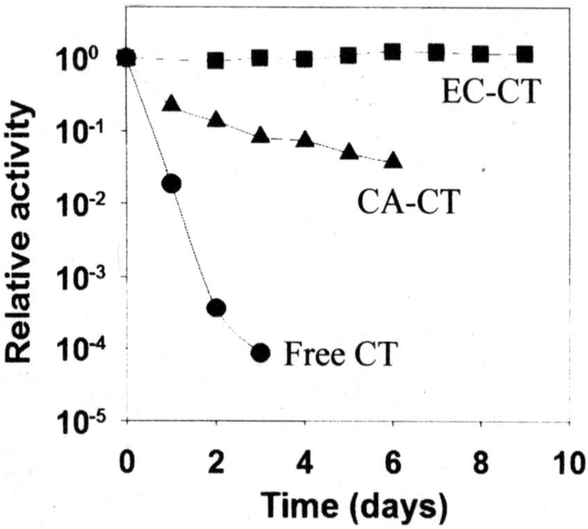

Figure 9. Stability of CA-CT and EC-CT nanofibers under 250 rpm shaking conditions in phosphate buffer. The deactivation of free CT under the same reaction conditions is also shown for comparison.

In addition to the improved activity, the EC-CT system shows a superior performance over the CA-CT system from the stability point of view (Figure 9). The EC-CT PS-PSMA nanofiber shows no loss of the CT activity for more than one month. The half-life of the CT activity could not be calculated due to the insufficient loss of activity over this period of time. This dramatic stabilization of EC-CT can be explained by no leaching of the immobilized CT and the good stability of crosslinked enzyme aggregates (CLEAs) themselves (*19*). In addition to preventing leaching of the enzyme molecules, the inter and intramolecular covalent cross-linking of the three dimensional protein structure is known to impart enhanced stability to the protein against thermal and organic solvent denaturation. It is also noteworthy that enzyme aggregation prevents the autolysis of CT molecules.

Dispersion of Hydrophobic Nanofibers in Aqueous Solution

In order to maximize the enzymatic activity, it is very important to disperse the enzyme-loaded supporting materials in the reaction media. We have devised a simple method for dispersion of PS-PSMA nanofibers in water without using conventional surfactant molecules (*20*). First, the electrospun nanofiber mat is dispersed fully in the aqueous alcohol solution (20% by volume). The nanofibers are then thoroughly washed *without drying* using a copious amount of distilled water until alcohol is completely removed from the solution phase. Figure 10

Figure 10. Pictures of as-spun and alcohol-treated PS-PSMA nanofiber bundles immersed in water.

compares the as-spun and alcohol-treated PS-PSMA nanofibers immersed in water. Due to the hydrophobicity of polystyrene, the as-spun PS-PSMA nanofibers form a tightly aggregated clump in aqueous solution. Even though the density of polystyrene (1.05 g/cm3) is higher than water, the nanofiber aggregates float on the water. This might indicate that the nanofiber aggregate contains tiny air bubbles inside. In contrast, the PS-PSMA nanofiber mat pre-treated with the alcohol solution is fully expanded in water, indicating that the nanofibers are well dispersed. Moreover, the nanofibers sink in the water, as expected from the density of the polymer. Even after further rinsing with water for several days, the alcohol pre-treated PS-PSMA nanofibers remain fully dispersed in water. The expanded nanofiber mat does not break into pieces even after vigorous shaking since the nanofibers are physically entangled.

The alcohol solution pre-treatment does not alter the physical and chemical feature of the PS-PSMA nanofibers. Since short chain alcohols such as ethanol and propanol do not dissolve polystyrene, the diameter of the nanofibers does not change after the alcohol treatment. The maleic anhydride functional group of the PS-PSMA nanofiber remains intact (not hydrolyzed) throughout these pretreatment processes and still available for enzyme immobilization. As shown in Figure 11, the reaction of the maleic anhydride group of the dispersed PS-PSMA nanofiber with n-propyl amine in phosphate buffer shows the decrease of the maleic anhydride peak at 1780 and 1860 cm^{-1} and the growth of the amide bond peak at 1557 and 1642 cm^{-1}.

The enzyme immobilization to the PS-PSMA nanofibers and the reactivity of the enzyme-loaded nanofibers are tested with lipase (LP). The LP immobilization (CA-LP) can also accomplished by direct reaction between the free amine at the enzyme surface and the maleic anhydride group at the nanofiber surface. The total loading of the LP enzyme is significantly increased by the dispersion of the PS-PSMA nanofibers. The amount of the immobilized LP is 5.4 ± 2.2 μg per mg of nanofiber for the as-spun PS-PSMA system while 42.4 ± 18.5 mg per mg of nanofiber for the alcohol pre-treated PS-PSMA system. The increase of a factor of ~8 in the loading is attributed to the increase of the nanofiber surface area available for enzyme loading upon dispersion.

The CA-LP activity is obtained from the kinetic study of 4-nitrophenyl butyrate hydrolysis reaction in phosphate buffer (pH 6.5) at 35 °C. As shown in Table 3, the maximum activity, V_{max}, of CA-LP on the as-spun PS-PSMA nanofibers is 0.217 ± 0.049 U/mg nanofiber, while that of CA-LP on the dispersed PS-PSMA nanofibers is 1.51 ± 0.28 U/mg nanofiber. The difference in V_{max} is consistent with the total amount of the CA-LP per unit mass of the nanofibers. So, the specific enzyme activity of CA-LP is the same for both cases, regardless of the nanofiber dispersion. It is interesting to note that the Michaelis constant, K_m, is slightly lower for CA-LP on the as-spun hydrophobic nanofibers than for CA-LP on the alcohol-treated dispersed nanofibers, although the difference is not significantly larger than the experimental error range. This

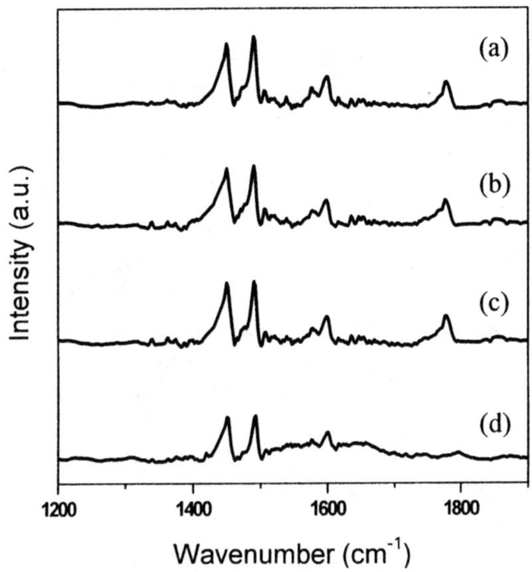

Figure 11. Infrared spectra of PS-PSMA nanofibers after various treatments. (a) as-spun nanofibers, (b) nanofibes washed with phosphate buffer, (c) nanofibers pre-treated with alcohol solution, and (d) nanofibers after alcohol pretrements and reaction with n-propylamine.

Table 3. Michaelis-Menten kinetic parameters for free lipase in solution (pH=6.5) and lipase immobilized on as-spun and dispersed (alcohol-treated) PS-PSMA nanofibers.

Sample	K_m (mM)	V_{max} (U/mg nanofiber)
Free LP	0.38 ± 0.11	-
CA-LP on as-spun nanofiber	0.34 ± 0.09	0.22 ± 0.05
CA-LP on dispersed nanofiber	0.48 ± 0.12	1.51 ± 0.28

might be caused by the preferential segregation of the hydrophobic substrate to the hydrophobic nanofiber surfaces in water or due to a different agitation behaviour for aggregated and dispersed nanofibers under the shaking condition.

Use of Hydrophilic Polymer Nanofibers

Nanofibers of hydrophilic polymers can also be utilized as a solid support. The hydrophilic nanofibers will be dispersed very well without any surface modification. However, the polymer chains should be cross-linked to prevent the dissolution of hydrophilic polymers in aqueous solution. This can be done after the fibers are spun or during the electrospinning process (*21,22*). In these processes, the degree of polymer cross-linking is not well controlled. If the croslinking is not complete, there is a chance to loose immobilized enymes through dissolution of non-cross-linked polymer chains in aqueous solution.

Conclusion

Electrospun polymer nanofibers are successfully used as a solid support material for enzyme immobilization. The large surface area of electrospun nanofibers allows high enzyme loadings and high apparent enzymatic activities. The sustainability of the enzymatic activity of biocatalytic nanofibers can be achieved by crosslinking enzyme molecules that prevents irreversible enzyme denaturation. Hydrophobic polymer nanofibers can be well dispersed in aqueous solution by a simple pre-treatment with aqueous alcohol solutions, which further improves enzyme loading and activity. With the improvement of enzyme loading, activity, and stability, the biocatalytic nanofibers will make efficient biocatalytic systems for bioconversion, bioremediation and biosensing. In addition, the easy control of polymer nanofiber structures can open up a new potential for more versatile and efficient use of biocatalytic nanofibers in various applications including drug delivery and tissue engineering.

Acknowledgements

This work was supported by the National Science Foundation (DMI-0210229), the Americal Chemical Society Petroleum Research Funds (40605-G5), the 3M Non-tenured Faculty Award, and and the Water Remediation & Environmental Research Center of Korea Institute of Science and Technology (KIST).

References

1. Schmid, A.; Dordick, J. S.; Hauer, B.; Kiener, A.; Witholt, B. *Nature* **2001**, *409*, 258.
2. Ball, P.; Ziemelis, K.; Allen, L. *Nature* **2001**, *409*, 225.
3. Koeller, K. M.; Wong, C. H. *Nature* **2001**, *409*, 232.
4. Zhao, X. S.; Bao, X. Y.; Guo, W.; Lee, F. Y *Materials Today* **2006**, *9*, 32.
5. Huang, Z. M; Zhang, Y. Z.; Kotaki, M.; Ramakrishna, S. *Composite Sci. Technol.* **2003**, *63*, 2223.
6. Li, D.; Wang, Y.; Xia, Y. *Adv. Mater.* **2001**, *409*, 258.
7. Theron, S. A.; Zussman, E.; Yarin, A. L. *Polymer* **2004**, *45*, 2017.
8. Megelski, S.; Stephens, J. S.; Chase, D. B.; Rabolt, J. F. *Macromolecules* **2002**, *35*, 8456.
9. Casper, C. L; Stephens, J. S.; Tassi, N. G.; Chase, D. B.; Rabolt, J. F. *Macromolecules* **2004**, *37*, 573.
10. Kim, B. C.; Nair, S.; Kim, J.; Kwak, J. H.; Grate, J. W.; Kim, S. H.; Gu, M. B. *Nanotechnology* **2005**, *16*, S382.
11. Herricks, T. E.; Kim, S.-H.; Kim, J.; Li, D.; Kwak, J. H.; Grate, J. W.; Kim S. H.; Xia, Y. *J. Mater. Chem.* **2005**, *15*, 3241.
12. Paradkar, V. M.; Dordick, J. S. *J. Am. Chem. Soc.* **1994**, *116*, 5009.
13. Wang, P.; Sergeeva, M. V.; Lim, L.; Dordick, J. S. *Nature Biotechnol.*, **1997**, *15*, 789.
14. Novick, S. J.; Dordick, J. S. *Chem. Mater.* **1998**, *10*, 955.
15. Kim, J.; Delio, R.; Dordick, J.S. *Biotechnol. Prog.* **2002**, *18*, 551.
16. Kim, J.; Grate, J. W. *Nano Lett.* **2003**, *3*, 1219.
17. Kim, J.; Kosto, T. J.; Manimala, J. C.; Nauman, E.B.; Dordick, J. S. *AICHE J.* **2001**, *47*, 240.
18 Jia, H.; Zhu, G.; Vugrinovich, B.; Kataphinan, W.; Reneker, D. H.; Wang, P. *Biotechnol. Prog.* **2002**, 18, 1027.
19. Roy, J. C.; Abraham, E. T. *Chem. Rev.* **2004**, *104*, 3705.
20. Nair, S.; Kim, J.; Kim, S. H. *Biomacromolecules.* (in press)
21. Xie, J.; Hsieh, Y.-L. *J. Mater. Sci.* **2003**, *38*, 2125
22. Kim, S. H.; Kim, S.-H.; Nair, S.; Moore, E. *Macromolecules* **2005**, *38*, 3719.

Chapter 8

Reactivity and Characterization of Bioengineered Metal Oxide Nanoparticles

Hazel-Ann Hosein, Sudeep Debnath, Gang Liu, and Daniel R. Strongin

Department of Chemistry, Temple University, Philadelphia, PA 19122

Nanoscale materials could potentially form the basis of a new generation of environmental remediation technologies that provide solutions to some of the challenging environmental cleanup problems. In this study, we report on the preparation of a series of supported iron and cobalt oxyhydroxide nanoparticle model surfaces and also investigated their reactivities toward a SO_2/O_2 mixture. Horse spleen ferritin was used to prepare 3 nm and 5 nm supported ferrihydrite nanoparticles and a ferritin like protein from Listeria innocua was used to prepare 3-4 nm cobalt oxide nanoparticles. Atomic Force Microscopy (AFM) was used to characterize the particles. Attenuated total reflection-Fourier Transform Infrared spectroscopy (ATR-FTIR) was used to study the in-situ oxidation of SO_2 on the nanoparticles.

INTRODUCTION

Metallic and metal oxide nanoparticles often have optical, magnetic, chemical and electronic properties that are very different from those of their bulk counterparts. This potential of nanoparticles for exhibiting unique chemical and physical properties, for example, has been a driving force for the many studies in the area of heterogeneous catalysis.[1-10] In contrast, fewer studies have dealt with the influence on particle size on the adsorption of pollutants in the gaseous phase. Atmospheric chemists, geochemists and environmentalists are all concerned about the reactions taking place on the surfaces of nanoparticles present in the atmosphere and in our ecosystems, since fundamental studies of these reactions can aid in the understanding of the mechanisms that lead to the problems of pollution.[11] The reactions of sulfur oxides with particulates in the atmosphere for example, are of chief concern, and this is primarily because sulfur oxides are known precursors to sulfuric acid, a major contributor to acid rain formation.[12] SO_2 is the principal component of sulfur-containing emissions in urban environments and is released in mass quantities into the atmosphere mainly from electric utilities, petroleum refineries, cement manufacturing and metal processing facilities. The health and environmental effects of SO_2 are widespread and range from respiratory effects to visibility impairments in humans to aesthetic damage to buildings and changes in plant and animal ecosystems.[13]

The adsorption of sulfur dioxide on different metal oxides (MgO, Al_2O_3, CeO_2, ZrO_2, TiO_2, CaO, CuO)[14-25] as well as on oxidized metal surfaces[26] has been studied extensively by a host of researchers using a variety of experimental techniques and varying reaction conditions. Early work was aimed principally at studying the removal of sulfur oxides from flue gases produced from combustion of fossil fuels and other industrial processes that treat sulfur-containing compounds, since the presence of these sulfur oxides in the atmosphere posed a severe air pollution problem.[27-29] A significant portion of these early studies was carried out primarily to determine the efficacy of the metal oxides as sorbents for SO_2. However, research in this area has expanded to include atmospheric studies on the reactions of sulfur dioxide with mineral dust[30-32] and a host of catalytic studies, which look at the poisoning of catalysts by SO_2.[32]

The adsorption and oxidation of SO_2 specifically on iron oxides at room temperature has received limited recent attention and few fundamental studies have concentrated on this area. A kinetic and conductivity study carried out by Kim *et al.*[33] looked at the mechanism of SO_2 oxidation on α-Fe_2O_3 at elevated temperatures but provided little information regarding the nature of the adsorbed species or the oxidation products of the reaction. In another study, researchers looked at the capacity of ferric oxide particles to oxidize SO_2 in air at room temperature and postulated a mechanism where O_2 was required for the conversion of SO_2 to SO_3, and also where sulfate was thought to be the final product.[34] It has also been reported that SO_2 can be oxidized by O_2 on semiconductor surfaces such as α–FeOOH and α-Fe_2O_3 in the atmosphere.[35,36] These semiconductor surfaces act as photocatalysts and readily oxidize SO_2 to SO_4^{2-} in the presence of water.

The present study investigates the chemistry of a SO_2/O_2 mixture on iron oxide nanoparticles as a function of particle size. The objective was to shed light on whether particle size affected the resulting chemistry. The protein, ferritin, was used to assemble ferrihydrite nanoparticles of two different sizes; 2-3 ± 0.5 nm and 5-6 ± 0.5 nm. In addition, cobalt oxide nanoparticles of approximately 2-3 ± 0.5 nm, assembled inside *Listeria innocua* ferritin like protein (LFLP), were investigated in this study.[37-39] The inorganic core (i.e., the iron and cobalt oxyhydroxide) was removed from the protein in an oxidizing environment and subsequently investigated in a mixture of SO_2 and O_2 with ATR-FTIR. We also show results in this contribution for the preparation of supported Mn(O)OH particles assembled by ferritin, but in this circumstance we did not investigate the reactivity of these particles toward SO_2/O_2.

METHODS FOR PREPARATION AND PARTICLE CHARACTERIZATION

Demineralization of Horse Spleen Ferritin

Ferritin is a protein that mineralizes and stores iron within its protein cage as a ferrihydrite nanoparticle. The protein cage consists of 24 subunits with an internal diameter of 8 nm and an outer diameter of 12 nm. The reactions to form the mineral particle include the oxidation of Fe^{2+} and its subsequent hydrolytic polymerization to form ferrihydrite. The Fe^{3+} rapidly forms a small mineral core within the protein shell, and this particle surface will itself catalyze the oxidation of Fe^{2+}.[37-39]

In the present study horse-spleen ferritin (Sigma Chemicals) was demineralized in a buffered solution with thioglycolic acid (TGA), which reduces Fe^{3+} to Fe^{2+} thereby making the mineral dissolve, and then removing the resulting aqueous iron through dialysis. The diluted protein solution in a dialysis bag was placed in a deoxygenated sodium acetate solution (pH 4.5), and aliquots of TGA were added and left for 2-3 hours in a deoxygenated environment. The solution was changed and the above steps were repeated until the protein solution was colorless. The solution was then dialyzed in sodium chloride solution and then in MES buffer pH 6.5. This protein solution contained the iron free apoferritin, that was then remineralized[37-39] to form the iron and manganese oxyhydroxide particles.

Preparation of Fe-loaded Horse Spleen Ferritin

The remineralization of apoferrtin was carried out using a deoxygenated ferrous ammonium sulfate (FAS) solution. Aliquots of the solution were added at regular intervals to the apoferritin solution in a MES buffer solution (pH 6.5) and stirred continuously in air. The volume of the FAS solution to be added was calculated to obtain the proper protein to ferrous ratio.[37,38]

Preparation of Mn-loaded Horse Spleen Ferritin

The remineralization of the manganese-loaded ferritin was similar to the Fe-loading procedure, except that the chemistry was carried out at pH 8.9. The solution used for Mn-loading (made from manganese chloride) was kept deoxygenated and aliquots of this solution were added to buffered apoferritin solution and stirred continuously in air.[37,38] The final solution after mineralization was dialyzed and kept in bis-tris buffer at pH 8.5. Each ferritin cage was loaded with approximately 1500 Mn atoms.

Reconstitution of cobalt bearing Listeria ferritin like protein

The cobalt oxide bearing Listeria ferritin-like protein samples were prepared by others[39](see acknowledgement). Unlike the horse spleen ferritin, Listeria ferritin-like protein is a small cage made of 12 subunits, with an inner diameter of 5 nm and an outer diameter of approximately 8.5 nm. Two cobalt oxide phases, Co(O)OH and Co_3O_4 could be constituted into the Listeria ferritin. In our study, we investigated the oxidation of SO_2 adsorbed on Co_3O_4. For details about the preparation of cobalt oxide bearing Listeria ferritin, refer to article 39.

AFM of nanoparticles

AFM was used to characterize the size of the inorganic core of the ferritin after removal of the protein shell. Individual solutions of the iron, cobalt, and manganese-loaded ferritin were dialyzed in deionized water to completely remove any buffer salts from the solution.[37-39] Samples used for AFM characterization were prepared by spreading individual iron, cobalt, or manganese diluted ferritin solution on either a ZnSe (for Fe-bearing ferritin) or silicon wafer surface (for Co and Mn-bearing ferritin) and then ozone treating the sample at 373 K to oxidize and remove the protein shell, as described elsewhere[37-39]. (The ZnSe surface was the same lens material used in the ATR-FTIR experiment detailed below). Figure 1 exhibits AFM images of particles resulting from 500 and 2000-Fe atom loaded ferritin. Analysis of the z-dimension of the protein-free particles showed sizes in the 3-4 ± 0.5 nm and 6-7 ± 0.5 nm range for the 500- and 2000-Fe loaded particles, respectively. Figure 2 exhibits images of the Co and Mn-bearing nanoparticles that exhibit sizes of 3 ± 0.5 and 6-7 ± 0.5 nm, respectively. Lateral dimensions were significantly larger, presumably due to tip convolution effects.

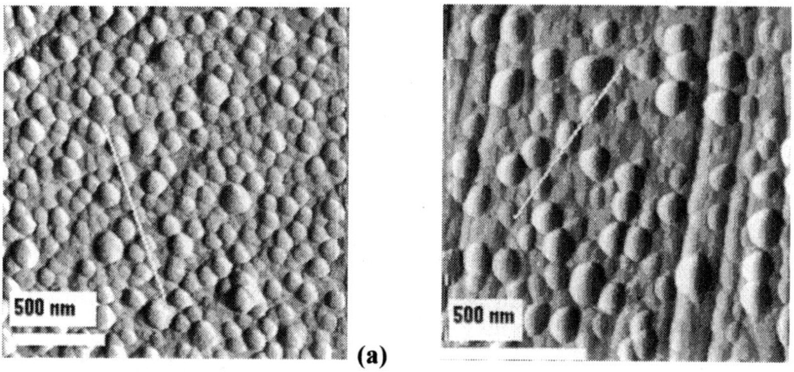

(a) (b)

Figure 1. AFM images of (a) 500 loaded Ferrihydrite particles and (b) 2000 loaded ferrihydrite particles deposited on a ZnSe ATR-FTIR crystal.

ATR-FTIR OF NANOPARTICLES WITH A SO_2/O_2 MIXTURE

ATR-FTIR spectra were recorded using a Nicolet Magna 560 IR spectrometer. A 45 ° horizontal ZnSe ATR crystal was used for all experiments. The samples were prepared by drying ferritin directly on the lens and then exposing the sample to reactive ozone to remove the protein shell. All spectra are referenced to a spectrum obtained immediately after exposing the sample to the SO_2/O_2 mixture (total pressure of 1 atm, approximately a 1:1 ratio of SO_2 and O_2). The total time of exposure was 24 h and all the data was obtained in situ.

ATR-FTIR spectra of the products formed following adsorption of SO_2 onto 6-7 nm sized ferrihydrite particles in the presence of molecular oxygen are shown in Figure 3a. The spectrum is dominated by an intense absorption in the region between 1250-1000 cm^{-1} with maximum spectral intensity centered about 1095 cm^{-1}. The spectrum taken after 24 h of reaction time shows most clearly the evolution of smaller bands appearing as shoulders at 910, 1030, and 1215 cm^{-1} that we attribute to HSO_4^-, bands at 984 and 1095 that are assigned to SO_4^{2-}, [36,40-42] and smaller peaks at 870, 890 and 965 cm^{-1} that are attributed to the presence of SO_3^{2-}.[43]

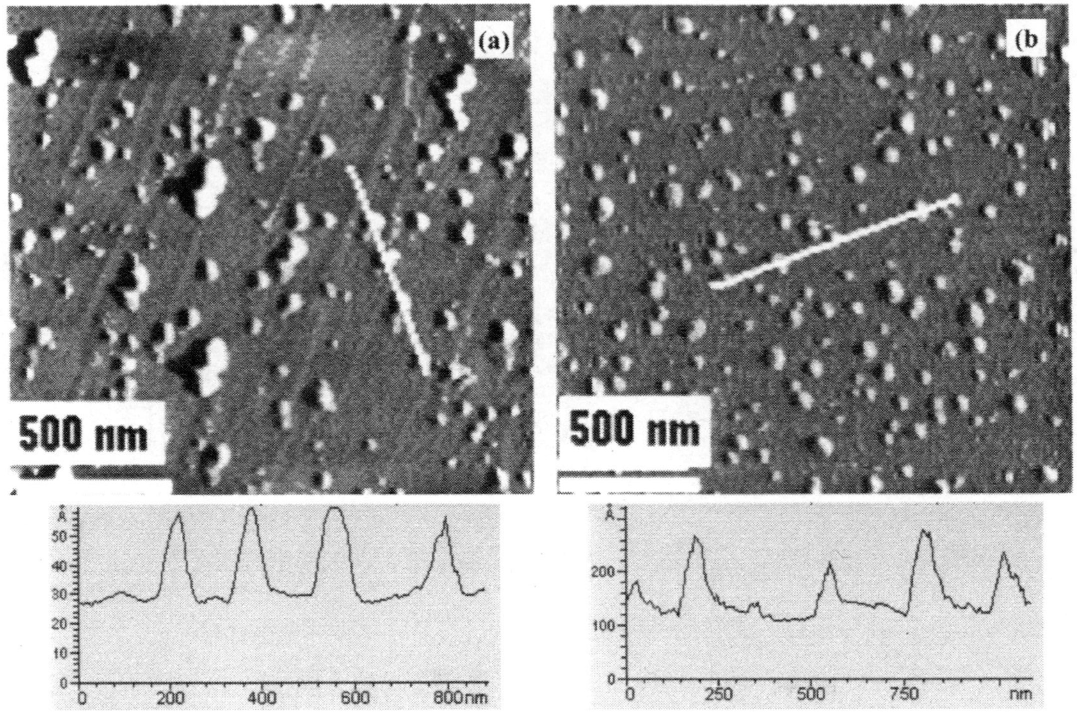

Figure 2. AFM images of (a) Cobalt oxide particles on silicon and (b) Mn oxide particles on silicon.

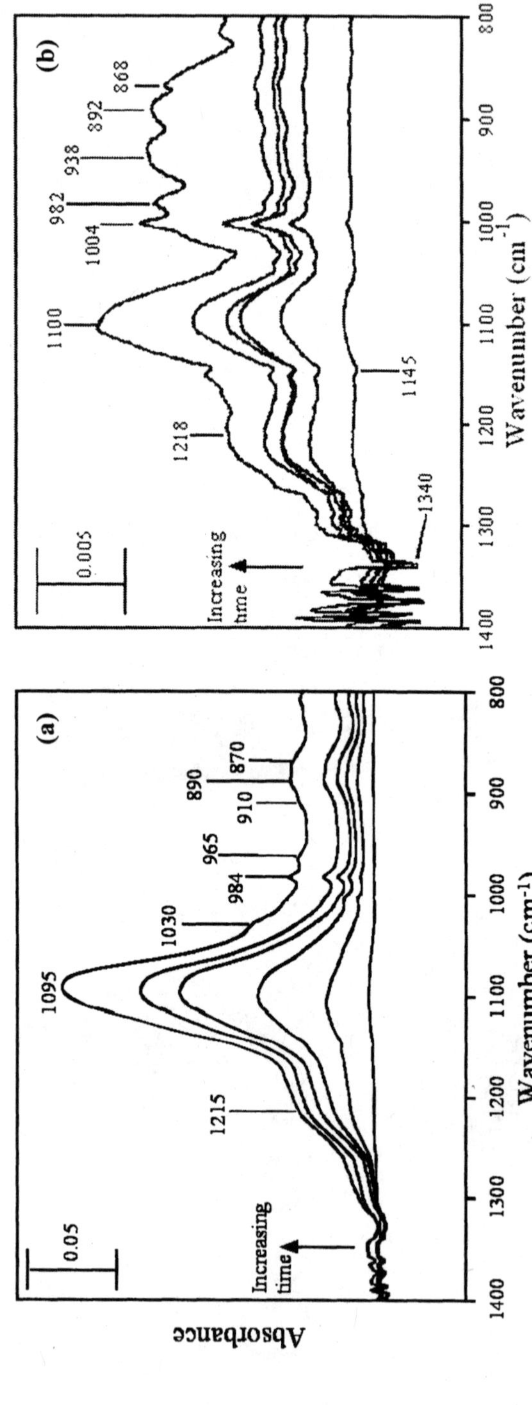
(a)
(b)
Absorbance
Increasing time
0.05
0.005
1095
1030
984
965
910
890
870
1215
1100
1218
1004
982
938
892
868
1145
1340
Wavenumber (cm-1)
1400
1300
1200
1100
1000
900
800

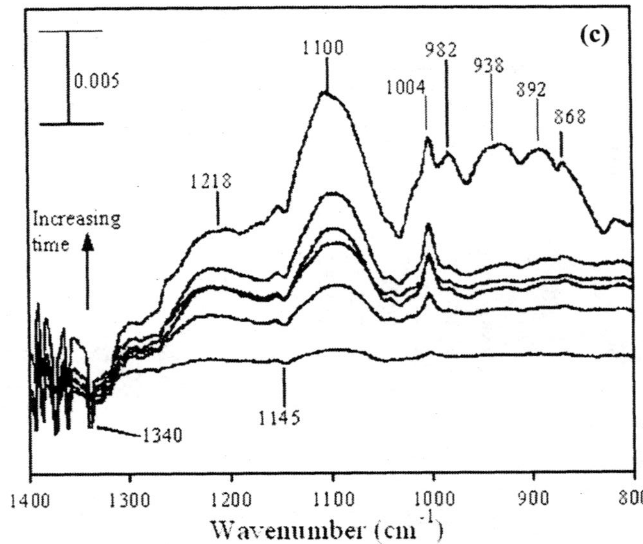

Figure 3. S-bearing product adsorbed on (a) 6 nm ferrihydrite, (b) 3 nm ferrihydrite, and on (c) on 3 nm cobalt oxide particles. The spectra were taken at regular intervals over a period of 24 h.

Figure 3b shows ATR-FTIR spectra of the products formed following adsorption of SO_2 onto 3-4 nm sized ferrihydrite particles in the presence of molecular oxygen. The disappearing peaks at 1340 and 1145 cm^{-1} can be ascribed to vibrational modes of physically adsorbed SO_2, that are converted to chemisorbed sulfur oxyanion species over time.[43] These spectra are complicated and it is difficult to conclusively make assignments of the peaks, but we attribute bands appearing at 1218, 1100, 1004, and 938cm^{-1} to SO_4^{2-} (and perhaps HSO_4^-),[43,44] and bands at 868, 892 and 982 cm^{-1} primarily to SO_3^{2-}.[40]

Figure 3c exhibits a spectrum associated with the reaction of 3-4 nm cobalt oxide (Co_3O_4) nanoparticles with SO_2 in the presence of molecular oxygen, which is similar to that of 3-4 nm ferrihydrite nanoparticles with SO_2 and O_2. Based on the experimental data, there is a mixture of sulfato and sulfite species with the 1113, 1071, 1025 and 1006 cm^{-1} peaks being assigned to a sulfate species[40,43] and the 964, 908 and 872 cm^{-1} peaks being assigned to a sulfite species.[43] The negative features at 1145 cm^{-1} are attributed to the depletion of adsorbed SO_2 over time.

Our results for the reactivity of the different nanoparticles in the SO_2/O_2 environment show some differences depending on the particle size. First, the dominating species in the 6 nm ferrihydrite particle is SO_4^{2-}, while on the smaller particles SO_3^{2-} shows a higher relative surface concentration, although SO_4^{2-} is still present. Perhaps, interestingly, the FTIR spectrum associated with the 3-4 nm Co oxyhydroxide particle is similar to that of the 3-4 nm ferrihydrite particle. Hence, the smaller particles show a reduced amount of sulfate, compared to the larger ferrihydrite particle. Differences in defect density between the large and small particles may be one reason. Also, the availability of surface sites may become low on the smaller particles, leading to incomplete oxidation. For example, the cross-sectional area of SO_2 is almost 2 nm^2,[45] so it may be that the co adsorption of SO_2 and O_2 on the smaller particles becomes difficult because of steric considerations. On the larger particles (6-7 nm) it may be that more complete oxidation to SO_4^{2-} is facile because of the increased availability of surface sites for the oxidation process to proceed. Certainly more detailed studies are needed to determine the origin of the size dependent chemistry. The results, however, do show that the adsorbed sulfur oxyanion product distribution is a sensitive function to the substrate particle size in the nano-regime.

SUMMARY

Ferrihydrite particles with nominal sizes 3nm and 6 nm were synthesized using ferritin as a precursor. The particle sizes were characterized using AFM. The reaction of SO_2 and O_2 adsorption on the particles led to the formation of primarily sulfate and sulfite. A higher relative surface concentration of SO_4^{2-} compared to SO_3^{2-} was found on the larger particles, while on the smaller particles a higher proportion of the surface monolayer was comprised of SO_3^{2-}. Small Co-bearing oxides (3 nm) showed a similar speciation to the 3 nm ferrihydrite particles. The results show the chemistry of SO_2 and O_2 is sensitive to the size of the reacting substrate.

ACKNOWLEDGEMENTS

D.R.S acknowledges the partial support of this research by the Center for Environmental Molecular Science (CEMS) at Stony Brook (NSF-CHE-0221934), U.S. Environmental Protection Agency (EPA), and the donors of the Petroleum Research Fund (PRF), administered by the American Chemical Society. Professor Trevor Douglas and Mark Allen at Montana State University are gratefully acknowledged for providing the Listeria ferritin-like protein bearing cobalt oxide.

REFERENCES

1. Valden, M.; Goodman, D. W. *Isr. J. Chem.* **1998**, *38*, 285.
2. Haruta, M.; Tsubota, S.; Kobayashi, T.; Kageyama, H.; Genet, M. J.; Delmon, B. *J. Catal.* **1993**, *144*, 175.
3. Haruta, M. *Stud. Surf. Sci. Catal.* **1997**, *110*, 123.
4. Zhang, J. Z. *Acc. Chem. Res.* **1997**, *30*, 423.
5. Zhang, Z.; Wang, C.; Zakaria, R.; Ying, J. Y. *J. Phys. Chem. B* **1998**, *102*, 10871.
6. Zhang, Q.; Gao, L. *Langmuir* **2004**, *20*, 9821.
7. Xu, X. *Catal. Lett.* **1994**, *24*, 31.
8. Xu, X.; Chen, P.; Goodman, D. W. *J. Phys. Chem.* **1994**, *98*, 9242.
9. Goodman, D. W. *Chem. Rev.* **1995**, *95*, 523.
10. Foger, K.; Anderson, J. R. *J. Catal.* **1979**, *59*, 325.
11. Usher, C. R.; Michel, A. E.; Grassian, V. H. *Chem. Rev.* **2003**, *103*, 4883.
12. Ohshiro, T.; Izumi, Y. *Biosci. Biotech. Biochem.* **1999**, *63*, 1.
13. Carnow, B. W.; Bouchard, E. *Illinois Inst. Nat. Res.* **1979**, *79*, 28.
14. Low, M. J. D.; Goodsel, A. J.; Takezawa, N. *Environ. Sci. Technol.* **1971**, *5*, 191.
15. Waqif, M.; Saad, A. M.; Bensitel, M.; Bachelier, J.; Saur, O.; Lavalley, J. C. *J. Chem. Soc. Faraday Trans.* **1992**, *88*, 2931.
16. Waqif, M.; Pieplu, A.; Saur, O.; Lavalley, J. C.; Blanchard, G. *Solid State Ionics* **1997**, *95*, 163.
17. Goodsel, A. J.; Low, M. J. D.; Takezawa, N. *Environ. Sci. Technol.* **1972**, *6*, 268

18. Datta, A.; Cavell, R. G.; Tower, R. W.; George, Z. M. *J. Phys. Chem.* **1985**, *89*, 443.
19. Stark, J. W.; Park, D. G.; Lagadic, I.; Klabunde, K. J. *J. Chem. Mater.* **1996**, *8*, 1904.
20. Bensitel, M.; Waqif, M.; Saur, O.; Lavalley, J.-C. *J. Phys. Chem.* **1989**, *93*, 6581.
21. Goodman, A. L.; Li, P.; Usher, C. R.; Grassian, V. H. *J. Phys. Chem. A* 93, 6581.
22. Waqif, M.; Saur, O.; Lavalley, J.-C. *J. Phys. Chem.* **1991**, *95*, 4051.
23. Schoonheydt, R. A.; Lunsford, J. H. *J. Catal.* **1972**, *26*, 261.
24. Matsumoto, A.; Kaneko, K. *Coll. Surf.* **1989**, *37*, 81.
25. Uysal, B. Z.; Aksahim, I.; Yucel, H. *Ind. Eng. Chem. Res.* **1988**, *27*, 434.
26. Pradier, C. M.; Jacqueson, E.; Dubot, P. *J. Phys. Chem.* **1999**, *103*, 5028.
27. Siddiqui, S.; Ahmad, A.; Hayat, S. *Pollut. Res.* **2004**, *23*, 327.
28. Gurtu, D.; Vaidya, M.; Gajghate, D. G. *Ind. J. Environ. Prot.* **2001**, *21*, 683.
29. Jonathan, C. P.; Ojha, K. G. *Pollut. Res.* **2004**, *23*, 333.
30. Ullerstam, M.; Vogt, R.; Langer, S.; Ljungström, E. *Phys. Chem. Chem. Phys.* **2002**, *41*, 4694.
31. Usher, C. R.; Al-Hosney, H.; Carlos-Cueller, S.; Grassian, V. H. *J. Geophys. Res.* **2002**, *107*, 4713.
32. Shang, J.; Zhu, Y.; Du, Y.; Xu, Z. *J. Solid State Chem.* **2002**, *166*, 395.
33. Kim, K. H.; Choi, J. S. *J. Phys. Chem.* **1981**, *85*, 2447.
34. Chun, K. C.; Quon, J. E. *Environ. Sci. Technol.* **1973**, *7*, 532.
35. Leland, J. K.; Bard, A. J. *J. Phys. Chem.* **1987**, *91*, 5083.
36. Faust, B. C.; Hoffmann, M. R.; Bahnemann, D. W. *J. Phys. Chem.* **1989**, *93*, 6371.
37. Hosein, H.; Strongin, D. R.; Allen, M.; Douglas, T. *Langmuir* **2004**, *20(23)*, 10283
38. Klem, M. T.; Young, M.; Douglas, T. *Materials Today* **2005**, *8(9)*, 28.
39. Allen, M.; Willits, D.; Young, M.; Douglas, T. *Inorganic Chemistry* **2003** *42(20)*, 6300.
40. Hug, S. J. *J. Coll. Interface Sci.* **1997**, *188*, 415.
41. Madejová, M. *Vib. Spectrosc.* **2003**, *31*, 1.
42. Wijnja, H.; Schulthess, C. P. *J. Coll. Interface Sci.* **2000**, *229*, 286.
43. Parry, D.; Samant, M. G.; Seki, H.; Philpott, M. R. *Langmuir* **1993**, *9*, 1878.
44. Yamaguchi, T.; Jin, T.; Tanabe, K. *J. Phys. Chem.* **1986**, *90*, 3148.
45. Klabunde, K. J.; Stark, J. W.; Koper, O.; Mohs, C.; Park, D. G.; Decker, S.; Jiang, Y.; Lagadic, I.; Zhang, D. *J. Phys. Chem.* **1996**, *100*, 12142.

I: Nanostructures for Biomolecular Catalysts

I-3: Nanobiocomposites

Chapter 9

Surfactant Aggregates as Matrix Nanocontainers for Proteins (Enzymes) Entrapment and Regulation

Natalia L. Klyachko and Andrey V. Levashov

Department of Chemical Enzymology, Faculty of Chemistry, M.V. Lomonosov Moscow State University, 119992 Moscow, Russia

The important issue of the use of surfactant or lipid aggregates formed in ternary systems of 'Surfactant (Lipid)-water-oil (organic solvent)' type for enzyme (protein) regulation is that the systems with bioactives wrapped tightly by nanosized matrix play an important role both on activity and conformational level. Peculiarities in system or protein manipulations as well as protein stability and enzyme activity regulations are shown on several examples.

Introduction

Surfactant or lipid aggregates of different structure are well-known and wide used carriers for proteins (enzymes) and other bioactive substances (1-8). There is a long history of using micelles (normal and reverse) (called also microemulsions or nanoemulsions) as well as liquid crystalline structures (lamellar, hexagonal, cubic, etc.) for enzyme regulation, protein modification applying the fundamental research to chemical synthesis, analysis, medical needs, and so on (3-5,8-15). According to the phase diagram of a ternary system 'Surfactant (Lipid)-water-oil (organic solvent)' (Fig.1), different ordered structures (varying by size and shape) can be formed spontaneously depending on the nature and component's ratio in the system (8). In diluted aqueous solutions, a surfactant forms up normal spherical micelles (L_I), the core of which is constituted of apolar chains (tails), while the outer shell contacting with water consists of polar (and often charged) heads. In oil or apolar organic solvents a surfactant also forms spherical – but inverted (or reverse) – micelles (L_2); here, the core consists of polar (charged) heads while non-polar tails are oriented outside into the solvent. Normal micelles are capable to solubilize apolar compounds while reverse micelles solubilize polar compounds, primarily water. Hydration of reverse micelles is accompanied by an increase in their sizes and formation of an independent aqueous phase inside the micelle.

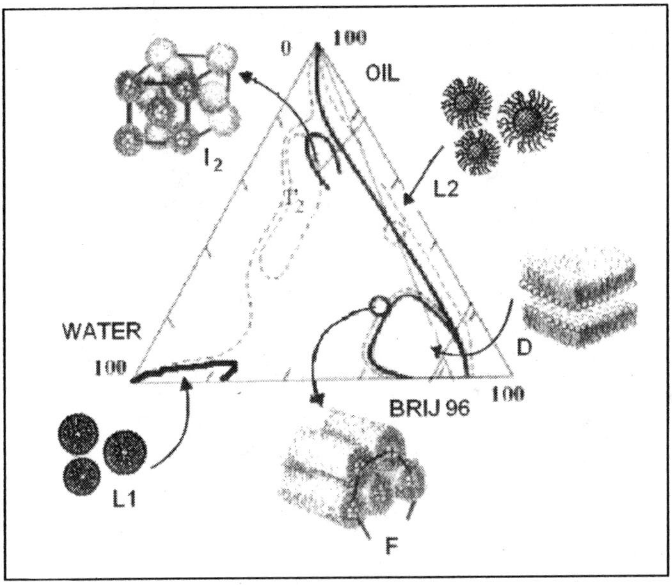

Figure 1. Phase diagram of ternary system Brij 96-water-oil and schematic representation of normal and reverse micelles (L_1 and L_2), lamellar (D), reverse hexagonal (F) and cubic (I_2) phases

Ternary systems of 'Surfactant (Lipid)-water-oil (organic solvent)' type can be considered as a model of membrane environment for enzyme functioning *in vivo* where enzymes mostly act either *on* or *near* the 'water/organic medium' interface contacting or interacting with other proteins, lipids or polysaccharides. Moreover, it is accepted now that lipid polymorphism, i.e. structural rearrangements of lipids from the bilayer to non-lamellar phases, either micelles, cubic or hexagonal phases, plays an important role in many processes occurring in the living cell, such as fusion and compartmentalization of membranes (16,17), protein-lipid interactions (18,19) and transport (20-24), regulation of enzyme activity (25-27) and oligomeric composition (28), etc. (see also reviews (29-33) and refs. therein).

In contrast to difficulties in studying dynamic intramembraneous lipidic particles formed between monolayers of lipid bilayer, surfactant or lipid hydrated reverse micelles exhibit a relatively ordered structure, characterized by a definite diameter with easy variable curvature, molecular weight (aggregation number), and packing density (see, for example, (34-37)). Surfactants, both low molecular mass and polymeric, lipids, polysaccharides and polyelectrolytes, all can be used as matrix building material, and the size and properties of the matrix can be easily varied by changing the nature and concentration of components. General methodology, techniques, and potentialities of using such systems have been analyzed in recent reviews (3,10-12,38). Micelles of o/w type from amphiphilic co-polymers have attracted much attention recently in the area of drug delivery as nanocontainers for poorly soluble drugs (see recent reviews, for example (38-40)). The size of such particles lies in nano-diapason, and recently the systems were 'rediscovered' as nanoemulsions becoming widely spread in new direction of drug delivery as drug carriers (for recent review see (41)).

Methods of protein (enzyme) incorporation into reverse micelles

Protein (enzyme) incorporation into reverse micelles of surfactant in organic solvents may be achieved by one of the following methods.

The first method proposed by our group (42) often referred as an "injection method" and is now most widely used. A small amount (the order of a few volume percent) of protein aqueous solution is introduced into the surfactant solution in organic solvent ("dry" or slightly hydrated). A ratio of water and organic solutions is defined by the conditions of the experiment, primarily, by the value of the desired surfactant hydration degree (water to surfactant molar ratio, $w_0 = [H_2O]/[surfactant]$). Thus obtained mixture is shaken up vigorously (seconds or tens of seconds) until formation of the optically transparent solution. This method is simple and effective.

In the second method proposed by Menger and Yamada (43), a desired amount of water (aqueous buffer solution) is first introduced into the solution of surfactant in organic solvent in order to achieve the desired value of hydration degree, w_0, after that a dry (for instance, lyophilized) protein preparation is dissolved in the obtained micellar solution and is energetically shaken (mixed). The time needed for solubilization of the dry protein is normally much longer than in the case of solubilization of aqueous solutions – it varies from a few minutes to a few hours. With this procedure a protein stays in contact with the surfactant and the organic solvent during a relatively long time and as a result its denaturation often occurs. However, in the end (at the expense of the loss of the protein) micellar solutions with much higher (often saturating) protein concentrations may be obtained compared to those in the first (injection) method.

The third method first described in the work of D. Hanahan (44) was intensively employed and developed in the works of the group of P-L. Luisi (45), see also review (2). The principle of this method lies in the spontaneous transfer (distribution) of the protein in a biphase system constituted of nearly equal volumes of protein aqueous solution and surfactant containing organic solvent (actually, a micellar system with the fixed hydration degree). Protein transfer occurs without stirring or with a slight stirring and lasts a relatively long time (from tens of minutes to one day).

Proteins purification, folding, aggregation

The use of reverse micelles for protein fractionation (separation and/or purification) is one of the aspects that make systems attractive tools in biotechnology (see review (46) and refs therein). Such parameters as pH and ionic strength regulating the processes of protein incorporation into the micelles and its withdrawal back into the water solution can easily be varied to find conditions for the selective extraction of the required protein from the mixture, for example. This approach allows extraction not only proteins (enzymes) but also other water-soluble components, DNA, for example (47).

Reverse micelles can also be considered as a convenient matrix whose shape favors folding extended biopolymers even metastable proteins (48,49). The problem of aggregation for folding of gene-engineered proteins that often form inclusion bodies exists. Attempts to extract proteins from inclusion bodies result in the formation of insoluble aggregates. One can overcome this problem using surfactant reverse micelles (48,49). Varying water content in the system, i.e. sizes of protein nanocontainers, it is possible to assemble both supramolecular complexes and particular subunits. For triosephosphate isomerase (50) and formate dehydrogenase (51), a possibility of obtaining (under specific conditions) folded monomeric or dimeric protein particles has been shown. Moreover, even polyelectrolytes incorporated into the reverse micelle can be folded to form compact particles, globules (52,53).

Manipulations with the system

Inside polar core of the nanocontainer (nanoemulsion of w/o type), water can be replaced with water-miscible organic solvent without destroying the system. Glycerol, 2,3-butanediol, DMSO, glucose and other sugars were used as water-miscible solvents (54-56) for enzyme activity and stability regulation.

Different polymeric materials such as polyelectrolytes, gelatin can be used to modify an inner hydrophilic protein microenvironment. The formation of 'nanogels' i.e. gel formation on the base of gelatin inside the water pools of nanodroplets has been studied in details (57-59). Further 'hardening' of nanocomposite enzyme-containing matrix based on gelatin gels has been made by in situ polymerization of tetraethoxysilane (60) that allowed obtaining dry enzyme-containing powder stable both in aqueous and organic solvents.

Different monomers capable to polymerization in the inner phase can be used, acrylamide and N,N'-methylenebisacrylamide (61) or ethyl 2-cyanoacrylate (62), for example. First example (61) was used to prepare nanogranules of stabilized chymotrypsin (the nanogranulated enzyme did operate in organic solvent at elevated temperatures (up to 70C), while the native enzyme was totally inactivated). Second example (62), used for preparation of insulin-containing nanocapsules by interfacial polymerization, represents one of the methods used in wide spread area of drug delivery.

Another way to manipulate with the protein or enzyme-containing nanoemulsions is to use surfactants with double bonds capable to polymerization (63-65).

Schematic representation of different manipulations with reverse micelles is given in Fig. 2.

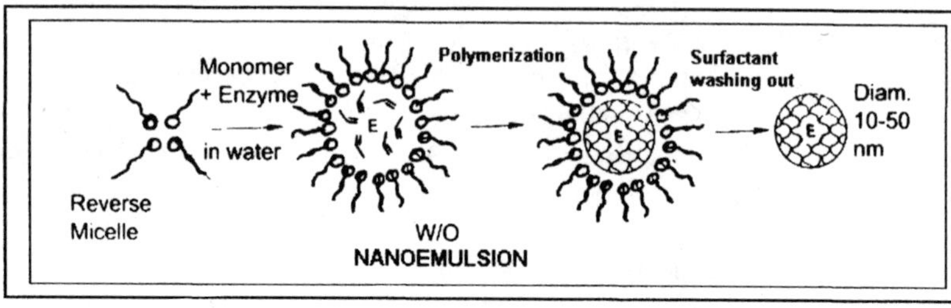

Figure 2. Schematic representation of different manipulations with reverse micelles: 1 - addition of water or/and aqueous solution of enzyme and monomer; 2 - polymerization of the content of an inner cavity; 3 - washing out the surfactant layer resulting in enzyme-containing nanogranules

Manipulations with the protein entrapped in surfactant nanocontainers

Reverse micelles are an excellent and even unique tool for various chemical manipulations with proteins (enzymes). Micellar systems are very convenient for the introduction hydrophobic poor water-soluble reagents into protein molecules in controlled quantities (first of all, of one-two hydrophobic residues per protein molecule), long-chain fatty acid residues (66) or phospholipids (67), for example. Such an elegant protein modification in mild conditions can bring new features to hydrophilic protein - the ability to membrane translocation (66).

Normally, simple proteins (enzymes) form protein-containing micelles in a stoichiometric ratio 1 to 1. In the case of oligomeric proteins (enzymes), variation of micelle sizes (water content in the system) allows obtaining a whole set of different oligomeric forms: monomers, dimers, tetramers and so on, for many individual enzymes and enzyme mixtures (see (68-69), for example, and reviews (3,9,12)). Employing micellar matrix it is possible to "construct" non-conventional (from the point of view of classical "aqueous" enzymology) protein-protein complexes, such as a compact non-covalent chymotrypsin dimer or a stable (dissociating in water only in presence of 8 M urea) non-covalent complex of chymotrypsin with peroxidase (70).

In a homogeneous solution, "cross-linking" of two macromolecules is a problem because of a vast number of intermolecular reactions resulting in the formation of the cross-linked block copolymer. In a micellar media it is possible to restrict a sphere of reaction by just a space of one micelle and to suppress (or to exclude when necessary) other reactions on the level of intermicellar interactions. In other words, employing classical linking reagents in micellar systems allows to link (to fix chemically) complexes in an "intramicellar" mode. This idea has been implemented in studies on the formation of protein and synthetic polymers conjugates of different stoichiometry (52). At low hydration degrees, when micelle sizes are small, the protein and polymer molecules are localized in different micelles (do not interacting virtually). After reaching a certain critical value of surfactant hydration degree (this value depends on the size of the complex), a complex is formed which may be linked chemically with a practically quantitative yield. Further increase of micelle sizes (surfactant hydration degree) may lead to formation of protein-polymer complexes of higher stoichiometry (52).

Enzyme (protein) stability regulation

The above-mentioned approaches to system and protein manipulations usually lead to the protein stabilization increasing both thermal and solvent stability. However, even without special additives and manipulations, a protein molecule is wrapped with surfactant molecules, representing 'immobilization' at molecular level and bringing enhanced stability to enzymes encapsulated in nanoemulsions at appropriate conditions. 'Close fitting cage' functions as a buffer (or dumper) of excessive spontaneous fluctuations that usually destroy (or disturb) the catalytic conformation of an enzyme in water. The illustrating scheme is shown in Fig. 3 (left upper picture). The general conclusion coming up: better fitting will bring higher protein stability, even unusual stabilization of oligomeric protein in monomeric form, as shown on the example in Fig. 3.

As an example, a comparative stability study was carried out for baker's yeast alcohol dehydrogenase (YADH) in water and reverse micelles (Fig. 3). The enzyme represents an oligomeric protein composed of four subunits working in water in its tetrameric form (for enzyme kinetics, see (71,72) and refs therein). First of all, it was found that the enzyme preferred to stay in micelles as monomer or dimer, or tetramer depending on micellar size (hydration degree, w_o). The results of the sedimentation analysis showing molecular masses of particles surrounded by the surfactant shell are presented in Fig.3 (left bottom). At w_o 11-20, only subunits (monomers) of YADH were found in the system, at w_o higher 21 – dimers were present (gray dots). Tetramers were appearing at high w_o only in the presence of NAD^+ (black rhombs).

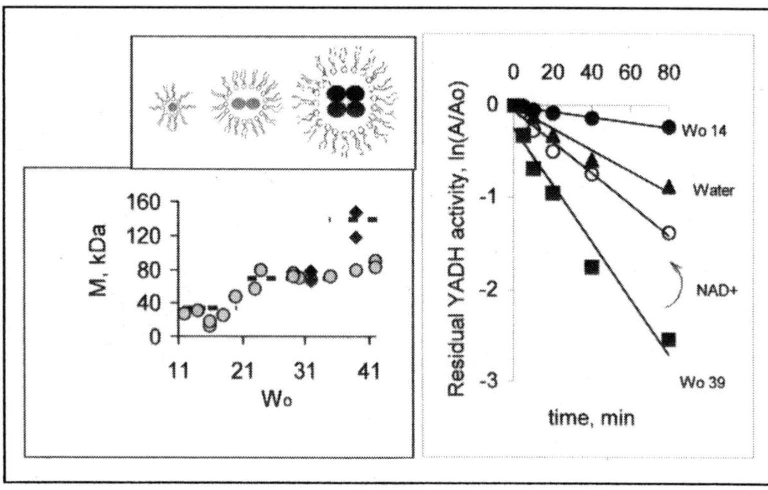

Figure 3. YADH in ternary AOT-water-octane system; Left bottom – molecular masses of protein found by sedimentation analysis; Left upper – schematic representation of the results of sedimentation analysis; Right - time dependence of the residual activity (ln(A/Ao)) of YADH in water and AOT- water- octane system at Wo 14 and 39, pH 8.6, 25°C. Arrow – the stability changes in the presence of NAD^+

As seen from Fig. 3 and known from previous works (see (71,72) and refs therein), the enzyme is denaturing quite fast in water at alkaline pH (black triangles). YADH stability can be regulated in wide range upon enzyme encapsulation in reverse micelles with different water content (different micellar size). As found, the enzyme in monomeric form (w_o 14) revealed higher stability than that in water probably because this form fits better to the spherical micelle (rhombs in Fig. 3). At w_o 39, the size of the micelle exceeds significantly the molecule of dimeric YADH (ellipsoid form) recognized by sedimentation analysis at this conditions. Possible additional unfavorable interactions with micellar matrix leaded to the situation when the enzyme was inactivated even faster being incubated at w_o 39 (squires in Fig. 3). Better fitting of tetramer formed in the presence of NAD^+ to this big micelle brought the enzyme stability improvement (white dots in the Fig. 3).

Enzyme activity regulation

An intriguing phenomenon observed for the enzymes entrapped into reverse micelles is a bell-shaped dependence of their catalytic activity on the hydration degree with the maximum observed at those w_0 where the size of a micelle is equal to that of a protein (1,3,8,9,11,12). In the case of oligomeric enzymes, the dependence of kinetic parameters on hydration degree shows a number of peaks corresponding to each oligomeric form of an enzyme (see, for example, 68-70).

Under conditions of such a complementarity's, there is a close contact of the enzyme molecule and the surrounding surfactant matrix. The enzyme in this case can be "squeezed" tightly by the micellar matrix, its rotational and vibrational motions can then be "freezed" leading in many cases to the fixation of the most catalytically active conformation, but usually in the expense of substrate binding parameters. The situation is illustrated in Fig. 4 by two examples, β-glucosidase (left) and lipoamide dehydrogenase (right). As found and confirmed by sedimentation analysis that allows calculating molecular masses of proteins encapsulated in nanoemulsion, sharp increase in enzyme efficiency is indeed observed in conditions when protein size fits to that of nanoemulsion inner cavity (top figures). As seen from the bottom figures, sharp increase in Km values, i.e. binding worsening occurs at the same conditions. It should be mentioned that two peaks seen for oligomeric lipoamide dehydrogenase reflect the formation of two catalytically competent enzyme forms, dimeric and tetrameric (69). One point should be stressed especially: normally, micellar concentration (or number of nanoemulsion particles) exceeds many orders of magnitude an enzyme concentration used. It means that, in principle, there are many empty containers in the system. The fact that enzyme molecules in some conditions 'prefer to stay together' in one micelle reflects that micelles are not forcing tools, they help to reveal natural potential of enzyme preferences.

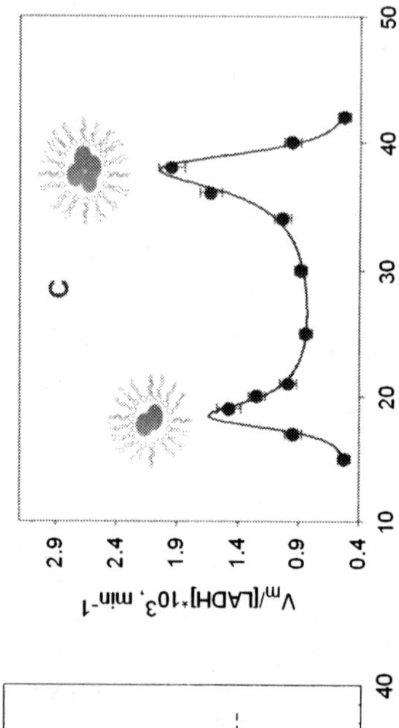
C
2.9
2.4
1.9
1.4
0.9
0.4
10
20
30
40
50
V_m/[LADH]*10^3, min^-1

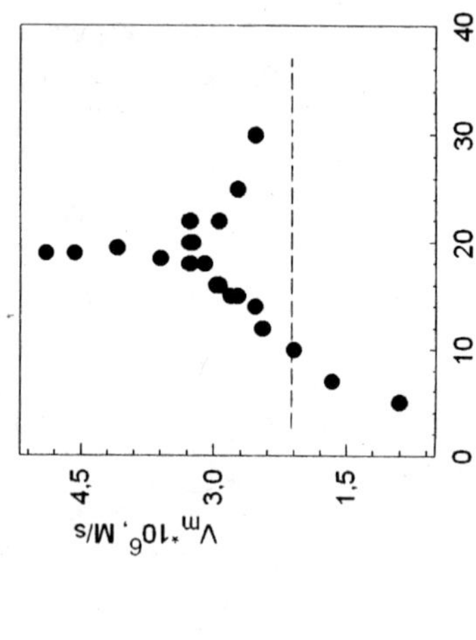
4,5
3,0
1,5
0
10
20
30
40
V_m*10^6, M/s

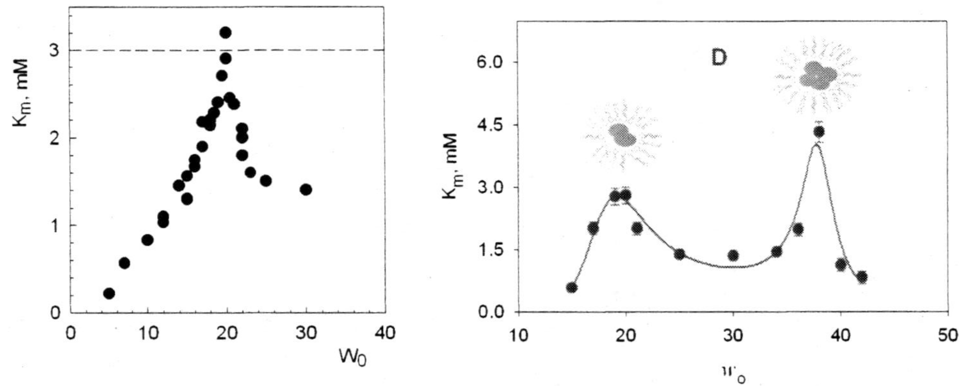

Figure 4. Regulation of kinetic parameters (Vm and Km) of sweet almond β-glucosidase (left) and pig heart lipoamide dehydrogenase (right) in AOT/water/octane reverse micelles; dashed line – kinetic parameters in water; 25°C; Substrates: o-nitrophenyl β-D-glucopyranoside; pH 4.5 (left); Lipoamide and NADH, pH 7.5 (right)

Conclusion

Protein- or enzyme-containing surfactant aggregates possess a whole set of unique advantages that makes them very perspective in using for both fundamental research and applied areas.

References

1. Martinek, K.; Klyachko, N.L.; Kabanov, A.V.; Khmelnitsky, Y.L.; Levashov, A.V. Micellar enzymology: its relation to membranology. *Biochim. Biophys. Acta* **1989**, *981*, 161-172.
2. Luisi, P.L.; Giomini, M.; Pileni, M.P.; Robinson, B.H. Reverse micelles as hosts for proteins and small molecules. *Biochim Biophys Acta* **1988**, *947*, 209-246.
3. Levashov, A.V.; Klyachko, N.L. Micellar enzymology: methods and technique. *Russ. Chem. Bulletin (Engl. Ed.)* **2001**, *50*, 1718-1732.
4. Orlich, B.; Schomacker, R. Enzyme catalysis in reverse micelles. *Adv. Biochem. Eng. Biotechnol.* **2002**, *75*, 185-208.
5. Savelli, G.; Spreti, N.; DI Profio, P. Enzyme activity and stability control by amphiphilic self-organizing systems in aqueous solutions. *Curr. Opin. Colloid Interface Sci.* **2000**, *5*, 111-117.
6. Walde, P. Enzymatic reactions in liposomes. *Curr. Opin. Colloid Interface Sci.* **1996**, *1*, 638-644.
7. Oberholzer, T.; Meyer, E.; Amato, I.; Lustig, A.; Monnard, P.A. Enzymatic reactions in liposomes using the detergent-induced loading method. *Biochim Biophys Acta* **1999**, *1416*, 57-68.
8. Klyachko, N.L.; Levashov, A.V.; Pshezhetsky, A.V.; Bogdanova, N.G.; Berezin, I.V.; Martinek, K. Catalysis by enzymes entrapped into hydrated aggregates having lamellar or cylindrical (hexagonal) or ball-shaped (cubic) structure in organic solvent. *Eur. J. Biochem.* **1986**, *161*, 149-154.
9. Klyachko, N.L.; Levashov, A.V.; Khmelnitsky, Y.L.; Martinek, K. *Catalysis by enzymes entrapped in hydrated surfactant aggregates having various structures in organic solvents*; Kinetics and Catalysis in Microheterogeneous systems, Gratzel, M.; Kalyanasundaram, K., Eds.; Marcel Dekker: New York, 1991, pp.135-181.
10. Torchilin, V.P. Recent advances with liposomes as pharmaceutical carriers. *Nat. Rev. Drug. Discov.* **2005**, *4*, 145-160.
11. Klyachko, N.L.; Levashov, A.V. Bioorganic synthesis in reverse micelles and related systems. *Curr. Opin. Colloid Int. Sci.* **2003**, *8*, 179-186.
12. Levashov, A.V.; Klyachko, N.L. *Enzymes in reverse micelles (microemulsions): theory and practice*; Interfacial catalysis; Volkov, A., Ed., Marcel Dekker: New-York, Basel, 2002, pp.355-376.

13. Oldfield, C. Enzymes in water-in-oil microemulsions ("reversed micelles"): principles and applications. *Biotechnol. Genetic Eng. Rev.* **1994**, *12*, 255-327.
14. Ono, T.; Goto, M. Application of reversed micelles in bioengineering. *Curr. Opin. Colloid Interface Sci.* **1997**, *2*, 397-401.
15. Chang, G.G.; Huang, T.M.; Hung, H.C. Reverse micelles as life-mimicking systems. *Proc. Natl. Sci. Counc. Repub. China B* **2000**, *24*, 89-100.
16. Ruiz-Arguello, M.B.; Basanez, G.; Goni, F.M.; Alonso, A. Different effects of enzyme-generated ceramides and diacylglycerols in phospholipid membrane fusion and leakage. *J. Biol. Chem.* **1996**, *271*, 26616-26621.
17. Ortiz, A.; Kilian, J.A.; Verkleij, A.J.; Wilschut, J. Membrane fusion and the lamellar-converted hexagonal phase transition in cardiolipin vesicle systems induced by divalent cations. *Biophys. J.* **1999**, *77*, 2003-2014.
18. Escriba, P.V.; Ozaita, A.; Ribas, C.; Miralles, A.; Fodor, E.; Farkas, T.; Garcia-Sevilla, A. Role of lipid polymorphism in G protein-membrane interactions: nonlamellar-prone phospholipids and peripheral protein binding to membranes. *Proc. Natl. Acad. Sci. USA* **1997**, *94*, 11375-11380.
19. Terrones, O.; Antonsson, B.; Yanaguchi, H.; Wang, H.-G.; Liu, J.; Lee, R.M.; Herrmann, A.; Basanez, G. Lipidic pore formation by the concerted action of pro-apoptotic BAX and tBID. *J. Biol. Chem.* **2004**, published online, manuscript M313420200.
20. Rietveld, A.G.; Chupin, V.V.; Koorengevel, M.C.; Hienk, H.L.J.; Dowhan, W.; de Kruijff, B. Regulation of lipid polymorphism is essential for the viability of phophatidylethanolamine-deficient Escherichia coli cells. *J. Biol. Chem.* **1994**, *269*, 28670-28675.
21. Rietveld, A.G.; Koorengevel, M.C.; de Kruijff, B. Non-bilayer lipids are required for efficient protein transport across the plasma membrane of Escherichia coli. *EMBO J.* **1995**, *14*, 5506-5513.
22. Ahn, T.; Oh, D.-B.; Kim, H.; Park, C. The phase property of membrane phospholipids is affected by the functionality of signal peptides from the Escherichia coli ribose-binding protein. *J. Biol. Chem.* **2002**, *277*, 26157-26162.
23. Epand, R.F.; Martinou, J.-C.; Fornallaz-Mulhauser, M.; Hughes, D.; Epand, R.M. The apoprotic protein tBid promotes leakage by altering membrane curvature. *J. Biol. Chem.* **2002**, *277*, 32632-32639.
24. Zazueta, C.; Ramirez, J.; Garcia, N.; Baeza, I. Cardiolipin regulates the activity of the reconstituted mitochondrial calcium uniporter by modifying the structure of the liposome bilayer. *J. Membrane Biol.* **2003**, *191*, 113-122.
25. Slater, S.J.; Kelly, M.B.; Taddeo, F.J.; Ho, C.; Rubin, E.; Stubbs, C.D. The modulation of protein kinase C activity by membrane lipid bilayer structure. *J. Biol. Chem.* **1994**, *269*, 4866-4871.

26. Li, L., Zheng, L.X., Yang, F.Y. Effect of propensity of hexagonal II phase formation on the activity of mitochondrial ubiquinol-cytochrome c reductase and H(+)-ATPase. *Chem. Phys. Lipids* **1995**, *76*, 135-144.
27. Van der Does, C.; Swaving, J.; Van Klompenburg, W.; Driessen, A.J.M. Non-bilayer lipids stimulate the activity of the reconstituted bacterial protein translocase. *J. Biol. Chem.* **2000**, *275*, 2472-2478.
28. Or, E.; Navon, A.; Rapoport, T. Dissociation of the dimeric SecA ATPase during protein translocation across the bacterial membrane. *EMBO J.* **2002**, *21*, 4470-4479.
29. De Kruijff, B. Lipid polymorphism and biomembrane function. *Curr. Opin. Chem. Biol.* **1997**, *1*, 564-569.
30. Luzzati, V. Biological significance of lipid polymorphism: the cubic phases. *Curr. Opin. Struct. Biol.* **1997**, *7*, 661-668.
31. Epand, R.M. Lipid polymorphism and protein-lipid interactions. *Biochim. Biophys. Acta* **1998**, *1376*, 353-368.
32. Rappolt, M., Hickel, A., Bringezu, F., Lohner, K. Mechanism of the lamellar/inverse hexagonal phase transition examined by high resolution X-ray diffraction. *Biophys. J.* **2003**, *84*, 3111-3122.
33. Pali, T., Garab, G., Horvath, L.I., Kota, Z. Functional significance of the lipid- protein interface in photosynthetic membranes. *Cell. Mol. Life Sci.* **2003**, *60*, 1591-1606.
34. Eicke, H.F. Surfactants in nonpolar solvents: aggregation and micellization. *Top.Curr.Chem.* **1980**, *87*, 85-145.
35. Kotlarchyk, M.; Huang, J.S.; Chen, S.-H. Structure of AOT reversed micelles determined by small-angle neutron scattering. *J. Phys. Chem.* **1985**, *89*, 4382-4386.
36. Moulik, S.P.; Paul, B.K. Structure, dynamics and transport properties of microemulsions. *Adv. Colloid Int. Sci.* **1998**, *78*, 99-195.
37. Kohling, R.; Woenckhaus, J.; Klyachko, N.L.; Winter, R. Small-angle neutron scattering study of the effect of pressure on AOT-n-octane-water mesophases and the effect of chymotrypsin incorporation. *Langmuir* **2002**, *18*, 8626-8632.
38. Kabanov, A.V.; Alakhov, V.YU. Pluronic block copolymers in drug delivery: from micellar nanocontainers to biological response modifiers. *Critical Rev. Therapeutic Drug Carrier Systems* **2002**, *19*, 1-73.
39. Lukyanov, A.N.; Torchilin, V.P. Micelles from lipid derivatives of water-soluble polymers as delivery systems for poorly soluble drugs. *Advanced Drug Delivery Rev.* **2004**, *56*, 1273-1289.
40. Torchilin, V.P. Lipid-core micelles for targeted drug delivery. *Curr. Drug Deliv.* **2005**, *2*, 319-327.
41. Sarker, D.K. Engineering of nanoemulsions for drug delivery. *Curr. Drug Deliv.* **2005**, *2*, 297-310.

42. Martinek, K; Levashov, A.V.; Klyachko, N.L.; Berezin, I.V. *Proc Acad Sci USSR (Russ)* **1977**, *236*, 920-923 (Eng. Ed. 1978, 236, 951-953).
43. Menger, F.M.; Yamada, K. *J. Amer. Chem. Soc.* **1979**, *191*, 6731-6734.
44. Hanahan, D.J. *J. Biol. Chem.* **1952,** *195*, 199-206.
45. Luisi, P.L.; Henningen, F.; Joppich, M. *Biochem. Biophys. Res. Commun.* **1977**, *74*, 1384-1389.
46. Harada, M.; Adachi, M.; Shioi, A. Tailoring of reverse micellar systems for separation of proteins. *Current Topics in Colloid Int. Sci.* **1997**, *2*, 95-109.
47. Goto, M.; Ono, T.; Horiuchi, A.; Furusaki, S. Extraction of DNA by reversed micelles. *J. Chem. Eng.* **1999**, *32*, 123-125.
48. Vinogradov, A.A.; Kudryashova, E.V.; Levashov, A.V.; van Dongen, W.M.A.M. Stabilization and refolding of inclusion body proteins in reverse micelles. *Analytical Biochem.* **2003**, *320*, 234-238.
49. Peterson, R.W.; Anbalagan, K.; Tommos, C.; Wand, A.J. Forced folding and structural analysis of metastable proteins. *J. Amer. Chem. Soc.* **2004**, *126*, 9498-9499.
50. Fernandez-Velasco, D.A.; Sepulveda-Becerra, M.; Galina, A.; Darszon, A.; Tuena de Gomez-Puyou, M.; Gomez-Puyou. A. *Biochemistry* **1995**, *24*, 361-369.
51. Uskova, M.A.; Klyachko. N.L. *Refolding of formate dehydrogenase in aqueous solution and reverse micelles*; Biocatalysis-2000: Moscow, RU, 2000; p.169.
52. Kabanov, A.V.; Levashov, A.V.; Khrutskaya, M.M.; Kabanov, V.A. *Macromol. Chem.* **1990**, *33*, 379-391.
53. Budker, V.G.; Slattum, P.M., Monahan, S.D., Wolff, J.A. Entrapment and condensation of DNA in neutral reverse micelles. *Biophys. J.* **2002**, *82*, 1570-1579.
54. Fletcher, P.D.I.; Galal, M.F.; Robinson, B.H. *J. Chem. Soc., Faraday Trans. I* **1985**, *81*, 2053-2065.
55. Klyachko, N.L.; Bogdanova, N.G.; Levashov, A.V.; Martinek, K. Micellar enzymology: superactivity of enzymes in reversed micelles of surfactants solvated by water/organic cosolvent mixtures. *Collect. Czesh. Chem. Commun.* **1992**, *57*, 625-640.
56. Klyachko, N.L.; Dulkis, Yu.K.; Sukhoruchenko, T.A.; Levashov, A.V. Stability and stabilization of recombinant peroxidase in reverse micelles. *Biochem.* (Russ.) **1997**, *62*, 394-399 (Engl.Ed. 337-341).
57. Quellet, C.; Eicke, H.-F. Mutual Gelation of gelatin and water-in-oil microemulsions. *Chimia* **1986**, *40*, 233-238.
58. Haering, G.; Luisi, P.L. Hydrocarbon gels from water-in-oil microemulsions. *J. Phys. Chem.* **1986**, *90*, 5892-5895.
59. Rees, G. D.; Robinson, B.H.; Stephenson; G.R. Preparative-scale kinetic resolutions catalysed by microbial lipases immobilized in AOT-stabilised

microemulsion-based organogels: cryoenzymology as a tool for improving enantioselectivity. *Biochim. Biophys. Acta* **1995**, *1259*, 73-81.
60. Schuleit, M.; Luisi, P.L. Enzyme immobilization in silica-hardened organogels. *Biotechnol. Bioeng.* **2001**, *72*, 249-253.
61. Abakumova, E.G.; Levashov, A.V.; Berezin, I.V.; Martinek, K. *Proc. Acad. Sci. USSR* **1985**, *283*, 136-139.
62. Watnasirichaikul, S.; Davis. N.M.; Rades, T.; Tucker, I.G. Preparation of biodegradable insulin nanocapsules from biocompatible microemulsions. *Pharm. Res.* **2000**, *17*, 683-689.
63. Khmelnitsky, Y.L.; Neverova, I.N.; Momcheva, R.; Yaropolov, A.I.; Belova, A.B.; Levashov, A.V.; Martinek, K. *Biotechnol. Tech.* **1989**, *3*, 275-280.
64. Gupta, A.; Nagarajan, R.; Kilara, A. *Biotechnol. Progr.* **1991**, *7*, 348-354.
65. Shapiro, Y.E.; Pykhteeva, E.G.; Levashov, A.V. ^{1}H NMR self-diffusion in polymer-surfactant nanocapsules and cryogels with enzyme. *J.Colloid Int. Sci.* **1998**, *206*, 168-176.
66. Levashov, A.V.; Kabanov, A.V.; Berezin, I.V.; Martinek, K. *Proc. Acad. Sci. USSR* **1984**, *270*, 246-248.
67. Lukyanov, A.N.; Klibanov, A.L., Kabanov, A.V.; Torchilin, V.P.; Levashov, A.V.; Martinek, K. *Bioorg. Chem. (Russ.)* **1988**, *14*, 670-673.
68. Kabanov, A.V.; Klyachko, N.L.; Nametkin, S.N.; Merker, S.; Zaroza, A.V.; Bunik, V.I.; Ivanov, M.V.; Levashov, A.V. *Protein Eng.* **1991**, *4*, 1009-1017.
69. Klyachko, N.L.; Shchedrina, V.A.; Efimov, A.V.; Kazakov, S.V.; Gazaryan, I.G.; Kristal, B.S.; Brown, A.M. pH-Dependent substrate preference of pig heart lipoamide dehydrogenase varies with oligomeric state: response to mitochondrial matrix acidification. *J.Biol.Chem.* **2005**, *280*, 16106-16114.
70. Rariy, R. V.; Klyachko, N. L.; Borisova, E. A.; Cortes –Penagos, C. J.; Levashov, A. V. Lectin-like center in the molecule of α-chymotrypsin: formation of complexes with peroxidase and artificially glycosylated α-chymotrypsin. *Biochem. Mol. Biol. Int.* **1995**, *36*, 31-37.
71. Ganzborn, A.J.; Green, D.W.; Hershey, A.D.; Gould, R.M.; Plapp, B.V. Kinetic characterization of yeast alcohol dehydrogenase. *J. Biol. Chem.* **1987**, *262*, 3754-3761.
72. Northrop, D.B.; Cho, Y.-K. Effects of high pressure on solvent isotope effects of yeast alcohol dehydrogenase. *Biophys. J.* **2000**, *79*, 1621-1628.

Chapter 10

Biomimetic Silica Encapsulation of Nanoparticles and Enzymes

Melanie M. Tomczak, Morley O. Stone, and Rajesh R. Naik*

Biotechnology Group, Materials and Manufacturing Directorate, Air Force Research Laboratory, Wright-Patterson Air Force Base, OH 45433
***Corresponding author: rajesh.naik@wpafb.af.mil**

In nature, several organisms elegantly deposit silica into shell and spicule structures. These include the unicellular algae family of diatoms (*1, 2*) as well as the marine sponges (*3, 4*). The basis for silica deposition in these organisms has been elucidated over the last several years. Families of proteins, termed silaffins in diatoms and silicateins in marine sponges, have been isolated from the silacious components of the respective organisms (*1, 3*). These protein families do not resemble each other in terms of sequence or isoelectric point, but the purified forms of each precipitate silica from silicic acid precursors (*1, 4*). The *in vitro* protein-mediated silica condensation reactions were the proof-of-concept that these organisms use proteins to deposit and/or template their intricate silica structures. Interestingly, these silica precipitating templates become entrapped with the silica matrix. The composite protein-silica structure found in nature was the impetus for encapsulating enzymes in silica using peptides and proteins as the silica template. This chapter will focus on the encapsulation of nanoparticles and enzymes using peptide-templated silica.

Silica Biomineralization in Nature

Silicateins are the protein family found in spicules of the marine sponge, *Tethya aurantia.* The silicateins are composed of three subunits: α, β, and γ that form ordered filaments, which constitute the core of the silica spicules and which can be observed by X-ray diffraction of intact spicules (*3*). The subunits are similar to each other and homologous to the cathepsin L family of cysteine proteases. Morse and co-workers propose that the protease active site corresponds to the catalytic site of the silicateins for condensation of silica although the active cysteine from cathepsin is replaced by a serine in α-silicatein (*4, 5*). Isolated silicatein filaments and recombinant α-silicatein were shown to mineralize silica using triethoxysilane (TEOS) as the silicic acid precursor but did not function after they were thermally denatured, indicating that the secondary structure of the proteins is important for activity (*4*).

The silaffin family of proteins in the diatom *Cylindrotheca fusiformis* is composed of 3 members: silaffin 1A, silaffin 1B and silaffin 2, which together make up the majority of the protein in the diatom cell walls (*1*). In contrast to silicateins, silaffins are heavily post-translationally modified with polyamine groups and phosphorylated side chains (*1*). In order to determine the role of the post-translational modifications, synthetic peptides of individual repeating units have been used. In particular the fifth repeating unit of the silaffin protein sil1p, which is termed R5, has been well studied (*1, 5, 6*). R5 is 19 amino acids long (MW = 2,013 Da) with the amino acid sequence: SSKKSGSYSGSKGSKRRIL resulting in a pI = 11.22 . Rapid condensation of silica as spherical particles occurs with R5 in the presence of the precursor silicic acid and phosphate ions (Figure 1; *1*). This occurs through the process of floculation (*1*) that results in the formation of a network of silica particles. In the presence of the peptide, silicic acid undergoes rapid condensation into silica compared with abiotic silica formation from the same precursor (*1, 7*). This results in a composite material of silica and protein in the diatoms (*1*), a point which spurred the idea for using R5 as a template for enzyme encapsulation in silica.

Biomimetic Silica Formation

R5-precipitated silica spheres have been well characterized. They are porous structures with a diameter of 400-600 nm, and have been extensively analyzed with SEM and TEM (*1, 6*). The majority of R5 peptide used in the templating reaction is associated with the silica rather than the remaining supernatant, as was shown by dissolving the silica with NaOH and interrogating the resulting material by SDS-PAGE, indicating that the peptide is intrinsic to the silica matrix (*8*). Thermogravimetric analysis (TGA) revealed that 15 %/wt

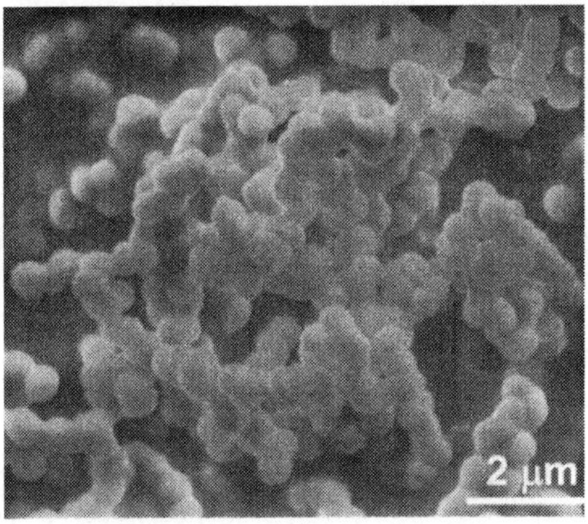

Figure 1. Scanning electron micrograph of R5-templated silica.

of the spheres is organic, 80 %/wt is silica and nitrogen absorption data indicate that the pore size is between 1.6 - 2.0 nm (unpublished results). Using this knowledge, we hypothesized that it would be possible to encapsulate other proteins or even other particles inside the silica matrix using R5-mediated condensation.

Several groups have identified peptides and proteins, other than those natural and recombinant proteins from diatoms and marine sponges, that catalyze the silica condensation reaction. Peptide phage display was used to select for twelve amino acid peptides that specifically bind to silica (*9*). These peptides were tested for condensation of silica using hydrolyzed tetramethoxysilane (TMOS) as the silicic acid precursor. In these experiments it was observed that some peptides are good at binding silica but not catalyzing the condensation reaction, and that some peptides are excellent precipitants of silica. For example, the peptide termed Si 4-1 is rich in histidine and is effective at both binding and precipitating silica however Si 4-10, which is arginine rich, is only effective at binding silica and does not precipitate significant amounts of silica (*9*). Poly(L-lysine) has been shown to be an effective catalyst of silica condensation due to its positively charged character (*5, 10-15*). Additionally, the "flexible" nature of the poly(L-lysine) secondary structure (*16*) results in different shapes of silica being formed under different conditions and when using different molecular weights of poly(L-lysine) (*5, 14, 15*). The silica templated by poly(L-lysine) has pores approximately ~18 Å in diameter (C. Perry, personal communication), making the resultant silica structures suitable for enzyme encapsulation as a substrate would be able to enter the silica matrix. Coradin and Livage (*17*) have shown that poly-amino acids, including poly(L-lysine), favor silica formation but that the same amino acids free in solution only have a small positive affect on the precipitation of silica compared to the buffer control. Lysozyme also is an effective precipitant of silica nanoparticles (*18, 19*).

In addition to proteins and peptides, small organic molecules containing nucleophilic groups, notably cysteamine (which contains both an –SH and an –NH_2 group), were shown to catalyze the precipitation of silica (*20*). As well, amine-terminated dendrimers that similar to the post-translational modifications made to silaffin lysines have been shown to be effective templates for silica formation (*21*). Clearly, there are many effective and diverse templates for the room temperature, neutral pH condensation of silica.

Biomimetic Enzyme Encapsulation

The biomimetic approach for silica precipitation and enzyme encapsulation has several important differences from the sol-gel method. First, the condensation reaction reaches completion within 5 – 15 minutes whereas with the sol-gel method the condensation occurs over a 24 hr period. Second, the biomimetic condensation reaction occurs preferentially at room temperature and neutral pH in aqueous solution, while the sol-gel method traditionally occurred in organic solvents. The hydrolysis of TMOS generates an alcohol side-product but, as the silica condensation reaction occurs within fifteen minutes, the residual alcohol can be removed from the silica-encapsulated enzymes quickly and further alcohol byproducts are not generated. These points allows the alcohol produced as a side product of the reaction to be quickly removed from the system, as compared with the much slower sol-gel method. This is a great advantage when considering that alcohol can denature enzymes, so the biomimetic silica may allow for superior stability of encapsulated enzymes compared to traditional sol-gel. Third, a more consistent and uniform size distribution of silica particles can be achieved using a peptide template than is possible with the sol-gel method. Biomimetic enzyme encapsulation using the R5 peptide-mediated silicification reaction was first shown by Luckarift and co-workers (*6*). This paper showed that butyrylcholinesterase could be immobilized into R5-templated silica at the efficient rate of approximately 90 % using a 2 mg/mL starting enzyme concentration. This yields approximately 220 mg enzyme/ g silica or ~20 % (w/w) compared to only 0.1 – 0.5 % (w/w) enzyme incorporation with traditional sol-gel methods (*22*). Subsequently, horseradish peroxidase, catalase, luciferase, blue and green fluorescent protein (BFP and GFP, respectively) and β-galactosidase all have been encapsulated in biomimetically-templated silica. The incorporation rate of the different enzymes depends on the starting solution concentration of the enzyme however, all reported incorporation rates have been greater than 50 %.

The initial activity of R5-mediated silica-encapsulated butyrylcholinesterase was equal to that of free enzyme, and that activity was maintained at about 100

% after storage for over 25 d at 25 °C in aqueous solution. In comparison, the activity of free butyrylcholinesterase without added antibiotic was less than 10 % after storage for 8 d under the same conditions (Figure 2; *6*). The same characterisitics of increased stability were shown to extend to heat stability of silica-encapsulated enzymes. After 1 hr at 65 °C, 100 % of the encapsulated butyrylcholinesterase activity was retained compared with only 15 % activity for the free enzyme (*6*).

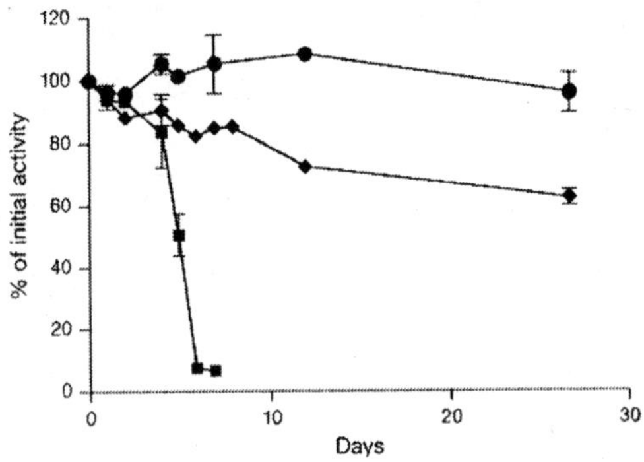

Figure 2. Butyrylcholinesterase activity at 25 °C in free and biosilica-encapsulated systems. Butyrylchoinesterase was assayed and stored at 25 °C for the indicated time as a free enzyme (squares), free enzyme in an antibiotic (diamonds), or biosilica encapsulated enzyme (circles). Used with permission from (6).

Catalase and horseradish peroxidase (HRP) also have been encapsulated in silica using an R5 template, which results in 65–85 % incorporation rates of the starting enzyme into the silica matrix. The biomimetically-encapsulated catalase and HRP are equally active as their free counterparts (Figure 3a; *23*). Both catalase and HRP could be dried after encapsulation, stored at room temperature for several days, and recover full activity when re-hydrated and assayed, similar to the results with butyrylcholinesterase (Figure 3b; *23*).

There are several reports that β-galactosidase has been successfully encapsulated in silica, using sol-gel and silica-coated alginate microcapsules for silica condensation, and retained activity while increasing stability (*24, 25, 26*) however using R5- mediated silica encapsulation of β-galactosidase we were unable to detect any activity (Tomczak and Naik, unpublished results). There are several possible explanations for this: 1) β-galactosidase is a tetramer in its active form and may not have been encapsulated as such; and/ or 2) the substrate could not effectively enter the silica matrix to be accessible by the encapsulated enzyme.

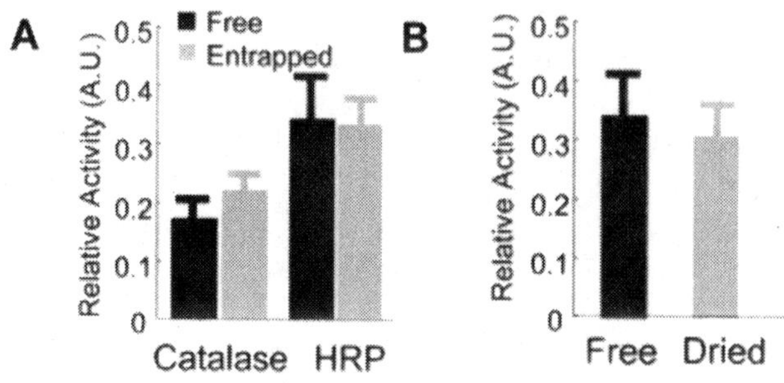

Figure 3. Catalase and HRP are active and stabilized when encapsulated in R5-templated silica. Catalase and HRP were encapsulated in R5-templated silica and assayed for activity before drying (panel A) and after drying and rehydrating catalase (panel B). The free catalase control (never dried) is shown in panel B for comparison of activity. (Reproduced with permission from reference 23. Copyright 1998.)

Molecular crowding of proteins has been studied using sol-gel matrices to mimic the effects of confinement on proteins that would occur in a cell, for example (*27, 28*). These studies also have shed light on the effects on the structure of sol-gel encapsulated enzymes. Generally, these studies have shown that the native structures of encapsulated enzymes are preserved, but that the encapsulated enzymes do come into contact with the solvent system. This is consistent with the findings that encapsulated enzymes can act on externally added substrates. Hydrogen bonding between the protein surfaces and the silica surface hydroxides in the matrix clearly plays a role in structure stabilization. Once the enzymes are incorporated into the silica matrix, the network of hydrogen bonding may mimic the water network that would normally stabilize the protein in solution. In this case, the native or active protein structure would be maintained inside the silica matrix even when the system were dehydrated or heated, which is what has been observed in the cases described here.

The kinetics of encapsulated enzymes has been measured in both stirred and unstirred systems. In a stirred system, the kinetics of encapsulated butyrylcholinesterase were similar to the free enzyme kinetics (H.R. Luckarift, personal communication). However when encapsulated HRP was tested in an unstirred system, the kinetics were considerably slower than those of the free enzyme (Figure 4). This decrease in kinetic rate could be an advantage in some cases when the slow release of an enzyme reaction product is desired. For example, in the slow release of insulin for diabetics or for the slow release of a drug therapy. The utility of having both similar and reduced enzyme kinetics is another advantage of the biomimetically-encapsulated enzyme system.

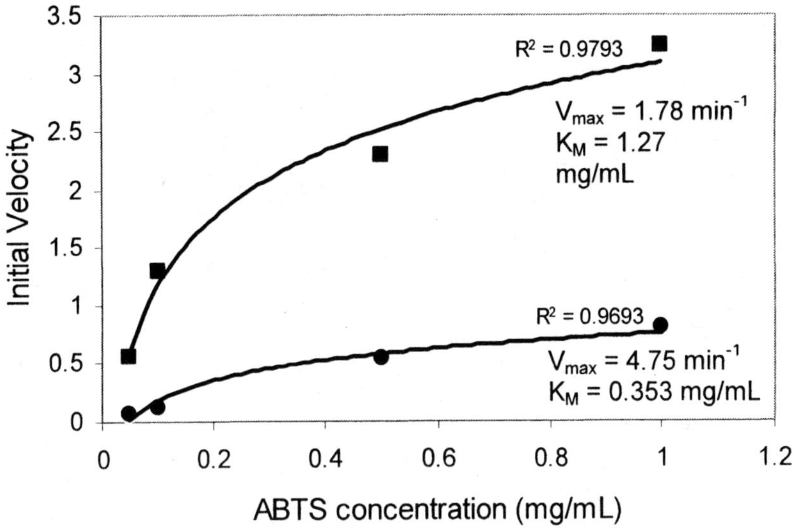

Figure 4. Kinetics of unstirred encapsulated HRP vs. free HRP. The V_{max} and K_M values were calculated based on the initial velocity of HRP acting on its substrate, ABTS at increasing concentrations in unstirred systems. Squares represent free enzyme, circles represent encapsulated enzyme.

Small Molecule Templates for Silica Encapsulation

In addition to protein catalysts for encapsulating enzyme in silica, rationally selected small molecules have proved useful as well. For example, the bifunctional small organic molecule cysteamine has been shown to effectively encapsulate two different enzymes, both of which retained activity. Luciferase is an enzyme that catalyzes the production of bioluminescence after oxidizing its substrate luciferin, a reaction that requires ATP (*29*). Morse and co-workers showed that encapsulated luciferase remained active when luciferin and ATP were added externally to the silica (*20*). These results show that luciferin and ATP entered the porous matrix and that luciferase acted on them, even though the enzyme was embedded in the silica matrix. Clearly, the porosity of the silica matrix allows for such enzyme reactions to occur while maintaining enzyme stability, however the details of how these reactions occur spatially within the silica matrix are still unknown.

Green fluorescent protein and blue fluorescent protein (GFP and BFP, respectively) also were encapsulated in silica using cysteamine (*20*). Both proteins retain their fluorescent character when encapsulated, with no evidence of fluorescent quenching by the silica. These experiments were expanded to show that live *E. coli* expressing GFP could be encapsulated using the same catalyst. *E. coli* cells were mixed with silica precursor and then patterned using a PDMS stamp on a silicon wafer. The GFP fluorescence was restricted to the area where silica formed, indicating that the cells were encapsulated in the silica matrix. The GFP fluorescence from *E. coli* was maintained for seven days (*20*). These results could have a significant impact on cell and tissue engineering for medical uses. If cells could be encapsulated in silica and then inserted while still live into an organism, they could synthesize and deliver chemotherapy or other drug therapies -- potentially without the antigenic problems that would be associated with the introduction of foreign cells. Additionally, encapsulating cells in silica could stabilize them for storage, which is an area of considerable research.

Multi-Functional Biomimetic Silica

There has been some interesting recent work on the use of lysozyme as a catalyst for silica condenstation. Lysozyme is an antibacterial enzyme that functions by hydrolyzing peptidoglycan linkages in Gram-positive bacterial cell walls (*30*). Lysozyme is a basic protein with a pI = 10.5 and its sequence contains many hydroxy- and imidazole- containing amino acids as well as basic amino acids (serine, threonine, histidine, aspartic acid, glutamic acid, lysine and arginine), characteristics which have been shown to be important for silica biomineralization. Lysozyme was shown to be an effective catalyst of silica condentation (*18*). This group found that lysozyme caused the formation of fused-spheres silica with average pore size of 30 nm using water glass (27 % SiO_2, 10% NaOH) as the silica precursor. They propose that lysozyme induces silica polymerization through electrostatic interactions, similar to the findings with poly(L-lysine)-templated silica. Luckarift and co-workers extended these findings to show that lysozyme as the silica template was encapsulated efficiently (~85 % incorporation) into the matrix in its antibacterial active form (*19*). When assayed against *Micrococcus lysodeikticus*, the silica-encapsulated lysozyme had as much lytic activity as the free enzyme and that activity was maintained after heating to 75 °C for an hour, while the free enzyme had less than 10 % activity after the same treatment (*19*). They proceeded by incorporating another enzyme, butyrylcholinesterase, into the silica matrix along with the lysozyme. These experiments showed that both the lysozyme and the butyrylcholinesterase maintained their activity when co-encapsulated, demonstrating that multi-functional materials can be made simply in a "one pot" synthesis.

Clearly, co-encapsulation of two enzymes is an important step for useful applications of biomimetic enzyme encapsulation. It is also important to have a facile method for separating the silica particles out of a solution. To this end, we have incorporated magnetic nanoparticles into the R5-templated silica condenstation reaction and found that they are effectively incorporated as well. TEM analysis shows that cobalt-platinum (CoPt) nanoparticles were encapsulated into the silica matrix (Figure 5a). Additionally, it was possible to co-encapsulate an enzyme with magnetic nanoparticles in order to separate them from the reaction. In doing this, we were able to show that by using this method of separation we could reuse the enzyme over several cycles while retaining 100 % enzyme activity (Figure 5b).

We have also shown that quantum dots can be encapsulated into a silica matrix using an R5-template. Quantum dots are reknowned for their stable and bright fluorescence character, and these characteristics were not inhibited when the quantum dots were encapsulated into a silica matrix (Figure 6). Silica-encapsulated quantum dots may be useful in directing and observing co-encapsulated enzyme activity in a cell, for instance.

Another area of encapsulation that has been developing over the last several years is alginate microcapsules that are coated with silica. These alginate microcapsules can be used for encapsulating something as simple as a protein to something as complex as a cell (reviewed in (*31*)). The alginate microspheres are coated with poly(L-lysine), which buffers the negatively charged alginate surface and acts as a template and catalyst for silica condensation on the microsphere (*26*). These coatings were permeable to myoglobin (~ 20 kDa) but not to Bovine Serum Albumin (~ 70 kDa), indicating that they would be effective at uptake or release of small molecules and proteins if used in therapeutic applications.

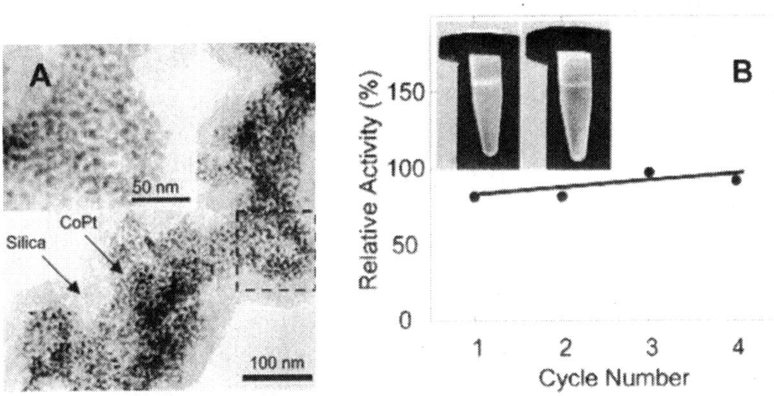

Figure 5. Biomimetic silica encapsulation of magnetic particles and catalase Panel A: TEM micrograph of CoPt nanoparticles encapsulated in biomimetically-templated silica. Panel B: Catalase was co-encapsulated with iron oxide nanoparticles in R5-templated silica and catalase activity was assayed after separating the silica-encapsulation matrix from the solution using a magnetic stand (shown) over several cycles.

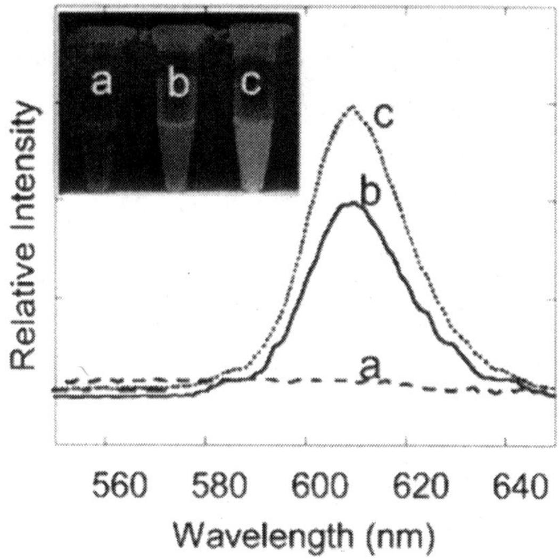

Figure 6. Fluorescence of silica-encapsulated quantum dots. Quantum dots were encapsulated in silica using R5 as a template. Tubes and curves labeled: (a) represent R5-templated silica alone, (b) represent free quantum dots in solution, and (c) represent quantum dots encapsulated in R5-templated silica.

Applications of Silica-Encapsulated Nanoparticles and Enzymes

There are many potential applications for encapsulated enzymes, and they will be described in detail elsewhere in this book (*32*). First, the increased stability of encapsulated enzymes is a clear advantage over free enzymes. The increase in temperature stability could lead to the use of the enzymes in some industrial processes that require a relatively mild heat treatment, where free enzyme would be degraded. The stability against dehydration stress is also a great advantage over free enzyme. This would allow for room temperature dry storage or shipment of enzyme, expanding the range of use of the enzymes and greatly reducing shipping and storage costs. Dry storage would certainly be an advantage for military applications, where an emphasis has been put on low-cost lightweight technologies.

Second, the reusablility of the encapsulated enzymes is a major advantage over free enzyme. In a free enzyme system, it is laborious to separate out very small amounts of enzymes from a reaction mixture. These processes certainly lead to loss of total enzyme amount as well. With silica encapsulation, magnetic nanoparticles can be co-encapsulated and be exploited as effective separation agents simply by applying a magnetic field to the sample. This separation method could be used on research (microliters) to industrial (kiloliters) scales, broadening the usefulness of the technology to many different environments and applications.

Third, co-encapsulation of two or more different enzymes or of nanoparticles and enzymes in the same silica matrix is an important advance. This would allow for physical proximity of enzymes that form a degradation pathway, for example, to make the process more efficient. As well, co-encapsulation of disparate enzymes would make the silica a multi-functional material through a simple, room temperature synthesis.

Combining all of these advantages together, it is clear that biomimetic silica encapsulation of enzymes could be used for many applications. For example, enzymes for decontamination of a water supply could be encapsulated along with a magnetic nanoparticle. This system could be transported in the dry state to the site where it is needed, and then deployed. Once the solution was decontaminated the particles could be sequestered at the side the of vessel using a magnetic field, then collected for future use. Additionally, incorporation of enzymes that produce therapeutic agents could be used in biomedical applications as implants or injections for slow release of the therapies. The applications for silica-encapsulated enzymes stretch from research to industrial, from military to biomedical.

References

1. Kröger, N.; Deutzmann, R.; Sumper, M. Science **1999**, *286*, 1129-1132.
2. Sumper, M.; Brunner, E. Adv. Funct. Mater. **2006**, *16*, 17-26.
3. Shimizu, K.; Cha, J.; Stucky, G.D.; Morse, D.E. Proc. Natl. Acad. Sci. USA **1998**, *95*, 6234-6238.
4. Cha, J.N.; Shimizu, K.; Zhou, Y.; Christiansen, S.C.; Chmelka, B.F.; Stucky, G.D.; Morse, D.E. Proc. Natl. Acad. Sci. USA **1999**, *96*, 361-365.
5. Rodríguez, F.; Glawe, D.D.; Naik, R.R.; Hallinan, K.P.; Stone, M.O. Biomacromolecules **2004**, *5*, 261-265.
6. Luckarift, H.R.; Spain, J.C.; Naik, R.R.; Stone, M.O. Nat. Biotechnol. **2004**, *22*, 211-213.
7. Gordon, R.; Drum, R.W. Int. Rev. Cytol. **1994**, *150*, 243-372.
8. Naik, R.R.; Whitlock, P.W.; Rodríguez, F.; Brott, L.L.; Glawe, D.D.; Clarson, S.J.; Stone, M.O. Chem Commun. **2003**, 238-239.
9. Naik, R.R.; Brott, L.L.; Clarson, S.J.; Stone, M.O. J. Nanosci. Nanotechol. **2002**, *2*, 95-100.
10. Patwardhan, S.V.; Mukherjee, N.; Steinitz-Kannan, M.; Clarson, S.J. Chem. Commun. **2003**, 1122-1123.

11. Patwardhan, S.V.; Clarson, S.J.; Perry, C.C. Chem Commun. **2005**, 1113-1121.
12. Patwardhan, S.V.; Maheshwari, R.; Mukherjee, N.; Kiick, K.L.; Clarson, S.J. Biomacromolecules **2006**, *7*, 491-497.
13. Hawkins, K.M.; Wang, S.S.-S.; Ford, D.M.; Shantz, D.F. J. Amer. Chem. Soc. **2004**, *126*, 9112-9119.
14. Glawe, D.D.; Rodríguez, F.; Stone, M.O.; Naik, R.R. Langmuir **2005**, *21*, 717-720.
15. Tomczak; M.M.; Glawe, D.D; Drummy, L.F.; Lawrence, C.G.; Stone, M.O.; Perry, C.C.; Pochan, D.J.; Deming, T.J.; Naik, R.R. J. Amer. Chem. Soc. **2005**, *127*, 12577-12582.
16. Fändrich, M.; Dobson, C.M. EMBO J. **2002**, *21*, 5682-5690.
17. Coradin, T.; Livage, J. Colloids Surf. B **2001**, *21*, 329-336.
18. Coradin, T.; Coupé, A.; Livage, J. Colloids Surf. B. **2003**, *29*, 189-196.
19. Luckarift, H.R.; Dickerson, M.B.; Sandhage, K.H.; Spain, J.C. Small **2006**, *2*, 640-643.
20. Roth, K.M.; Zhou, Y.; Yang, W.; Morse, D.E. J. Amer. Chem. Soc. **2005**, *127*, 325-330.
21. Knecht, M.R.; Wright, D.W. Langmuir **2004**, *20*, 4728-4732.
22. Gill, I.; Ballesteros, A. J. Am. Chem. Soc. **1998**, *120*, 8587-8598.
23. Naik, R.R.; Tomczak, M.M.; Luckarift, H.R.; Spain, J.C.; Stone, M.O. Chem Commun. **2004**, 1684-1685.
24. Fukushima, Y.; Okamura, K.; Imai, K.; Motai, H. Biotechnol. Bioeng. **1988**, *32*, 584-594.
25. Heichal-Segal, O.; Rappoport, S.; Braun, S. Bio/technology **1995**, *13*, 798-800.
26. Coradin, T.; Mercey, E.; Lisnard, L.; Livage, J. Chem Commun. **2001**, 2496-2497.
27. Eggers, D.K.; Valentine, J.S. J. Molec. Biol. **2001**, *314*, 911-922.
28. Eggers, D.K.; Valentine, J.S. Protein Sci. **2001**, *10*, 250-261
29. DeLuca, M. Adv. Enzymol. **1976**, *44*, 37-68.
30. Pellegrini, A.; Thomas, U.; von Fellenberg, R.; Wild, P. J. Appl. Bacteriol. **1992**, *72*, 180-187.
31. Coradin, T.; Nassif, N.; Livage, J. Appl. Microbiol. Biotechnol. **2003**, *61*, 429-434.
32. Luckarift, H.R; Spain, J.C. Nano-scale Science and Technology in Biomolecular Catalysis; American Chemical Society: Washington, 2006.

Chapter 11

A Biomimetic Process for Enzyme Immobilization within Polycation-Templated Silica

C. Chang, D. E. Ward, B. Kelemen, and J. C. McAuliffe

Danisco Genencor, 925 Page Mill Road, Palo Alto, CA 94304

A simple and scalable procedure for enzyme entrapment, termed silica co-precipitation, has been developed. The process was initially inspired by the ability of diatoms and other marine organisms to rapidly mineralize silica through the action of polycationic peptides termed silaffins. A mixture of an enzyme and a polycationic polymer such as polyethylenimine (PEI) is added to a phosphate-buffered silicate solution (PBSi) and results in the rapid formation of a colloidal co-precipitate consisting of the enzyme, the polycation within a hydrated, amorphous silica matrix. The process has been applied to a series of enzymes and the properties of the resulting composites characterized.

The use of silicate based matrices for the immobilization of biomolecules has a long history beginning with the simple adsorption of enzymes onto diatomaceous earth and other silicate based inorganics, followed by the development of sol-gel technology and more recently, to the use of templated mesoporous silicates (1,2). One of the key limitations to the greater adoption of these technologies by industry relates to the cost of preparation. We have developed a concerted and economical method for the immobilization of a variety of enzymes within a polycation-templated silicate matrix (3,4). The method, termed silicate coprecipitation was developed during studies into the mechanisms through which diatoms and other marine organisms produce biosilicates under ambient conditions. The procedure involves the addition of a mixture of an enzyme and a polycation such as polyethylenimine, to a buffered silicate solution. A colloidal precipitate forms within seconds that contains a significant proportion of the added enzyme. The method appears to be comparable to alternative methods for enzyme immobilization with regard to retention of enzymatic activity, and has the potential to be optimized according to the enzyme of interest.

Industrial Enzyme Immobilization

Immobilized enzymes are used in a wide variety of commercial processes including fine chemical manufacture, food production, air filtration, cleaning, and bioanalytical applications (5). Immobilization imparts several advantages over the free, soluble enzymes including ease of removal from the process stream, improved chemical and thermal stability, enhanced activity in organic solvents, and suitability for continuous use over extended periods of time (6). There are however, several disadvantages inherent to many methods commonly employed for enzyme immobilization. These include loss of activity upon immobilization, diffusional limitations reducing catalytic rate, sub-optimal mechanical properties of the supports used for immobilization and most importantly the cost of immobilization itself. Typically this cost is equal to or greater than the cost of the enzyme itself, which in turn limits the number of applications practiced on an industrial scale (Table I) (5,7).

Table I. Commercial Applications of Immobilized Enzymes

Enzyme	*Application*
Glucose isomerase	High fructose corn syrup
Lipases	Resolution of Chiral intermediates
Aminoacylases	Amino acid synthesis
Lipase	Transesterification of fats
Penicillin acylase G	Antibiotic modification
Glucose oxidase	Glucose biosensors

An 'ideal' enzyme immobilization process would enable one to rapidly render a given enzyme into a preferred physical form without loss of activity through a cheap and scalable process. In addition, the immobilized form of the enzyme would show improved kinetic properties and enhanced stability relative to the soluble form. In practice, most methods for enzyme immobilization fall well short of this goal. Our research has aimed to apply the principles of biomineralization demonstrated in Nature, to the development improved methods for enzyme immobilization, with an emphasis on biosilicates (8).

Biomimetic Silicate Formation

The formation of silica structures by living organisms is known as biosilicification and occurs in a wide range of marine organisms such as sponges and diatoms, as well as higher plants. Formation of this amorphous opal-like silica occurs at ambient temperatures and neutral pH in contrast to the harsh conditions employed in man-made procedures. Several classes of silica-templating molecule have been described in the literature (9).

The silicateins α, β and γ were identified subunits of a protein filament isolated from the silica spicules of *Tethya aurantia*, a marine sponge. The silicateins catalyze the polymerization of silicate to biosilica (10). Interestingly, they have a high homology to the cysteine protease cathepsin L. In addition, silicatein α possesses unique tandem arrays of hydroxyl functions that may influence the morphology of biosilica. The silicateins also polymerize silicon alkoxides such as $Si(OEt)_4$ to silica and alkyl siloxanes to silicone polymers. Other proteins also possess the ability to form siloxane bonds. For example, Brandstadt and colleagues determined that Bovine Trypsin catalyzes the polymerization of both silicates and organosiloxanes and has potential application for the synthesis of novel silicon-based materials (11).

Other silica-templating molecules act in a non-catalytic fashion and become incorporated into the biosilica network. A series of polycationic peptides, the silaffins, were isolated from diatom cell walls by Kroger and coworkers (12). Characteristic features of these peptides include modified lysine residues featuring *N*-methyl-propylamine repeats. Solutions of silicic acid are rapidly precipitated by silaffin peptides within seconds producing silica nanoparticles. The modified lysine residues are essential to activity. Peptides with silica precipitating activity have also been discovered through the panning of bacteriophage peptide libraries (13). In general, these peptides are rich in basic amino acid residues including lysine, histidine and arginine. The ability of organic polymers to precipitate silica is not unique to polypeptides; in fact a range of synthetic polycationic polymers have this ability including polyethylenimine, polyallylamine and linear polyamines such as triethylene-tetramine (9,14). The mechanism is thought to involve the complexation of an organic cationic polymer with negatively charged silicate anions and their subsequent condensation to form silica colloids. This condensation reaction is most likely driven by the ability of the cationic polymer to act as both a general acid and base (Figure 1) (15).

The morphology of the silicates obtained by polyamine templating is profoundly affected by the hydrodynamics of the process. For example, the use of a peptide related to silaffin-1A_1 to precipitate silica gave rise to spherical silica composites in a static system, but braided fiber-like structures when the reaction was carried out in a tube under shear (16). Given the known interplay between the activity of an immobilized enzyme and the morphology of the

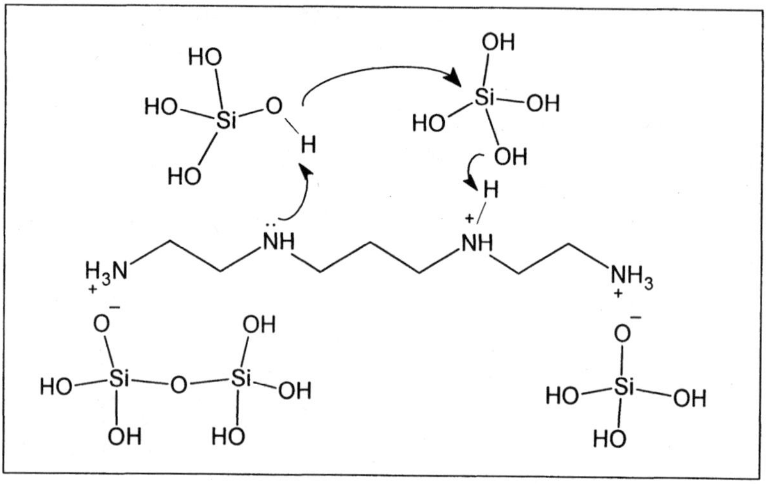

Figure 1. Polyamine-catalyzed silicate condensation is driven by the electrostatic interaction between silicate and cationic amines, combined with acid/base catalysis.

associated matrix (17), the use of silica precipitating molecules offers a potentially 'tunable' system for enzyme immobilization whereby the structure of the silicate matrix could be optimized around a chosen biocatalyst.

Silica Co-precipitation

Our early studies focused on modifying synthetic peptide templates related to silaffin peptides by covalent attachment of various functional groups including fluorophores, lipids, enzyme inhibitors and enzymes. Addition of these modified templates at initial concentrations of 0.1 to 10 mg/mL to alkoxysilane-derived silica sols (ca. 100 mM silicate) resulted in the precipitation of nanocomposites composed of silica and the peptide templates (18). Our initial attempts to extend this approach to enzyme immobilization involved the expression of a chimeric subtilisin protease containing one silaffin motif per protein molecule. Unfortunately, these fusion proteins did not promote silica precipitation upon exposure to a silicate sol. These observations led to a more practical approach where the protein to be immobilized was simply mixed with an excess of a silaffin peptide or alternative template molecule and allowed to associate in a non-covalent fashion, followed by mixing with a soluble silicate.

Silica co-precipitation with peptide templates

Immobilization of a subtilisin protease (GG36, Genencor) was carried out by first mixing the protease (0.1 mg) with the R5 silaffin peptide (1 mg, SSKKSGSYSGSKGSKRRIL) in Tris buffer (900 μL of 10 mM) (3). A silicate sol derived from acid-hydrolyzed tetraethoxysilane (TEOS) (100 μL of 1 M) was added to the buffered enzyme and resulted in the immediate formation of a colloidal white solid. The solid was collected by centrifugation and the supernatant was decanted for enzyme assay. The solid was resuspended and washed twice with deionized water. The reaction performed at pH 7 resulted in the formation of a gel (smaller particles), and the reaction performed at pH 8 gave a precipitate (larger and denser particles). The subtilisin enzyme alone did not produce a precipitate when treated with a silicate sol.

Enzyme assays were performed upon both the supernatant and the solids from the reaction using the substrate *N*-succinyl-Pro-Ala-Ala-Phe *p*-nitroanilide (suc-AAPF-pNA). The extent of enzyme immobilization was over 90%, with an expressed activity in the solid of 20-30% of the total initial enzyme activity. Subsequent experiments found that the degree of enzyme capture was a function of the enzyme to silaffin ratio. In addition, larger amounts of silicate resulted in greater enzyme capture but lower specific activities of the resulting solids. The incorporation of organosilanes such as phenyltriethoxysilane (PTES) seemed to reduce the specific activity of the resulting silica composites, possibly owing to reduced porosity of the resulting solids. Alternate peptide sequences could also be used as template molecules, for example poly-L-lysine and arginine oligomers. The protease/silica composites retained enzymatic activity as shown by a standard colorimetric assay even after repeated washes with brine and buffer solutions.

A similar approach has been described by Luckarift *et al.* for the immobilization of enzymes using silica-precipitating peptides (19). The authors reported a high level of enzyme capture and a large gain in stability as compared to the free enzyme in buffer. The kinetics of the immobilized enzyme, in this case butyrylcholine esterase, were also similar to that in solution. These experiments demonstrated that it was possible to immobilize enzymes through the use of silaffins and other silica precipitating peptides, in combination with silicate sols derived from alkoxysilanes.

Improved co-precipitation process

The approach described above was not thought to be scalable owing to the expense of synthetic peptides and the potential hazards involved with handling large amounts of alkoxysilanes. Accordingly, two modifications were made to the procedure. Firstly, cationic polymers such as polyethylenimine (PEI) were used instead of peptides as silica-precipitating agents. The second modification involved the use of a phosphate-buffered silicate (PBSi) solution, made by the partial neutralization of sodium silicate with phosphoric acid. The results of a titration of sodium silicate with phosphoric acid are shown in Figure 3. It can be seen that phosphate-buffered silicate has a good buffering capacity in the pH 6 to 8 range and as such is compatible with most enzymes.

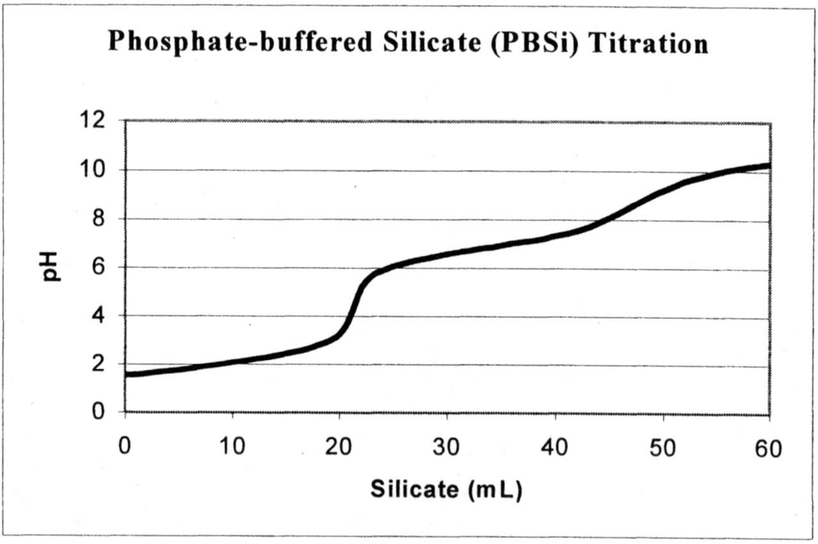

Figure 3. Titration of phosphoric acid (60 mL of 166mM) with sodium silicate (620mM silica) produces phosphate-buffered silicate (PBSi) solutions.

This improved process, termed *silica co-precipitation*, was applied to a range of enzymes with the aim of determining the generality of the method. A typical experiment involved first mixing a solution of an enzyme (1-10 mg/mL) with a 10 to 20-fold excess (w/w) of polyethylenimine (PEI, M_n = 60,000, M_w = 750,000, Sigma). An enzyme assay is usually carried out at this point in order to determine of enzyme inhibition has occurred. This solution is rapidly added to a 5 to 10-fold excess (v/v) of a freshly made PBSi solution (100-200 mM silicate, pH 6 to 8) and results in the instantaneous formation of a colloidal white precipitate. Collection of the solid can be achieved by centrifigation or filtration through a depth filter such as cheesecloth. Following washing, the solid is dried either by freeze drying or by desiccation. The results are discussed below.

Enzyme Capture and Activity

The determination of the activity of immobilized enzymes is not entirely straightforward as the apparent activity of these materials is often influenced by the manner in which the assays are performed. The amount of solid used and the degree of agitation are important parameters. In most cases, assays were performed with solid loadings of under 0.1% w/v, combined with rapid agitation. If the turbidity of the assay solution was too high for a direct spectrophotometric reading, then an aliquot was withdrawn and the immobilized enzyme quickly removed by centrifugation. The efficiency of enzyme immobilization was assessed by the following metrics;

- *Enzyme Capture* refers to the percentage of the initial enzyme that is immobilized. It is determined indirectly by the measuring the difference between the initial and final enzyme activities in solution.
- *Relative Specific Activity* (RSA) is the percent activity of the immobilized enzyme relative to an equivalent amount of the free, unimmobilized enzyme under excess substrate conditions. It is influenced by mass transfer (diffusional effects of substrate or product) and/or denaturation of the enzyme.
- *Total Expressed Activity* (TEA) is the enzyme activity of the solid containing the immobilized enzyme expressed in units (U).

Table II. Enzymes Immobilized by Silica Coprecipitation

Enzyme[1,2]	*Capture*	*Relative Specific Activity*
β-glucosidase	40%	20-40%
Glucose isomerase	75%	35%
Lactase	20%	57%
P. alcaligenes lipase	85%	20%
Acyltransferase	46%	24%
Glucose oxidase[2,3]	19%	15-25%

[1] Initial protein concentrations were in the 1-10 mg/mL range.

[2] Enzymes produced by Genencor International, Inc., Palo Alto, CA, U.S.A.

[3] Immobilization of GOx (OxyGo5000™) via the sol-gel method gave 94% capture and 54% RSA, but freeze drying resulted in the loss of most of the activity.

The results with regard to enzyme capture and retained activity varied according to the nature of the enzyme employed and the initial protein concentration prior to immobilization (Table II). For example, immobilization of glucose isomerase resulted in the capture of 75% of the protein and retention of 35% of the specific activity relative to the unimmobilized enzyme. In contrast, *P. alcaligenes* lipase (LIPOMAX™) showed much lower relative specific activity (<20%) despite excellent capture (>90%). For a given enzyme, the extent of immobilization was fairly reproducible, and was related to the initial enzyme concentration. Higher enzyme loadings typically resulted in a lower percentage capture. In contrast, the relative specific activity of the silica co-precipitates seemed to be somewhat variable, perhaps an indication of the polydispersity of the matrix created by this process. Inclusion of organosilanes in the composites was also pursued as it has been shown by Reetz and others to dramatically enhance the activity of sol-gel immobilized lipases (20). The addition of organosilanes such as sodium methyl siliconate (SMS) to the PBSi solution seemed to significantly reduce both the surface area and yield of the co-precipitated material. Overall, the silica coprecipitation process seemed to offer a convenient and general means for enzyme immobilization, however a better understanding of the underlying chemistry of the process was needed.

Physical Chemistry of Silica co-precipitation

One unique aspect of the co-precipitation approach is that enzyme immobilization and formation of the mesoporous silicate solid are concerted and occur within several seconds (Figure 3). This contrasts to sol–gel approaches where a mixture of a silicate sol and an enzyme forms a gel over minutes to hours. Several modes of immobilization are possible including physical entrapment of the enzyme within the silicate matrix, ionic pairing with the polyelectrolyte, and other non-covalent interactions (H-bonding, hydrophobic association with the template or organosilanes etc.). Obviously, it is important that the polycationic template used does not inhibit the enzyme and preferably plays a role in enhancing enzyme stability. Such a stabilizing effect has been demonstrated for the addition of PEI to proteases (21).

The silica biocomposites produced by this method take the form of an aggregated colloidal solid composed of amorphous, hydrated silica, the polycationic template polymer and the enzyme. Drying of the silica bio-composites by lyophilization gives a free flowing powder, whereas air drying gives an agglomerated solid. These solids were subjected to surface area (BET) and particle size analysis. Typical surface areas ranged from 70 to 180 m^2/g for silica-only composites. Pore sizes were in the 2-6 nanometer range and the size distribution was broad, in contrast to mesoporous silicates made by surfactant templating approaches (22). SEM images of the composites indicated that the material was polydisperse, consisting of aggregates of sub-micron particles. In solution, these aggregates tend to form loose agglomerates in the 10 to 30 μm range over the course of several minutes, which could be readily resuspended by sonication or vigorous shaking.

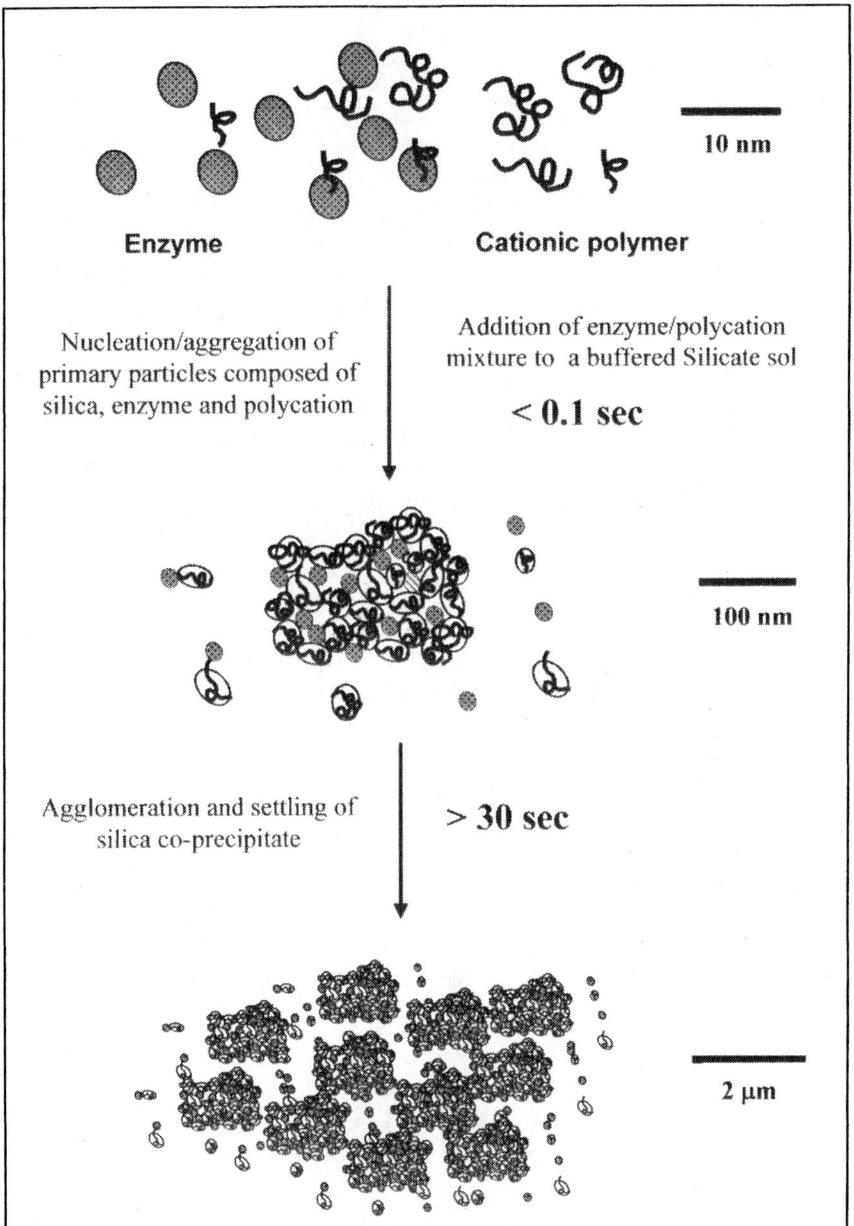

Figure 3. Model of the silica co-precipitation process.

Enzyme Stability

Encapsulation of enzymes often results in greater stability to denaturation due to exposure to solvents or elevated temperatures (6). The thermal activity profile of immobilized *P. alcaligenes* lipase (LIPOMAX™) was compared to that immobilized on Eupergit C resin and the free, unimmobilized enzyme using *p*-nitrophenyl butyrate as the substrate (23). Thermal activity refers to the absolute activity of an enzyme as a function of temperature. It should be noted that such activity is a function of both of the turnover number per protein molecule, in addition to the degree of denaturation of the enzyme. Thus faster turnover rates will compensate for increased denaturation up to a certain point, after which thermal activity will decline.

The results are depicted in Figure 4 and indicate that the silica-immobilized lipase has a higher thermal activity than both the Eupergit-immobilized and the free, unimmobilized enzyme. In this case the greater thermal activity at elevated temperatures is likely a function of greater thermal stability, as opposed to faster turnover relative to the two other forms of the enzyme.

Activity in Organic Solvents

The lipase from *Psuedomonas alcaligenes* (LIPOMAX™, Genencor) was immobilized using 2% PEI and 100mM phosphate-buffered silicate. The final

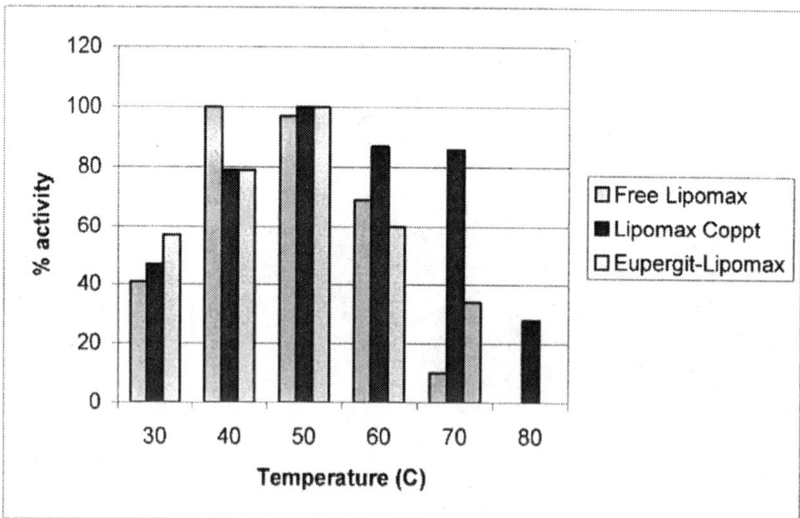

Figure 4. Thermal activity of a LIPOMAX™ silica coprecipitate as compared to the soluble enzyme and that immobilized on to EUPERGIT C® resin.

enzyme loading was calculated to be around 1% w/w following freeze-drying of the resulting silica co-precipitate. The immobilized lipase was studied for the ability to promote ester formation between 1-heptanol and oleic acid (each at 50mM) in isooctane at 50°C over 48 hours. For comparative purposes, the enzyme was also immobilized on Eupergit C, an epoxy-functional resin (24). Additional controls included lyophilized enzyme powder and enzyme lyophilized in the presence of KCl. Enzyme granules consisting of 2.5% w/w protein were also used.

The extent of reaction was analyzed by GC/MS, the results are shown as a function of activity relative to the amount of protein (Figure 5). The lipase immobilized by silica coprecipitation had the greatest specific activity, followed by that immobilized on Eupergit C resin. When the data was reanalyzed according to total mass of catalyst (including support), the silica-immobilized enzyme was slightly better than the enzyme granules.

Enzyme Kinetics

The kinetics of an immobilized glucose oxidase (GOx) coprecipitate were determined and compared to the soluble, unimmobilized form of the enzyme

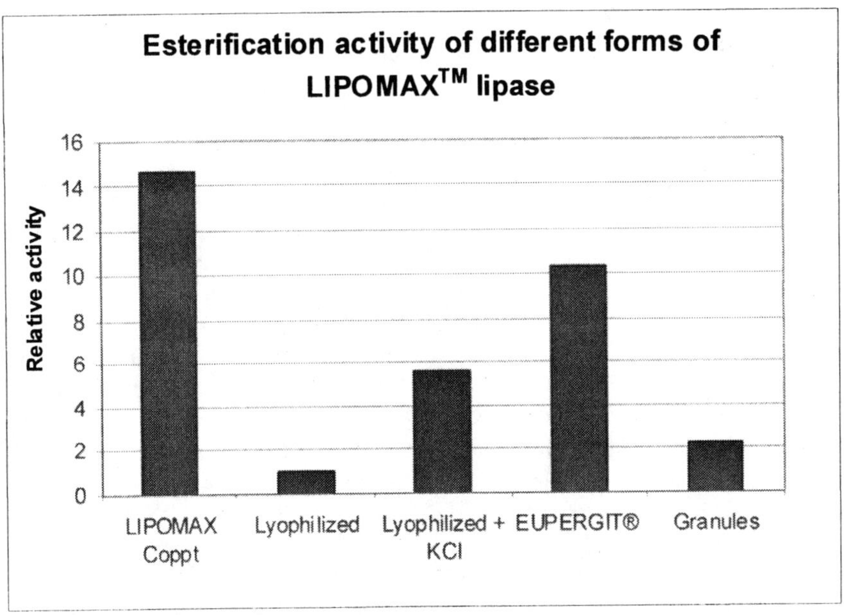

Figure 5. Relative activity of different forms of LIPOMAX™ lipase for the esterification of 1-heptanol and oleic acid in isooctane at 50°C.

(Table III). The reduced efficiency of immobilized GOx relative to the free enzyme is most likely related to hindered access to the active sites of the encapsulated enzyme (6,7). Efforts to improve the activity of glucose oxidase immobilized by this method are described below.

Table III. Kinetic Parameters of Free and Immobilized Glucose Oxidase[1]

Enzyme	V_{max} (mM/sec)	K_m (mM)
Soluble GOx	0.256	24.0
Immobilized GOx	0.084	36.8

[1] OxyGo5000™ Glucose Oxidase, Genencor International, Inc.

Optimization of enzyme capture and activity

One of the potential advantages of silica coprecipitation is the ability to fine-tune the process by adjusting one or more of the many parameters that influence the properties of the final product. A series of experiments were conducted with the goal of improving the immobilization efficiency of glucose oxidase (GOx, EC 1.1.3.4). The following parameters were tested;

- Nature and amount of polycation
- Order of addition
- Mixing dynamics
- Cross-linking agents (gluteraldehyde or sodium periodate)
- Drying method

Enzyme capture varied widely according to the method used for immobilization (Table IV). The manner in which the enzyme/PEI solution was added to the PBSi sol seemed to affect both the amount of enzyme captured and the relative specific activity of the immobilized enzyme. For example, when the silicate solution was vortexed during addition of the enzyme/PEI mixture, more enzyme was captured, although the RSA was reduced relative to the control. Increasing the PEI to enzyme ratio and/or using a 20% PEI solution also improved enzyme capture, but reduced the specific activity of the entrapped enzyme. A similar trend was noted inclusion of cross-linking agents such as gluteraldehyde and sodium periodate. Alternate polyamines, such as triethylenetetramine, did not greatly alter the degree of enzyme capture and activity. Overall, it seemed that while capture of the enzyme could be greatly improved, the specific activity of immobilized GOx was significantly less than that for the free enzyme.

The process for immobilizing the acyltransferase (AcT) from *M. smegmatis* was also studied in greater detail (25). In this case however, it was possible to improve both enzyme capture and relative specific activity through the addition of a small amount of gluteraldehyde. Increasing the enzyme loading 10-fold resulted in a lowering of both enzyme capture and specific activity (Table V).

Table IV. Effect of Process Parameters on GOx Immobilization

Condition	*Capture (%)*	*RSA (%)*	*TEA (U)*
Control[1]	19	15-25	65
Unstirred	15	25	19
Vortexed	39	38	75
20 mg PEI[2]	35	17	29
40 mg PEI[2]	64	17	55
80 mg PEI[2]	73	18	67
Gluteraldehyde[3]	89	14	63
10mM $NaIO_4$	100	12	62

[1] GOx (2 mg, 500 U) was mixed with 2% PEI (40 mg PEI), then added to a PBSi sol. Assays were performed with 5% glucose in 100mM citrate buffer at pH 5.0 with detection of hydrogen peroxide formation by ABTS/HRP.

[2] PEI was added as a 20% solution.

[3] Gluteraldehyde (1.25 mg) was added to a GOx/PEI mixture.

Table V. Effect of Process Parameters on AcT Immobilization

Condition	*Capture (%)*	*RSA (%)*	*TEA (U)*	
			Before Drying	*After Drying*
Control[1,2]	46	24	14.8	4.7
2 mg AcT + Glut[2]	81	35	38.1	6.0
20 mg AcT + Glut[2]	26	17	59.6	12.9
20 mg AcT + Glut[3]	32	15	63.1	40.9
2 mg AcT in sol-gel[3]	100	28	34.7	0

[1] Control condition utilized AcT (2 mg, 124 U) and PEI (40mg) in 1 mL of water, added to 9 mL of 100mM PBSi, followed by centrifugation and washing of the resulting solid. Assays were performed with 1mM *p*-nitrophenyl butyrate in 100 mM Tris HCl, 0.005% Tween 80, pH 8.6, with determination of ΔA at 410 nm.

[2] Air-drying of the resulting solid. [3] Freeze-drying of resulting solid.

The way in which the co-precipitates were dried had a large effect on the activity of the solid, with the freeze-dried material maintaining a greater activity as compared to air-dried material. In contrast, the sol-gel immobilized enzyme lost all activity upon freeze-drying, perhaps as a result of pore shrinkage. The addition of humectants such as glycerol and PEG is currently under study as a means of further reducing the loss of enzyme activity upon drying.

Conclusion

We have demonstrated that enzymes can be readily immobilized in polycation-precipitated silicates by the simple addition of polyamine/enzyme mixtures to buffered solutions of sodium silicate. The product takes the form a colloidal silicate hydrogel consisting of the enzyme, the polyamine and amorphous silica. This process, termed silica co-precipitation, was successfully applied to a range of enzymes with modest to good enzyme capture and retention of catalytic activity in most cases. Fine-tuning of the process to suit a particular enzyme can be achieved by adjusting parameters such as the nature and concentration of the polyamine, enzyme loading, the rate and manner of addition, cross-linking agents and drying method. Optimization of the immobilization of two enzymes, glucose oxidase and acyltransferase was achieved in this fashion. In addition, the use of readily available and non-toxic starting materials makes silica co-precipitation amenable to scaled up production. Further improvement of silica co-precipitation and a detailed characterization of the kinetic parameters of enzymes entrapped by this method is ongoing. One of the major goals of future work is to gain greater control over the structure of the materials produced, particularly with regard to the degree of porosity of the silica matrix.

In the longer term, it should be possible to design enzymes that have the ability to create their own microenvironment through the mineralization of suitable inorganics. Approaches might include the modification of the surface residues of a desired protein, or alternatively using the catalytic activity of the protein to generate an inorganic matrix. Further development of silica co-precipitation might lead to the realization of this and other improvements for the industrial immobilization of enzymes.

References

1. Gill, I. Bio-doped Nanocomposite Polymers: Sol-Gel Bioencapsulates. *Chem. Mater.* **2001**, *13*, 3404-3421.

2. Lee, J. *et al.* Simple Synthesis of Hierarchically Ordered Mesocellular Mesoporous Silica Materials Hosting Crosslinked Enzyme Aggregates. *Small*, **2005**, *1*, 744-753.
3. Bond, R; Jewett, M. C.; McAuliffe, J. C.; Ward, D. E.; *U.S. Pat. Appl.* 20050158837 A1, **2005**.
4. McAuliffe, J. C.; Smith, W. C.; Bond, R.; Zimmerman, J.; Ward, D.; Sanford, K.; Lane, T. H. Template-driven enzyme immobilization: development of a rapid and practical process inspired by diatoms. *PMSE Preprints* **2005**, *92*, 641.
5. Buchholz, K., Kasche, V., Bornscheuer U.T., *Biocatalysts and Enzyme Technology,* Ch. 1, Wiley-VCH, Weinheim, 2005.
6. Kennedy, J. F. *Principles of Immobilization of Enzymes;* in *Handbook of Enzyme Biotechnology;* Wiseman, A., Ed.; Ch 4, p.147; Wiley: Chinchester.
7. Straathof A.J.J. and Aldercreutz P. (Eds.), *Applied Biocatalysis*, 2nd edn. Harwood, 2001.
8. *Biomineralization;* Baeuerlein, E.; Ed.; Wiley-VCH: Weinheim, 2000.
9. Patwardhan, S.V.; Clarson, S.J.; Perry, C.C. On the Role(s) of Additives in Bioinspired Silicification. *Chem. Commun.* **2005**, 1113-1121.
10. Cha, J. N.; Shimizu, K.; Zhou, Y.; Christiansen, S. C.; Chmelka, B. F Stucky, G. D.; Morse D. E. *Proc. Natl. Acad. Sci. USA* **1999**, *96*, 361-365.
11. Bassindale, AR.; Brandstadt, K.F.; Lane, T.H.; Taylor, P.G. *J. Inorg. Biochem.* **2003**, *96*, 401-406.
12. Kröger, N.; Deutzmann, R.; Sumper, M. *Science,* **1999**, *286*, 1129.
13. Naik, R. R.; Brott L. L.; Clarson, S. J.; Stone, M. O. *J. Nanosci. Nanotech.* **2002**, *2*, 95-100.
14. Mizutani, T.; Nagase, H.; Fujiwara, N.; Ogosh, H. Silicic acid polymerization catalyzed by amines and polyamine *Bull. Chem. Soc. Jpn.* **1998**, *71*, 2017.
15. Coradin, T.; Livage, J. Effect of some amino acids and peptides on silicic acid polymerization *Coll. Surf. B: Biointerfaces*, **2001**, *21*, 329-336.
16. Rodriguez, F.; Glawe, D. D.; Naik, R. R.; Hallinan, K. P.; Stone, M. O. *Biomacromolecules* **2004**, 5, *261-265.*
17. Lei, C. Shin, Y. Liu, J.and Ackerman E. J. Entrapping Enzyme in a Functionalized Nanoporous Support. *J Am. Chem. Soc.* **2002**, *124*, 11242.
18. McAuliffe, J. C., Bond, R. L., Cuevas, W. A., *US Pat. Appl.* 20040039179 A1, **2004**.
19. Luckarift, H. R.; Spain, J. C.; Naik, R. R.; Stone, M. O. *Nature Biotechnol.* **2004**, *22*, 211-213.
20. Reetz, M. T.; Wenkel R.; Avnir D. Entrapment of lipases in hydrophobic sol-gel matrices: Efficient heterogeneous Biocatalysts in aqueous Medium *Synthesis*, 2000, 781-783

21. Andersson, M. M.; Hatti-Kaul, R. Protein stabilizing effect of polyethyleneimine *J. Biotechnol.*, **1999**, *72*, 21-31.
22. Asefa, T.; Ozin, G. A.; Grondey, H.; Kruk, H.; Jaroniek, M.; Recent developments in the synthesis and chemistry of periodic mesoporous organosilicas. *Stud. Surf. Sci Catal.* **2002**, *141*, 1-26.
23. Gerritse, G., Hommes, R. W. J., Quax, W. J. Development of a lipase fermentation process that uses a recombinant *Pseudomonas alcaligenes* strain *Appl. Environ. Microbiol.* **1998**, *64*, 2644-2651.
24. Boller, T.; Meier, C.; Menzler S. *Org. Proc. Res. Dev.* **2002**, *6*, 509-519.
25. Amin, N.S. *et al.* WO2005056782, **2005**.

Chapter 12

Nanoporous Sol–Gel Supports Enzymatic Hydrolysis of Chlorophyll in Organic Media

Yunyu Yi[1], Selim Kermasha[2], and Ronald Neufeld[1,*]

[1]Department of Chemical Engineering, Queen's University, Kingston, Ontario K7L 3N6, Canada
[2]Department of Food Science and Agricultural Chemistry, McGill University, 21111 Lakeshore Road, Ste-Anne de Bellevue, Quebec H9X 3V9, Canada

Entrapment of membrane proteins is a challenging task compared to that involving soluble proteins. Chlorophyllase, a membrane protein, was entrapped in nanoporous tetramethoxysilane (TMOS)-derived sol-gel, demonstrating higher protein mass and activity yield than that in polysaccharide gel and hybrid gel. Both sol-gel formulation process and chemistry may affect apparent activity of sol-gel entrapped chlorophyllase, thus need to be taken into account to improve entrapped chlorophyllase activity. Even though the external transfer of chlorophyll to the sol-gel matrix can be minimized by increasing mixing rate up to 150 rpm, internal diffusion of chlorophyll in nanoporous sol-gel was largely restricted, confirmed by the measured diffusion coefficient of 10^{-14} m^2/s. The apparent activity was also affected by the availability of hydrophobic chlorophyll substrate within the hydrophilic gel, which was controlled by partitioning of chlorophyll between a water/organic solvent medium and the gel phase.

Introduction

Chlorophyllase (EC 3.1.1.14) is a membrane glycoprotein, present in the photosynthetic membranes of higher plants and algae (1). *In vitro*, chlorophyllase catalyzes the hydrolysis of hydrophobic chlorophyll (2) suggesting application in the enzymatic removal of chlorophyll, coextracted as a contaminant into Canola oil. Hydrolysis of chlorophyll in contaminated oil by chlorophyllase enzyme has not as yet been demonstrated due to the fact that chlorophyllase is not produced commercially, and both enzyme and substrate are highly water immiscible (3). Thus the enzymatic hydrolysis of chlorophyll, a water-insoluble substrate, was carried out in non-aqueous media to solubilize chlorophyll and increase access of chlorophyll to the enzyme. Thus far, a wide variety of nonaqueous media have been developed for chlorophyllase-catalyzed reactions aimed at optimizing chlorophyllase biocatalysis, including water/miscible organic solvent (4), aqueous/hexane biphasic organic solvent (5), biphasic system (6), and ternary micellar systems containing polysorbates (7) and nonionic surfactant (8). The biocatalysis of chlorophyllase was also studied in ternary micellar systems but with pheophytin as substrate (9,10).

Successful immobilization would allow the reuse of enzyme, which is costly and difficult to produce. Meanwhile, immobilized enzymes demonstrate enhanced stability against harsh environments, such as that of Canola oil and organic solvent, compared to the free enzyme counterpart. Hence, chlorophyllase was entrappd in polysaccharide gel, hybrid gel and tetramethoxysilane (TMOS)-based sol-gel in earlier investigations (11,12). Chlorophyllase demonstrated higher mass and activity yield upon entrapment in TMOS-based sol-gel. This paper is then focused on optimizing catalytic performance of nanoporous sol-gel entrapped chlorophyllase in a water/acetone reaction medium.

Sol-gel derived silica was reported as entrapment matrix for a wide-range of enzymes (13, 14), whereas only a few reports describe the encapsulation of membrane proteins in sol-gel (12, 15-20). As the field is relatively new, the sol-gel immobilization process is not well understood (21), and some results are difficult to understand. For example, encapsulation of the same class of enzyme, but derived from alternate sources can result in different activity yields, even though the same sol-gel process has been followed (22). These differences may be due to sol-gel reactions being highly complex, and thus there is little effective control over the structure of the gel at a molecular level (21).

Improving entrapped enzyme activity is always one of the objectives while developing enzyme preparation for industrial application. Apparent activity of entrapped enzyme depends on the intrinsic activity of enzyme in the specific microenvironment of entrapment matrix, such as pH, ionic strength, hydrophobic or electrostatic interaction between enzyme molecule and matrix. Also substrate diffusion within a nanoporous matrix to reach the enzyme is largely hindered. For example, translational diffusion of small molecules, such as acetone, in pores with average diameter of 3 nm is reduced by a factor of 5 (23). Any factor that leads to small pore size of sol-gel leads to greater restriction. Apparent activity of entrapped enzyme would also depend on the accessibility of enzyme. For sol-gel entrapped enzyme, not all the enzyme entrapped is readily available to substrate (13,22), as some protein may be deeply entrapped in enclosed pores as a more cross-linked gel network forms during aging.

For entrapping chlorophyllase in a nanoporous TMOS-derived sol-gel, the factors that may affect activities are (a) gelation time, defined as the time between the addition of protein to pre-gel sol and the loss of sol fluidity (24); (b) acidic pH when using HCl as catalyst for hydrolysis of sol-gel precursor (25); (c) alcohol produced as a result of hydrolysis of sol-gel precursor (25); (d) gel morphology and thus the accessibility of substrate (26); and (e) drying stress (15). Sol-gel chemistry during formulation of the pre-gel sol would affect the activity of entrapped chlorophyllase through the first four factors. Sol-gel process involving aging and drying conditions would affect the activity of entrapped chlorophyllase through the last two factors. Hence, this paper outlines the sol-gel process and chemical parameters that need to be considered to improve entrapped chlorophyllase activity in a water/acetone organic solvent system (15,26). Some other factors, including diffusion/partition coefficient, of substrate within or to the nanoporous sol-gel support, may decrease the apparent activity of chlorophyllase entrapped in nanoporous sol-gel, and thus is also presented (26). This main objective then was to establish the ground work for industrial chlorophyll removal from Canola oil by an enzymatic approach.

Optimization of Sol-gel Process on Catalytic Performance of Entrapped Chlorophyllase

As shown in Figure 1, the entrapment process starts with a pre-gel sol consisting of partially hydrolysed precursor containing the active encapsulant, in this case chlorophyllase. Sol transforms to a gel over a few minutes or hours (t_g, gelation time), depending on composition. Fresh gel is aged at 4°C to develop a more chemically condensed gel network. Gel is then dried to remove water/alcohol from the porous matrix. Both aging and drying conditions likely affect chlorophyllase activity through altering enzyme hydration, or through changes in gel morphology affecting accessibility of the substrate chlorophyll to the enzyme chlorophyllase.

Aging Conditions

Stronger gel structures develop with longer aging times, reducing gel shrinkage during drying, thus preserving the conformation of entrapped enzyme

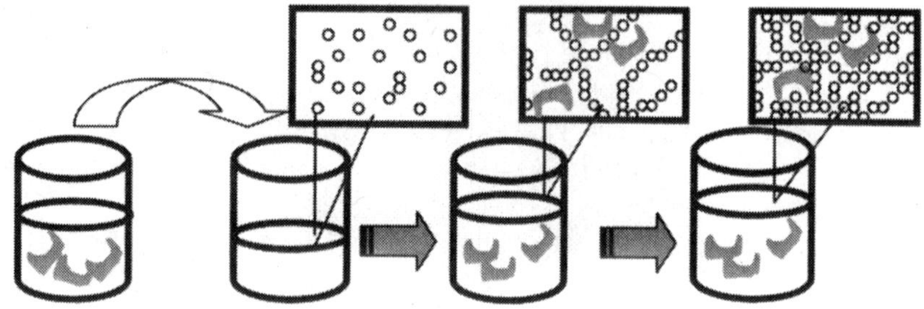

Figure 1. Schematic illustration of the process of entrapping chlorophyllase in TMOS-derived nanoporous sol-gel.

to a larger extent [27]. However, longer aging time may lead to a small pore size, thus decreasing the apparent activity of sol-gel entrapped enzyme. An example would be the apparent activity of lipase decreased upon aging for 25 days at 4°C (28). Aging time ranging from 1 day to 1 week was studied with chlorophyllase. An increase in apparent activity was found with aging time (15), indicating that aging for up to 1 week did not lead to a significant decrease of matrix pore size. Various researchers adopted aging times for entrapped enzymes, ranging from hours to weeks or months, however for practical reasons, a 24 h period was adopted toward an optimal formulation, even though stronger gel networks are preferable.

Aging temperature is another factor that may affect the gel structure, as well as entrapped enzyme activity. Aging at 4°C is preferable for maintaining enzyme activity, and thus is most commonly used.

Drying Conditions

The natural environment for enzyme is water, as an essential water layer (bound water) surrounds the enzyme molecule, maintaining activity or correct conformation (29). Drying or dehydration thus may denature enzyme if the essential water layer is removed (30). During drying, free water is likely be removed during early stages of drying, thus the enzyme is likely to retain activity to a large degree as long as the bound water remains associated. However, the removal of bound water during later stages of drying may cause detrimental conformational changes to proteins. For example, antibody denatured during 100 h of drying (31) and lipase completely lost its activity when air-dried to less than 50% of initial water, which was not restored by water replacement (32). Hence, extensive drying is generally detrimental to proteins. In the case of chlorophyllase, air-drying for 6 h would still lead to a highly active chlorophyllase preparation, whereas negligible activity was observed when drying time exceeded 12 h, as seen in Figure 2 (15).

The decrease of entrapped chlorophyllase activity could also be due to drying stress as a result of high capillary forces developed during evaporation of solvents from gel nanopores. The drying approach that eliminates the capillary force is likely to reduce the drying stress imposed. This hypothesis was supported by a higher specific activity observed with freeze-dried gel than that of air-dried, vacuum-dried and solvent-dried gel (15) and is consistent with that reported by Kamakami and Yoshida (32) with sol-gel entrapped lipase. Even though freeze-drying is preferable to obtain a more active chlorophyllase preparation, air-drying for 3 h is more convenient for practical reasons, and thus used toward optimization.

Sol-gel Chemistry on Catalytical Performance of Entrapped Chlorophyllase

Sol-gel chemistry or formulation determines the gelation time (24), and microenvironment of the sol-gel matrix. It may also affect pore morphology of sol-gel through altering the relative hydrolysis and condensation rate (33). Hence, sol-gel chemistry can affect the catalytical performance of sol-gel entrapped chlorophyllase either directly through altering intrinsic activity of chlorophyllase, or indirectly through altering the gel morphology and thus accessibility of chlorophyllase.

Water/silane Molar Ratio

Water/silane molar ratio in the starting sol can affect the relative hydrolysis and condensation rate, impacting on pore morphology of sol-gel, and thus the accessibility of chlorophyllase to substrate. There is always an optimum water/silane molar ratio for a specific enzyme. An optimum chlorophyllase activity was observed at an R-value of 30, with lower activities obtained at both lower and higher R (15). Chen and Lin (34) and Reetz *et al.* (22) reported similar findings with sol-gel entrapped lipase. At low R value, reduced lipase activity was explained by enzyme aggregation and alcohol denaturation in the pre-gel sol. Sol with R-values over 24 were thought to form sol-gel through alternate gelation behavior, resulting in increasing amounts of residual enzyme in the aqueous supernatant, lowering entrapped lipase activity. In the present study, a change in gelation behavior was not observed. In contrast, over 94% encapsulation yields were observed for all levels of R examined. An examination

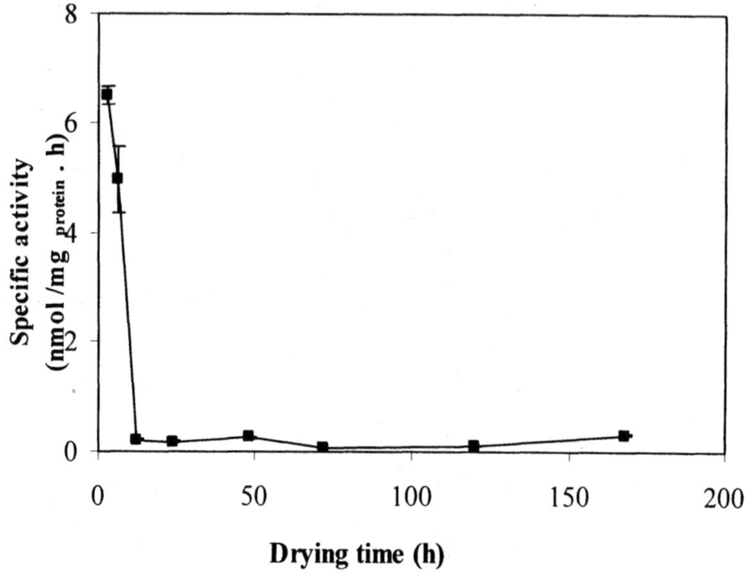

Figure 2. Activities of sol-gel entrapped chlorophyllase air-dried at different time periods (reproduced from reference 15 with permission of Springer Science. Copyright 2006.)

of pore morphology by nitrogen absorption-desorption analysis shows that the pore volume and average pore size of TMOS-derived gel increased with increasing water/silane molar ratio (15). The increased pore volume and average pore size may facilitate the transfer of substrate within sol-gel, thus resulting in the higher measured activity.

Enzyme Loading

Enzyme loading is another factor that needs to be considered in order to improve the activity of sol-gel entrapped enzyme. The general trend is that the specific activity of enzyme decreases at high loading, explained by intraparticle diffusional limitations and/or aggregation of enzyme during sol-gel preparation (35,36). The optimum enzyme loading may vary from enzyme type, sources and support materials, For chlorophyllase, the optimum enzyme loading is 0.257 $mg_{protein}/g_{gel}$. (15). Other researchers reported optimum lipase (35) and trypsin (36) loading in sol-gel as 6.25 and 0.1% respectively. It is possible that the substrate chlorophyll has better access to the enzyme at higher enzyme loading, yet beyond an optimum point, further increase in enzyme loading creates a diffusional limitation to chlorophyll diffusion. High concentrations of enzyme may also trigger aggregation, reducing the availability of enzyme throughout the gel particle matrix (22).

Catalyst / Precursor Type

Catalyst employed to facilitate the hydrolysis and condensation of sol-gel precursor can be classified into basic and acidic catalyst. Base catalyst favors condensation reaction over hydrolysis reaction, thus leading to a denser sol-gel network, whereas acidic catalyst tends to form a branched sol-gel network, as a result of favoring hydrolysis reaction versus condensation reaction (37). The intrinsic activity of chlorophyllase is likely to be reduced in an acidic pH environment when HCl is employed as catalyst for hydrolysis of sol-gel precursor (25). The activity of sol-gel entrapped chlorophyllase was improved by a factor of 2 by using basic catalyst NaF, instead of acidic catalyst HCl (12,15).

Most studies on sol-gel entrapment used TMOS as precursor via the alkoxide route, producing a high content of alcohol as a result of hydrolysis and condensation of TMOS. It will be a problem for molecules that are not stable in alcohol (38). To eliminate the denaturing effect of alcohol, aqueous sol-gel protocols have been developed using sol-gel precursors that do not produce alcohol, including poly(glyceryl silicate) (14), diglycerylsilane (39), monosoribital silane (39), and sodium silicate (38). Alternatively, the denaturing effect of alcohol can be minimized by doping additives along with enzyme in the sol-gel, with some additional advantages as described in the section below.

Microenvironment

Microenvironment of enzyme in nanoporous sol-gel could be altered by doping additives. The additive would protect enzyme from denaturing effects by: forming a shield between protein and the environment (13); by reducing the shrinkage of gels (40) and thus the stress on protein conformation; by acting as a template to alter the gel physicochemical properties (41); or by effectively protecting enzyme from dehydration (13). An extensive study on the effect of a wide range of additives on entrapped chlorophyllase catalytical activity, as well as the properties of sol-gel materials has been conducted, including charged or uncharged polymeric additives, uncharged additives with small molecular weight, and novel or uncommon additives – lipase, lipids (24). It was found that different additives affect activity of entrapped chlorophyllase in different ways. In general, additives with smaller molecular weight, such as glucose or glycerol, better preserve chlorophyllase activity than those with high molecular weight, such as polyethylene glycol, poly (vinyl alcohol), polyethyleneimine, or lipase.

Among the additives investigated, lipids demonstrate better performance on improving chlorophyllase activity upon sol-gel entrapment compared to its additive-free control (Table I, Figure 3). This might be due to the fact that lipids are essential to maintain the activity of chlorophyllase - a membrane protein (3, 9,10). When chlorophyllase was entrapped in sol-gel in the form of a liposome, the lipid provided a shield between chlorophyllase and the denaturing water/alcohol environment. Employing lipids as additive to protect enzyme during the sol-gel process has not been previously reported.

Other Factors Affecting Apparent Chlorophyllase Activity in Nanoporous Sol-gel Matrix

Even though the synthesis process is optimized to preserve the intrinsic activity of enzyme as described in the two previous sections, the apparent enzyme activity in nanoporous sol-gel is often reduced by some other factor, including internal/external diffusion resistance, partition coefficient of substrate between reaction medium and gel phase, as well as the components of the reaction medium.

Internal/external Diffusion Resistance

The external mass transfer of substrate – chlorophyll to the gel matrix can be minimized by increasing the mixing rate up to 150 rpm (26). However, the internal diffusion of chlorophyll in the nanoporous gel was largely restricted, which was reflected by the measured diffusion coefficient on the order of 10^{-14} m^2/s (26). This value is orders of magnitudes lower than that of hydrophilic substrates such as glucose diffusing within a hydrogel matrix (42). The hydrophobic nature of chlorophyll, its higher molecular weight, and the smaller pore size of silica gel, would explain the large differences in diffusion coefficient between the widely different systems. The restricted diffusion of chlorophyll within nanoporous gel explained the lag period observed in chlorophyll hydrolysis catalyzed by entrapped chlorophyllase (26),

Partitioning of Chlorophyll between Gel Phase and Reaction Media

While catalyzing a reaction with hydrophobic substrate in a water/organic solvent system, the apparent activity of enzyme entrapped in hydrophilic nanoporous sol-gel is likely to be reduced by the availability of substrate, as well as the denaturing effect of the organic solvent in the reaction medium. The hydrophobic substrate has a lower tendency to partition into the hydrophilic

Table I. Single Lipid or Lipid Mixture Used to Form Chlorophyllase-Lipid Associate

Formulation	*MGDG (µg)*	*PG (µg)*
MGDG 1	600	-
MGDG 2	1200	-
PG 1	-	600
PG 2	-	1200
Lipid mix 1	300	300
Lipid mix 2	600	600

NOTE: MGDG refers to Monogalactosyl diglyceride;

PG represents phosphatidyl glycerol

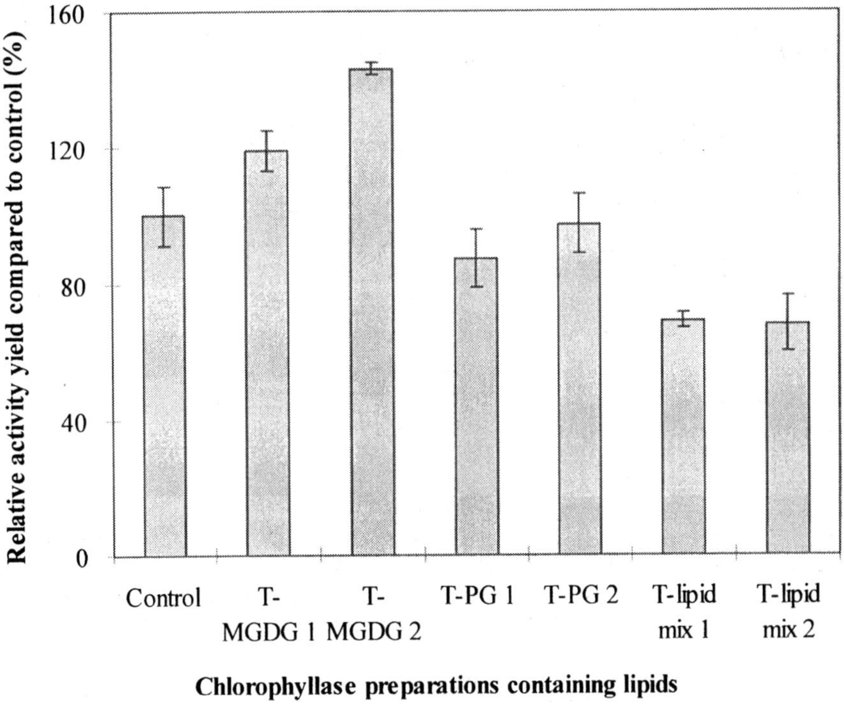

Figure 3. Relative activity yield of chlorophyllase entrapped in TMOS-derived sol-gel containing various types and combination of lipids (see Table I for detailed formulation).

support from the water/organic solvent system as a result of hydrophobic interaction. The partition coefficient of hydrophobic substrate increases when enzyme was entrapped in a more hydrophobic matrix (11).

As chlorophyll is a hydrophobic substrate, acetone was used as a solublizing solvent. Acetone serves as a co-solvent facilitating transfer/partition of chlorophyll to and within the hydrophilic sol-gel matrix (26). A strong catalytic dependence was found over a wide range of acetone concentration (26) while examining catalytic performance of chlorophyllase entrapped in nanoporous sol-gel in a water/acetone medium. Maximum activity was observed at 20% acetone, with decreasing activity at both lower and higher concentrations. As the partition coefficient of chlorophyll to the gel phase increases with increasing acetone concentration, the reduced activity in reaction medium with above 20% acetone arises from the capability of acetone to extract essential water from the enzyme, possibly resulting in denaturation (43).

Catalytic performance of chlorophyllase entrapped in nanoporous support was also examined in a medium consisting of water, acetone, and Canola oil. The Canola oil content was varied from 0 to 30% while holding acetone concentration fixed at 20% (26). The overall decrease in activity is 80% in the presence of 30% Canola oil, compared to a media without oil present (Figure 4). Canola oil differs from acetone in that it is hydrophobic. A 75% reduction in activity of the free enzyme was observed in a water/miscible organic solvent containing 30% oil (44). Reduced chlorophyllase activity in oil may be due to conformational modifications disrupting tertiary structure and denaturing the enzyme (45). Alternatively, the oil as hydrophobic solvent may reduce availability of the substrate chlorophyll to the enzyme, due to substrate partitioning into the oil phase. As conformational change is more likely restricted in a sol-gel matrix, the partitioning effect, reducing chlorophyll accessibility to within the gel matrix, is a more likely explanation for reduced activity with increasing oil content.

Summary and Conclusions

Table II summarizes the parameters of sol-gel process and chemistry before and after optimization. The optimized sol-gel protocols improve entrapped chlorophyllase activity by 80%.

TMOS- derived nanoporous sol-gel effectively entrapped the membrane protein chlorophyllase. Catalytical performance of entrapped chlorophyllase could be manipulated through varying sol-gel process parameters and sol-gel chemistry. External diffusion resistance of chlorophyll was minimized in water/acetone medium at mixing rates over 150 rpm. The internal transfer of chlorophyll within the nanoporous gel was significantly hindered, as a diffusion coefficient on the order of 10^{-14} m^2/s was obtained, which is 4 orders of magnitude lower than that of solutes such as glucose in polysaccharide hydrogel.

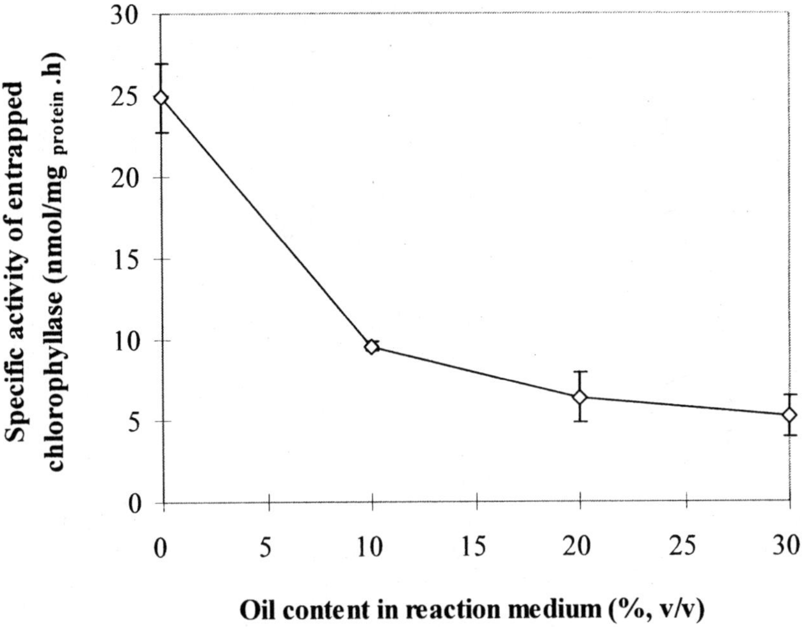

Figure 4. Effect of Canola oil content in reaction media on the activity of entrapped chlorophyllase (reproduced from reference 26 with agreement of Wiley Publishers. Copyright 2006.)

Activity of nanoporous sol-gel entrapped chlorophyllase was examined in an acetone/water organic system with or without the presence of Canola oil. The activity of entrapped chlorophyllase drops to a similar degree as that of free chlorophyllase at the same Canola oil level (30%,v/v). Enhanced stability of sol-gel entrapped chlorophyllase was compromised by the restricted diffusion and partition in such a solvent system. Considering that enzyme immobilization facilitates enzyme recovery, the nanoporous sol-gel entrapped chlorophyllase has the potential for industrial application in the upgrading of Canola oils.

To overcome the observed limitation of the current sol-gel protocol, further study shall be carried out to develop a hydrophobic matrix with bimodal pores, which can be achieved by using sol-gel precursor with non-hydrolysable alkyl chains, and doping the matrix with high molecular weight polymer such as PEG to induce phase separation. The challenge to develop such an matrix would be the fact that phase separation can only occur within a narrow range of PEG concentration. Neverthless, successful development of such an matrix would facilitate the mass transfer of substrate to and within the matrix, due to the micropores and hydrophobic character of the matrix. Meanwhile, the nanopores of the matrix are effective in retaining enzyme.

Table II. Comparison of Original and Optimized Sol-gel Protocol

Parameters	*Parameters optimized*	*Original Protocol*	*Approaches to optimize protocol*	*Optimized protocol*	*Ref*
Sol-gel process	Aging time	4°C for 24 h	4 °C for 24 – 120 h	4°C for 24 h	12,15
	Drying time, temperature	6 h at RT[c]	Air-drying at RT for 3~168 h	3 h at 4 °C	12,15
	Drying approach	Air-drying	Air,vacuum, solvent, freeze-drying	Air-drying	15
Sol gel chemistry	Water silane ratio R	20	7-40	30	15
	Precursor type	TMOS	TMOS, MTMS [a], PTMS[b]	TMOS	12,15
	Catalyst type	HCl	HCl, NaF	NaF	12,15
	NaF concentration (mM)	21.5	2.2-21.5	4.3	15
	Enzyme loading ($mg_{protein}$/g gel)	0.129	0.065~0.428	0.257	15
	Additives	none	Polymer, protein, lipid, sugar	lipids	24

NOTE: a MTMS denotes methyl trimethoxysilne

b PTMS refers to propyl trimethoxysilane

c RT represents room temperature.

Acknowledgement

Authors acknowledge financial support from a Strategic Grant of the Natural Sciences and Engineering Research Council of Canada.

References

1. Terpstra, W. Chlorophyllase in *Phaeodactylum tricornutum* photosynthetic membrane. Extractability, small-scale purification and molecular weight determination by SDS-gel-electrophoresis. *Physiol. Plant* **1978**, *44*, 329-334.

2. Moll, W.; A. W.; DE Eit B.; Lutter, R. Chlorophyllase activity in developing leaves. *Planta* **1978**, *139*, 79-83.
3. Khalyfa, A.; Kermasha, S.; Marsot, P.; Goetghebeur, M. Purification and characterization of chlorophyllase from alga Phaeodactylum tricornutum by preparative native electrophoresis. *Appl. Biochem. Biotechnol.* **1995**, *53*, 11-27.
4. Khamessan, A.; Kermasha, S.; Khalyfa, A.; Marsot, P. Biocatalysis of chlorophyllase from the alga Phaeodactylum tricornnutum in a water/miscible-organic-solvent system. *Biotechnol. Appl. Biochem.* **1993**, *18*, 285-292.
5. Khamessan, A.; Kermasha, S.; Marsot, P. Effects of polar organic solvent on the biocatalysis of chlorophyllase in a biphasic organic system. *Biosci. Biotech. Biochem.* **1994**, *58*, 1947-1952.
6. Khamessan, A.; Kermasha, S.; Marsot, P.; Ismail, A. Biocatalysis of chlorophyllase from Phaeodactylum tricornutum in a biphasic organic system. *J. Chem. Technol. Biotechnol.* **1994**, *60*, 73-81.
7. Khamessan, A.; Kermasha, S.; Mollimard, L. Optimization of chlorophyllase-catalyzed hydrolytic activity in a micellar ternary system. *Biotech. Appl. Biochem.* **1995**, *22*, 327-335.
8. Khamessan, A.; Kermasha, S. Biocatalysis of chlorophyllase from Phaeodactylum tricornutm in micellar ternary system containing spans. *J. Biotechnol.* **1996**, *45*, 253-264.
9. Samaha, H.; Kermasha, S. Biocatalysis of chlorophyllase in ternary micellar system using pheophytins as substrates. *J. Chem. Technol. Biotechnol.* **1997**, *68*, 315-323
10. Samaha, H.; Kermasha, S. Biocatalysis of chlorophyllase in a ternary micellar system containing Span85 using purified and oxidized pheophytins as substrates. *J. Biotechnol.* **1997**, *55*, 181-191.
11. Yi, Y.; Kermasha, S.; L'Hocine, L.; Neufeld, R. J. Encapsulation of chlorophyllase in hydrophobically modified hydrogel. *J. Mol. Cata. B: Enzy.* **2002**, *19-20*, 319-325.
12. Yi, Y.; Kermasha, S.; Neufeld, R. J. Matrix physicochemical properties affect activity of entrapped chlorophyllase. *J. Chem. Technol. Biotechnol.* **2005**, *80*, 1395-1402.
13. Jin, W.; Brennan, J. D. Properties and applications of proteins encapsulated within sol-gel derived materials. *Anal. Chim. Acta* **2002**, *461*, 1-36.
14. Gill, I.; Ballestero, A. Bioencapsulation within synthetic polymers (part 1): sol-gel encapsulated biologicals. *Trends Biotechnol.* **2000**, *18*, 282-296.
15. Yi, Y.; Neufeld, R. J.; Kermasha, S. Optimization of tetramethoxysilane-derived sol gel entrapment protocol stabilizes active chlorophyllase. *J. Sol-Gel Sci. Technol.* **2006**, *38*, 251-259.
16. Weetall, H. H. Retention of bacteriorhodopsin activity in dried sol-gel glass. *Biosen. Bioelectr.* **1996**, *11*, 327-333.

17. Wu, S.; Ellerby, L.M.; Cohan, J.S.; Dunn, B.; El-Sayed, M.A.; Valentine, J.S.; Zink, J. I. Bacteriorhodopsin encapsulated in transparent sol-gel glass: A new biomaterial. *Chem. Mater.* **1993**, *5*, 115-120.
18. Luo, T.M.; Soong, R.; Lan, E.; Dunn, B.; Montemagno, C. Photo-induced proton gradients and ATP biosynthesis produced by vesicles encapsulated in a silica matrix. *Nat. Mater.* **2005**, *4*, 220-224.
19. Besanger, T.R.; Hodgson, R.J.; Guillon D.; Brennan, J.D. Monolithic membrane-receptor columns: Optimization of column materials for frontal affinity chromatography/mass spectrometry applications. *Anal. Chim. Acta* **2006**, *561*, 107-118.
20. Besanger, T. R.; Easwaramoorthy, B.; Brennan, J.D. Entrapment of highly sctive membrane-bound receptors in macroporous sol-gel derived materials. *Anal. Chem.* **2004**, *76*, 6470-6475.
21. Xu, J.; Dong, H.; Feng, Q.; Wei, Y. Direct immobilization of horseradish peroxidase in hybrid mesoporous sol-gel materials. *Poly. Prep.* **2000**, *41*, 1044-1045.
22. Reeze, M.T.; Zonta, A.; Simpelkamp, J. Efficient immobilization of lipase by entrapment in hydrophobic sol-gel materials. *Biotechnol. Bioeng.* **1996**, *49*, 527-534.
23. Koone, N.; Shao, Y.; Zerda, T.W. Diffusion of simple liquids in porous sol-gel glass. *J. Phys. Chem.* **1995**, *99*, 16976-16981.
24. Yi. Y. Ph.D Thesis, Queen's University, Kingston, Canada. **2004**.
25. Ellerby, L.M.; Clinton R.; Ninshida, F.; Stacey N.; Yamanaka, A.; Dunn, B.; Valentine, J. S.; Zink, J. I. Encapsulation of proteins in transparent porous silicate glasses prepared by the sol-gel method, *Science* **1992**, *255*, 1113-1115.
26. Yi, Y.; Neufeld, R. J.; Kermasha. S. Characterization of sol-gel entrapped chlorophyllase. *Biotech. Bioeng.* **2006**, in press.
27. *The Chemistry of Silica. Solubility, polymerization, colloid and surface properties, and biochemistry;* Iler, R. K., Eds.; Wiley-Interscience publications: New York, 1979; p 532.
28. Keeling-Tucker, T.; Rakic, M.; Spong, C.; Brennan, J. D. Controlling the material properties and biological activity of lipase within sol-gel derived bioglasses via organosilane and polymer doping, *Chem. Mater.* **2000**, *12*, 3695-3704.
29. Zaks, A.; Klibanov, A.M. The effect of water on enzyme action in organic media. *J. Biol. Chem.* **1988**, *263*, 8017-8021.
30. Depaz, R. A.; Dale, D. A.; Barnett, C. C.; Carpenter, J. F.; Gaertner, A. L.; Randolph, T. W. Effects of drying methods and additives on the structure, function, and storage stability of subtilisin: role of protein conformation and molecular mobility. *Enzyme Microb. Technol.* **2002**, *31*, 765-774.
31. Zuehlke, J.; Knopp, D.; Niessner, R. Sol-gel glass as a new support matrix in immunoaffinity chromatography. *Fresenius J. Anal. Chem.* **1995**, *352*, 654-659.

32. Kawakami, K.; Yoshida, Y. Entrapment of lipase in silica glass by the sol-gel method and its esterification activity in organic media. *Biotechnol. Tech.* **1994**, *8*, 441-446.
33. Sato, S., Murakata, T., Suzuki, T., Ohgawara, T. Control of pore size distribution of silica gel through sol-gel process using water soluble polymer as additives. *J. Mater. Sci.* **1990**, *25*, 4880-4885.
34. Chen, J.P.; Lin, W.S. Sol-gel powders and supported sol-gel polymers for immobilization of lipase in ester synthesis. *Enzyme Microb. Technol.* **2003**, *32*, 801-811.
35. van Unen, D.-J.; Engbersen, J.F.J.; Reinhoudt, D. N. Sol-gel immobilization of serine proteases for application in organic solvents. *Biotechnol. Bioeng.* **2001**, *75*, 154-158.
36. Avnir D.; Braun S.; Lev O.; Ottolenghi M. Enzymes and other proteins entrapped in sol-gel materials, *Chem. Mater.* **1994**, *6*, 1605-1614.
37. Brinker C. J.; Scherer, G. W. 1990, *Sol-gel Science. The physics and chemistry of sol-gel processing*, Chapter 5. Gelation. Academic Press, Inc. P309.
38. Bhatia, R.B.; Brinker, C.J.; Ashley, C.S.; Harris, T.M. Synthesis of sol-gel matrices for encapsulation of enzymes using an aqueous route, Mat. Res. Soc. Symp. Proc. **1998**, *519*, 183-188.
39. Rupcich, N.; Goldstein, A.; Brennan, J. Optimization of sol-gel formulations and surface treatments for the development of pin-printed protein microarrays, *Chem. Mater.* **2003**, *15*, 1803-1811
40. Wambolt, C.L.; Saavedra, S.S.; Iodide fluoresecence quenching of sol-gel immobilized BSA. *J. Sol-gel Sci. Technol.* **1996**, *7*, 53-57.
41. Dong, H.; Xu, J.; Qiu, K. Y.; Jansen, S.A.; Wei, Y. Synthesis of nonsurfactant-based hybrid mesoporous sol-gel materials. *Polym. Prep.* **2000**, *41*, 194-195.
42. Zhang, W.; Furusaki, S. On the evaluation of diffusivities in gels using diffusion cell technique. *Biochem. Eng. J.* **2001**, *9*, 73-82.
43. Gorman, L.A.S.; Dordick, J.S. Organic-solvents strip water off enzymes. *Biotechnol. Bioeng.* **1992**, *39*, 392–397.
44. Bitar, M. Masters Thesis, McGill University, Montreal, Canada. **2003**.
45. Buskov, S.; Sorensen, H.; Sorensen, S. Separation of chlorophylls and their degradation products using packed column supercritical fluid chramotography. *J High Res. Chromat.* **1985**, *22*, 339-342.

Chapter 13

Enzyme Encapsulation Using Nanoparticle-Assembled Capsules

Vinit S. Murthy[1] and Michael S. Wong[1,2,*]

Departments of [1]Chemical and Biomolecular Engineering and [2]Chemistry, Rice University, Houston, TX 77005

We discuss the synthesis and characterization of enzyme-containing nanoparticle-assembled capsules (NACs) using acid phosphatase as the model enzyme. NACs are a new capsular material that can be synthesized rapidly under ambient conditions and mild pH values. Unlike vesicles and other organic-based hollow sphere structures, NACs can be handled in the absence of water. The hollow forms spontaneously, allowing for non-destructive encapsulation. The enzyme is shown to preserve its activity within the capsules, and a maximum loading and encapsulation efficiency of ~18 wt% and ~70%, respectively, can be achieved. The encapsulated enzymes can be recovered and re-used through centrifugation.

Immobilization of enzymes in robust solid supports has captured the attention of material scientists and biochemists for almost three decades *(1,2)*. Broadly speaking, bioimmobilization involves either trapping the individual enzyme molecules within solid support materials or physically binding the enzymes on the surface of the supports in order to restrict their freedom of movement. There are a number of approaches for immobilizing enzymes: (1) covalent binding to the support, (2) physical adsorption, (3) microencapsulation, (4) physical entrapment, and (5) crosslinking *(1,3-5)*. No single one solves all immobilization issues, but the strategy of combining multiple approaches could.

We have developed a new enzyme immobilization route which combines physical adsorption, microencapsulation, and physical entrapment *(6)*. This route is based on the nanoparticle assembly synthesis of organic/inorganic hollow capsule structures. In this two-step synthesis, solutions of a polyamine and a multivalent anion are combined to form a suspension of polymer-salt aggregates. A nanoparticle (NP) sol is then added, leading to the formation of nanoparticle-assembled capsules ("NACs") *(6-8)*. The benign synthesis conditions (aqueous environment, room temperature, no organic solvents involved, near neutral pH) and rapid generation of NACs (from seconds to minutes) allow for the non-damaging encapsulation of enzymes and other sensitive molecules. The capsule material is structurally robust and does not collapse when removed from water.

Herein we discuss the synthesis and characterization of enzyme-encapsulated NACs, in which encapsulation simply involves the additional step of adding an enzyme solution to the polymer-salt aggregate suspension before the NP sol (Figure 1). Acid phosphatase was used as the model enzyme, and two substrates (florescein diphosphate and 4-nitrophenyl phosphate) were used to characterize catalytic activity. Recoverability and re-use of the encapsulated enzyme were demonstrated. Fluorescence imaging of NACs containing dye-conjugated protein molecules provided additional evidence for successful acid phosphatase encapsulation.

Experimental Section

Materials

Poly(*L*-lysine hydrobromide) (195,000 g/mol, bromide counterion, "PLL"), poly(*L*-lysine hydrobromide) conjugated with fluorescein isothiocyanate (68,600 g/mol, bromide counterion, "PLL-FITC"), acid phosphatase (from potato, 7 units/mg or U/mg of solid, "AP"), 4-nitrophenyl phosphate (disodium salt, "4-NPP"), citrate buffer solution (pH 4.8, 0.09 M), bovine serum albumin conjugated to fluorescein isothiocyanate (67,000 g/mol, "BSA-FITC"), protease solution (from *Bacillus* sp., ≥16 U/g) were all procured from Sigma-Aldrich and

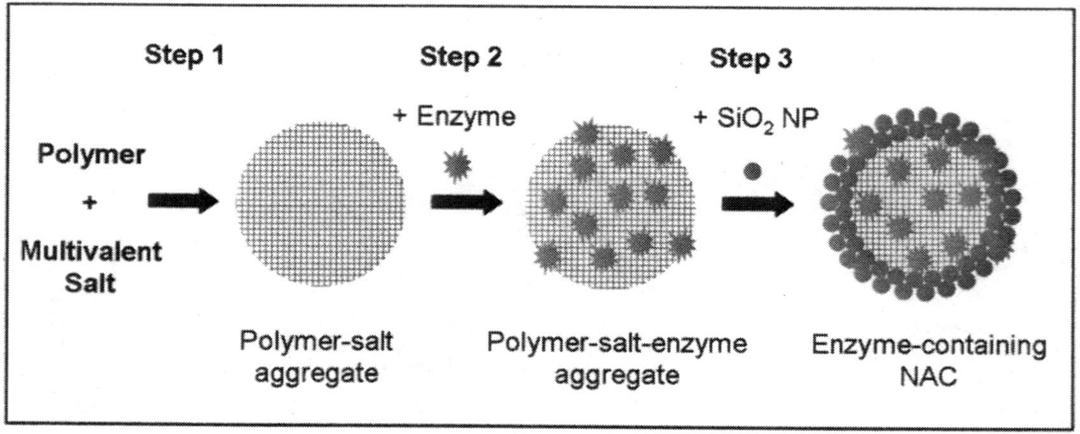

Figure 1. Conceptual diagram illustrating the synthesis of enzyme-NACs via a three-step mixing process.

were used without further purification. Stock solutions of BSA-FITC (20 mg/ml) and enzyme (10.6 mg/ml) were prepared in 0.01 wt% sodium azide aqueous solution to prevent microbial degradation. Trisodium citrate ("cit") was obtained from Fisher Scientific. A 10 mM solution of fluorescein diphosphate (purchased from Anaspec, "FDP") was prepared using citrate buffer solution (pH 4.8), divided into 100 µl aliquots, and frozen. Silica (SiO_2) NPs (Snowtex-O, Nissan Chemicals) were available as an aqueous colloidal suspension or sol (20.5 wt%, pH 3.4, ionic strength I = 16.9 mM). The NPs measured 13±3 nm in diameter, according to dynamic light scattering. Zeta potential of SiO_2 NPs in the original sol condition was calculated from its electrophoretic mobility (-1.4 (µm/sec)/V/cm)) to be -16 mV in the Henry's limit, indicating the particles were negatively charged.

Synthesis method

A typical synthesis protocol for enzyme-NACs was as follows: 21 µl of PLL (2 mg/ml in water) was mixed with 125 µl of cit (5.36 mM) by vortex mixing at low speeds ("5" speed on a scale from 1 to 10) for 10 seconds. The lightly turbid suspension was aged for 20 min without stirring or shaking. 50 µl of acid phosphatase solution (initial concentration varying from 0.1 to 1.6 mg/ml) was then added and the solution was mixed gently. After aging for 10 min, 125 µl of silica NP sol was added and rapidly vortex mixed for 20 seconds at "7" speed. Settling of the acid phosphatase-PLL-cit suspension was found to be noticeable after aging for longer than 10 min. The resulting enzyme-NACs were aged for 2 hrs and washed (through centrifugation and decantation) thoroughly with deionized water to remove excess enzyme and SiO_2 NPs. For BSA-encapsulated NACs, 50 µl of BSA-FITC solution (initial concentration of 20 mg/ml) was used in place of the acid phosphatase.

Enzyme content determination

The amount of encapsulated enzymes was determined by measuring the unencapsulated enzyme concentration with a standard AP activity assay *(9)*. The enzyme-containing NACs were washed two times with water. The supernatant from the NAC washes was added to 15 ml of 4-NPP (2 mM solution in citrate buffer, pH 4.8) and magnetically stirred at room temperature. 30 µl aliquots of the resulting solution were taken at different times (0, 2, 4, 6, 8, and 10 min) and added to 3 ml of NaOH solution (resulting in a pH of 10.0). Basic solutions of the nitrophenol product were colored (yellowish), allowing for quantification via UV-vis spectroscopy; the absorbance at 405 nm was measured. Enzyme amount was determined from the slope of the nitrophenol concentration-time curve, referenced to a free enzyme concentration-activity calibration curve. Proper control experiments were carried out to ensure that residual SiO_2 NPs, PLL, and cit did not affect enzyme activity.

The encapsulated enzyme content was directly quantified also, by breaking the enzyme-NACs using a high-concentration NaCl solution (2 M) to release the enzyme. The thus-freed enzyme concentration was then determined using the above technique. Control experiments indicated the salt solution did not affect enzymatic activity. Carried out on several enzyme-NACs samples, these measurements verified that the unencapsulated enzyme measurements (which was easier to perform) provided accurate encapsulated enzyme content information.

Encapsulation efficiency was defined as the percentage of enzyme molecules encapsulated in the NACs out of the total enzyme added, which was calculated from assay explained above. Enzyme loading was defined as the ratio of the weight of the enzyme encapsulated to the total weight of the enzyme-NACs.

Characterization

Confocal laser scanning microscopy.

Confocal images were captured with Carl Zeiss LSM 510 inverted microscope equipped with a 100× oil immersion objective (NA = 1.4). The laser excitation wavelength of 488 nm was chosen for FITC (λ_{ex} = 494.5 nm, λ_{em} = 519 nm). Samples were mounted on conventional glass slides and sealed under a cover slip to prevent drying. All samples were prepared 1-2 hrs prior to imaging.

Electrophoretic mobility measurement.

Electrophoretic mobility was measured using phase analysis light scattering (PALS), a variation of electrophoretic dynamic light scattering (DLS). Zeta potential measurements were calculated from electrophoretic mobility measurements using Henry's equation (*i.e.*, $0.1 \leq \kappa a \leq 100$, where κ is the Debye-Hückel parameter and a is the particle radius). A dip-in (Uzgiris type) electrode system with 4 ml polystyrene cuvettes was used.

UV-vis and fluorescence spectroscopies.

UV-vis spectroscopy was carried out using Shimadzu 2401-PC UV-vis spectrophotometer. Standard poly(methyl methacrylate) cuvettes with a path length of 1 cm were used. Fluorescence measurements were taken with Jobin-

Yvon Horiba Fluoromax 3 spectrophotometer. Standard quartz cells with a path length of 1 cm were used. Solutions were constantly stirred during analysis with a magnetic Teflon bar; stirring did not interfere with the fluorescence spectra.

Scanning electron microscopy.

Scanning electron microscopy (SEM) was carried out in JEOL 6500 field emission microscope equipped with in-lens thermal field emission electron gun. Secondary electron image (SEI) was taken at 15 kV electron beam with a working distance of 10 mm. The samples were loaded on aluminum stubs, dried and sputter-coated with gold for 1 min.

In situ fluorescence microscopy.

Fluorescence microscopy was performed on a Leica DM2500 upright microscope equipped with 100× oil immersion objective (NA 1.4). Excitation wavelength of 488 nm was chosen for fluorescein. FDP solution (0.1 mM solution in citrate buffer) was added to a suspension of enzyme-NACs. The suspension was then deposited on a poly-lysine coated glass slide. Differential interference contrast (DIC) image was obtained using the same microscope in transmitted light mode.

Thermogravimetric analysis.

Thermogravimetric analysis (TGA) was performed in the presence of air flow (100 cm^3/min) on DT 2960 from TA Instruments. The temperature was ramped at the rate of 5 °C/min at was kept constant at 1100 °C for 20 min.

Results and Discussion

Acid phosphatase (AP) encapsulation

The chemistry behind the NP assembly synthesis of the NAC material involves electrostatic interactions among the various precursors and allows for the encapsulation of charged molecules like AP. We chose AP as the model enzyme as it is well-studied and its dephosphorylation activity can be readily assayed. AP is ubiquitous in nature, and is found in plants (*e.g.*, potatoes and wheat germ) and animals (*e.g.*, the human prostrate gland). The potato-derived form has a molecular weight and isoelectric point (pI) of ~55 kDa and 5.2, respectively.

When the aqueous solutions of PLL and cit are combined, the positively-charged PLL form spherically shaped aggregates resulting from the cross-linking or salt-bridging action of the negatively charged citrate ions *(6,7)*. We hypothesized that zwitterionic AP molecules would bind to, and possibly incorporate into, the polymer aggregates after introducing the AP solution (Figure 1). Electostatic, as well as hydrogen-bonding and hydrophobic, interactions would be the responsible driving force. We further hypothesized that the SiO_2 NPs would lead to shell formation after addition to the enzyme-polymer-salt aggregates. The current model for shell formation is that the negatively charged NPs penetrate through the outer region of the polymer aggregates, and that the penetration depth determines the shell thickness *(6,7)*. For non-enzyme-containing NACs, the shell generally is 100-150 nm thick and comprised of the NPs and polymer.

Combining AP with PLL-cit aggregates resulted in micron-sized aggregates in Brownian motion (Figures 2a,b). A comparison of the brightfield and confocal images indicated that the PLL (observable in the latter image due to the fluorescent FITC) concentrated at the aggregate periphery after contact with AP molecules. Without the enzyme, the aggregates showed a homogeneous distribution of the polymer *(6)*. Noting the previous observation of similar PLL re-distribution behavior after contact with SiO_2 NPs, we concluded that the AP molecules (radius of gyration ~ 3 nm) interacted with the PLL-cit aggregates as charged NPs. Analogizing with the SiO_2 NP case, we inferred that the AP was located mostly near the aggregate surface at this synthesis stage.

The hypothesized electrostatic interactions were supported by pH and electrophoretic measurements. The pH values of the AP precursor solution and AP-PLL-cit suspension were measured to be ~6.5 and 6.8, respectively. They were higher than the pI of AP, indicating that the enzyme possessed a net negative charge before and after combining with the PLL-cit aggregates. The PLL-cit aggregates were positively charged, according to the electrophoretic mobility value of +0.7 (μm/s)/(V/cm) measured before combining with the enzyme.

The addition of SiO_2 NPs to the AP-PLL-cit aggregates led to the micron-sized AP-encapsulated NACs, as seen under brightfield microscopy (Figure 2c). The PLL clearly localized at the NAC shell, with little found in the interior in all but the smallest, observable NACs (Figure 2d). The same shell formation and hollowing were observed in PLL-cit-SiO_2 NAC synthesis *(6)*, with the latter attributed to the dis-assembly of the polymer aggregate during shell formation. In this AP-NAC case, the SiO_2 NPs and PLL were located in the shell but no information could be deduced about the enzyme location. Without the SiO_2 NPs, the AP-PLL-cit aggregate morphology changed after removal from water, forming a continuous film.

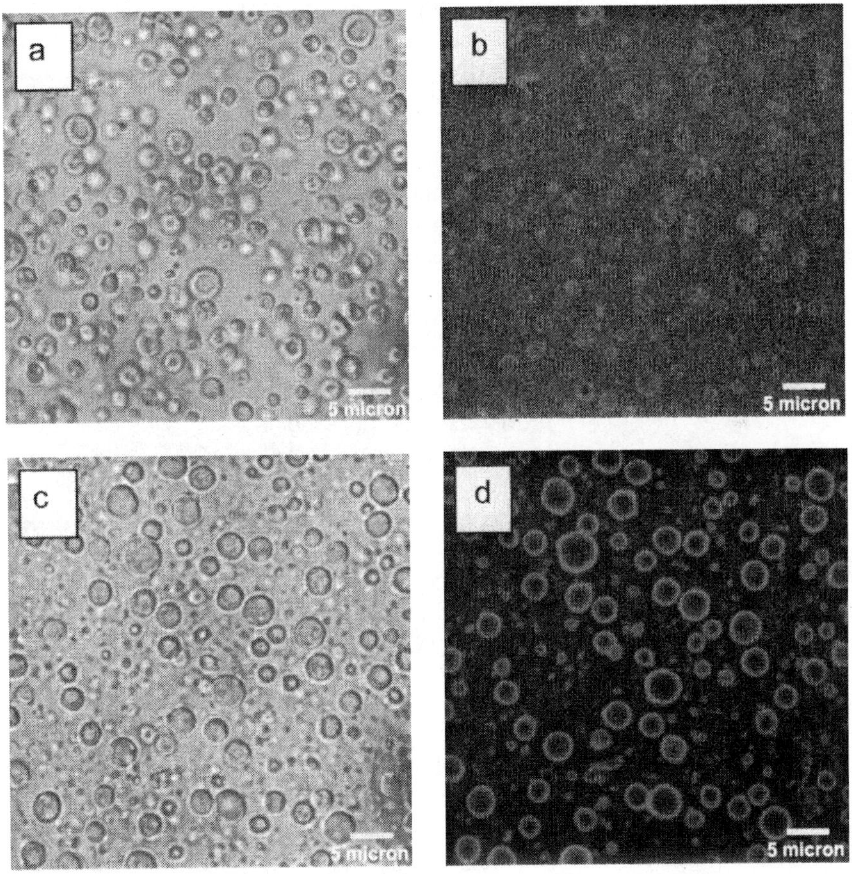

Figure 2. (a,c) Brightfield and (b,d) confocal microscopy images of (a,b) AP-PLL-cit aggregates and (c,d) AP-NACs. $[PLL]_{final}$ = 0.13 mg/ml; $[cit]_{final}$ = 2.1 mM; $[SiO_2\ NP]_{final}$ = 7.9 wt.%; $[AP]_{final}$ = 0.26 mg/ml. PLL was conjugated to FITC.

Control of AP loading

Typical loadings obtained with conventional sol-gel techniques are in the range of 0.5 to 5 wt%. Stone and co-workers recently reported a high 20 wt% butyrylcholinesterase enzyme loading using a biomimetic peptide/SiO_2 composite *(10)*. A high enzyme loading using our PLL/SiO_2 NP composite material of enzymes could be possible, and so we studied the effect of initial enzyme precursor concentration on the loading, encapsulation efficiency, and particle morphology of the AP-NACs.

Table 1 lists data on enzyme loading and encapsulation efficiency as a function of final enzyme concentration in the AP-PLL-cit aggregate suspension. High loading (16 wt%) and efficiency (70%) were achieved at the lowest enzyme concentration. With increasing enzyme concentrations, the loading remained roughly unchanged (within experimental error) and the encapsulation efficiency decreased. These trends suggested that there was an upper limit at which PLL-cit aggregate became fully saturated with the AP enzyme. Both the enzyme loading and efficiency dropped significantly at the highest AP concentration used. We attributed this to interference by the AP on the shell formation, leading to unstable AP-NACs.

Table 1. Enzyme loading and encapsulation efficiency as a function of acid phosphatase concentration.

$[AP]_{final}$, mg/ml	*Enzyme loading, wt%*	*Encapsulation efficiency, %*
0.019	16	70
0.034	15	52
0.10	18	20
0.26	9	4

The particle morphology of dried AP-NACs with varying enzyme concentrations was characterized via SEM (Figure 3). As the control sample, the "empty" (non-AP-containing) NACs were stable to drying (Figure 3a). With low AP loadings, the capsules were also stable, with a small fraction observed to have pinholes or small indentations in the shell wall (Figures 3b-d). At the highest loading, the size distribution became noticeably bimodal, with largest NACs mostly exhibiting shell damage (Figure 3e).

BSA-FITC encapsulation

The location of AP molecules within the NACs could not be discerned, as mentioned earlier (Figures 2c,d), though several locations could be possible: on the inner PLL/NP shell wall, within the wall itself, and/or in the core interior. To gain insights into the physical location of AP, we studied the encapsulation of a fluorescence-labeled protein of comparable size and pI. The BSA molecule used was ~7 nm in diameter with a pI of 4.7 *(11)*.

As expected, combining BSA with PLL-cit aggregates resulted in the formation of BSA-PLL-cit aggregates (Figures 4a,b). The BSA-FITC was negatively charged since the synthesis pH was measured to be 8.1, which was higher than the BSA pI. The brightfield image of the aggregates showed

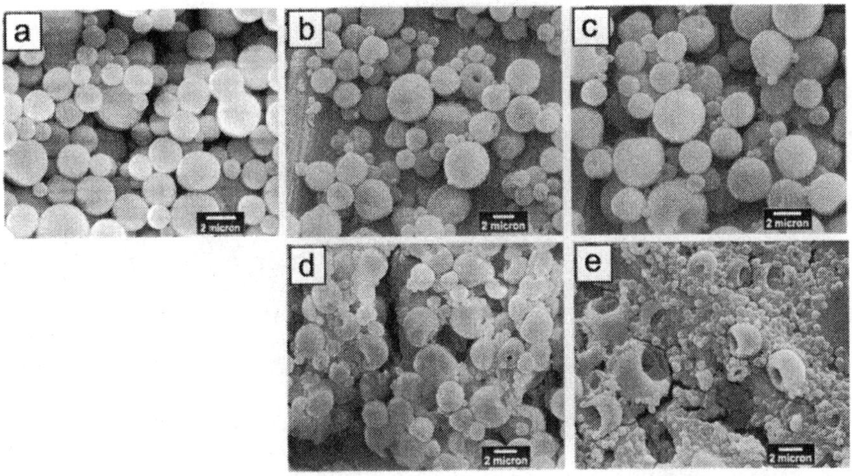

Figure 3. SEM images of AP-NACs with enzyme loadings of (a) 0 wt%, (b) 16 wt%, (c) 18 wt%, and 9 wt%.

optically opaque spheres, unlike the case for AP-PLL-cit aggregates (Figure 2a), suggesting the BSA-PLL-cit aggregates were more dense than the latter. The corresponding confocal image indicated the fluorescent BSA molecules were located throughout the aggregate, concentrating at the aggregate perimeter (Figure 4b and inset). We inferred that the BSA had not only adsorbed on the PLL-cit aggregates surface but also diffused inside the aggregates to form solid microspheres. Interestingly, the BSA-PLL-cit aggregates retained their particle morphology upon drying and subjection to SEM vacuum conditions (Figures 4c,d), consistent with their higher density than the AP-PLL-cit aggregates.

The addition of SiO_2 NPs led to BSA-containing NACs (Figure 5). The particle morphology and BSA spatial distribution were very similar to those of the BSA-PLL-cit aggregates. These BSA encapsulation results suggested the AP molecules encapsulated in the NACs were located in the core interior, with a higher concentration near the inner shell walls.

From fluorimetry measurements of free BSA-FITC in the supernatant and TGA measurements of the BSA-NAC product, the BSA loading of aggregates and NACs were found to be 33 wt% and 28 wt%, respectively, indicating the NPs contributed to the NAC weight-wise.

The SiO_2 NP shell wall could not be identified in the confocal and SEM images. So, we tested for the presence of the SiO_2/polymer shell by treating the BSA-PLL-cit aggregates and the BSA-containing capsules with protease. Proteases are enzymes which attack the peptide bonds in proteins in a process known as proteolytic cleavage *(3)*. The aggregate suspension was incubated with a bacterial protease solution, and after 15 hrs of aging, the aggregates

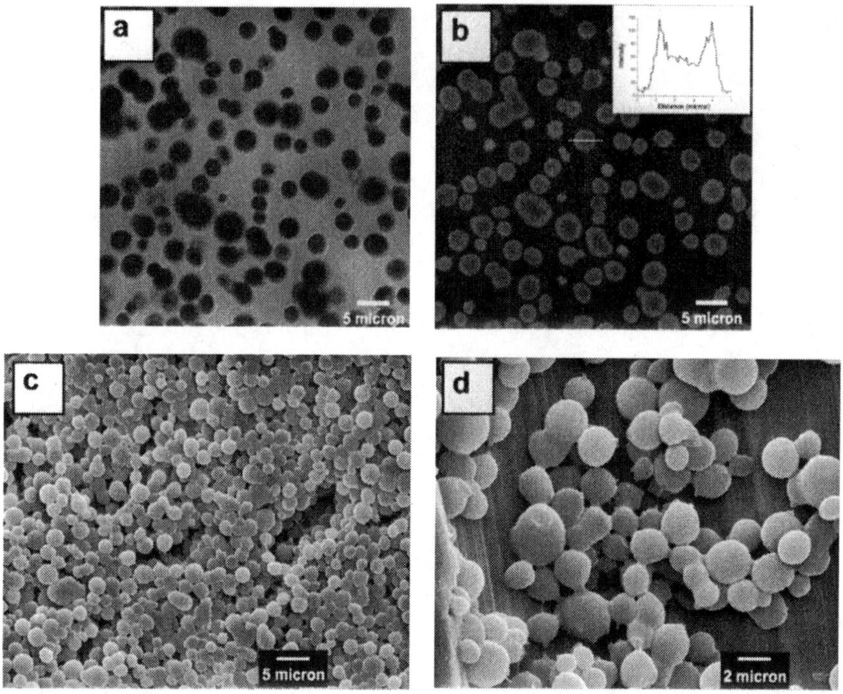

Figure 4. (a) Brightfield, (b) confocal, and (c,d) SEM images at two different magnifications of BSA-PLL-cit aggregates. Inset: Fluorescence intensity line profile of an aggregate. $[PLL]_{final} = 0.21$ *mg/ml;* $[cit]_{final} = 3.4$ *mM;* $[BSA]_{final} = 5.1$ *mg/ml. BSA was conjugated to FITC.*

disintegrated into a clear, fluorescent solution (Figures 6a,b). BSA-NACs treated in the same manner with the protease, on the other hand, were essentially unaffected (Figures 6c,d). Some loss in capsule shape was observed, suggesting structural stability differences among the BSA-containing NACs. The SiO_2 NPs provided a protective barrier against proteolytic attack via the formation of a SiO_2/polymer shell.

Enzymatic activity of AP-NACs

The enzyme-containing NACs can be used as tiny reaction vessels *(6)*. To more clearly illustrate this property of AP-NACs, we carried out a time-lapse fluorescence microscopy experiment in which a suspension of the AP-encapsulated NACs was diluted in a solution of FDP, a non-fluorescent

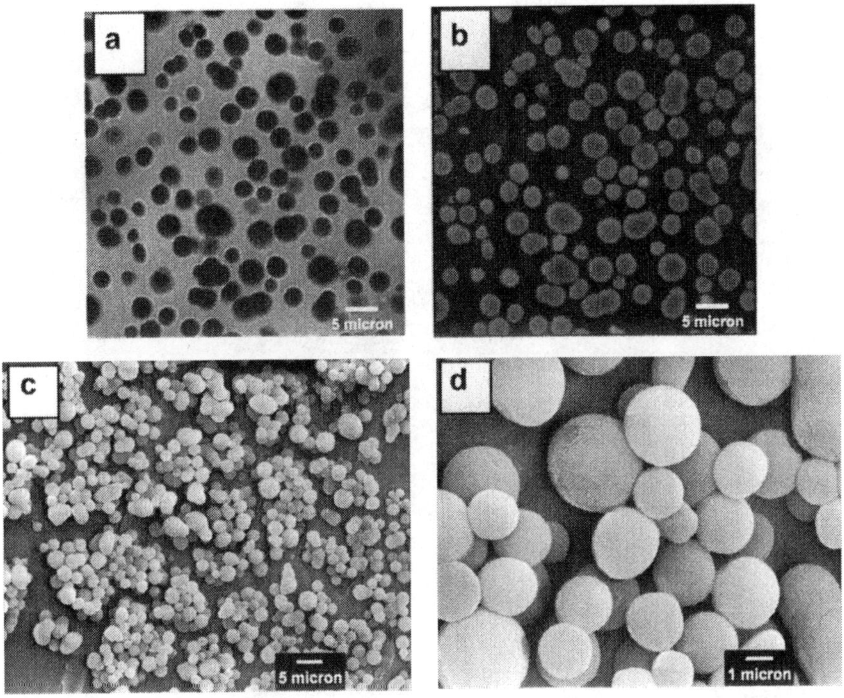

Figure 5. (a) Bright field, (b) confocal, and (c,d) SEM images at two different magnifications of BSA-NACs. $[PLL]_{final}$ = 0.13 mg/ml; $[cit]_{final}$ = 2.1 mM; $[BSA]_{final}$ = 3.1 mg/ml; $[SiO_2\ NP]_{final}$ = 7.9 wt%. BSA was conjugated to FITC.

substrate. Fluorescence emanated from the NAC shell wall and interior, due to the presence of the fluorescein product, and increased in intensity with time (Figures 7a-c). These images indicated the encapsulated enzyme was able to dephosphorylate FDP, which implied that the substrate was able to diffuse through the SiO_2/polymer shell wall. This was consistent with our observations of semipermeability of the NACs *(12)*. Image analysis indicated that the *in situ*-generated fluorescein accumulated within the shell wall and, to a lesser extent, in the core interior. The fluorescence of the aqueous medium surrounding NACs also grew in intensity, which was attributed to the diffusive release of fluorescein product resulting from the concentration difference between the shell wall and the exterior. The NACs did not degrade during the reaction after 30 min (Figure 7d) or even after 1 hr.

The possibility of AP adsorbing onto the external surface of NACs could not be discounted, in spite of multiple washings of the AP-NACs after encapsulation. We sought to remove any surface-bound enzymes by treating the

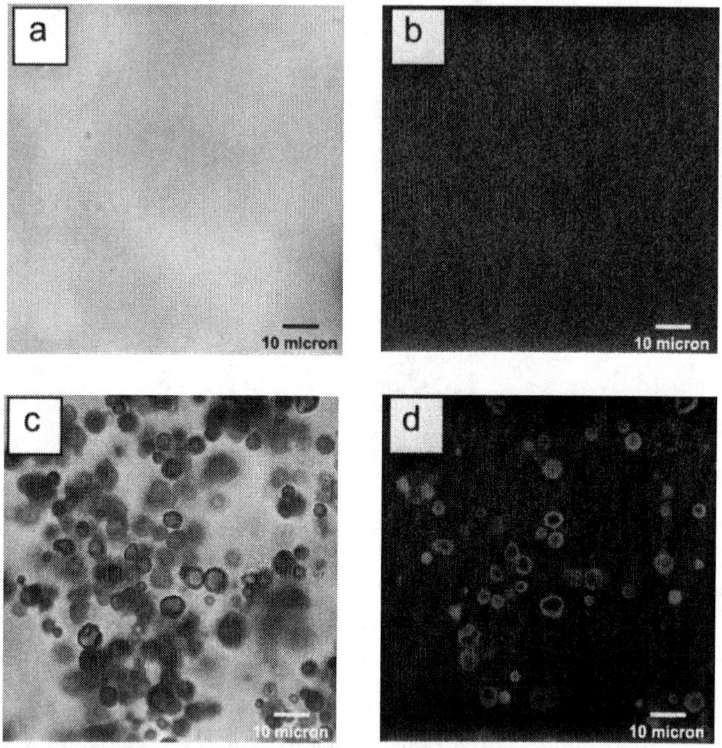

Figure 6. (a,c) Bright field and (b,d) confocal microscopy images of (a,b) disintegrated PLL-cit-BSA aggregates and (c,d) of BSA-containing SiO_2 NACs incubated in protease enzyme for 15 hr time period at room temperature.

NACs in a solution of protease, but during control experiments, we found that AP was resistant to proteolytic degradation *(13)*.

Recoverability and re-use of AP-NACs

Immobilized enzyme systems provides the advantage of recovering and re-using the enzyme multiple times. To show recoverability and reuse for our system, we used 4-NPP as the substrate (the AP activity assay used the same substrate to quantify enzyme concentration).

The rudimentary testing was performed as follows. NACs were washed two times with water to remove any un-encapsulated AP, and were suspended in a 4-NPP solution (Figure 8). After 30 min, the reaction vial was centrifuged at 6000

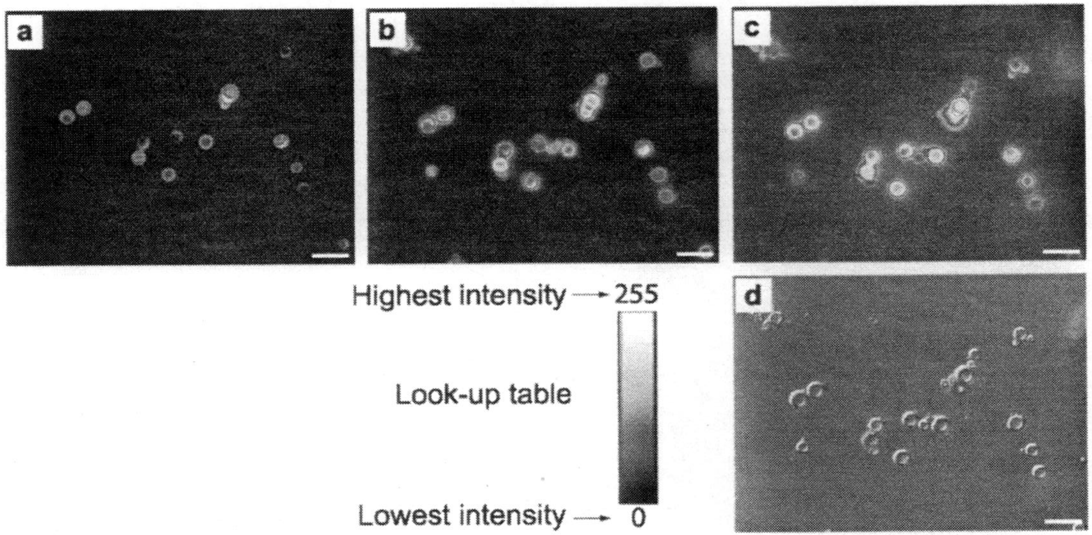

Figure 7. Series of time-lapse fluorescence microscopy images of AP-containing NACs suspended in an aqueous solution of FDP after (a) 3 min, (b) 15 min, and (c) 30 min. The images were pseudo-colored for visualizing the fluorescein product concentration gradient. (d) DIC image of AP-NACs after 30 min of reaction time, corresponding to panel (c). Scale bars: 10 μm.

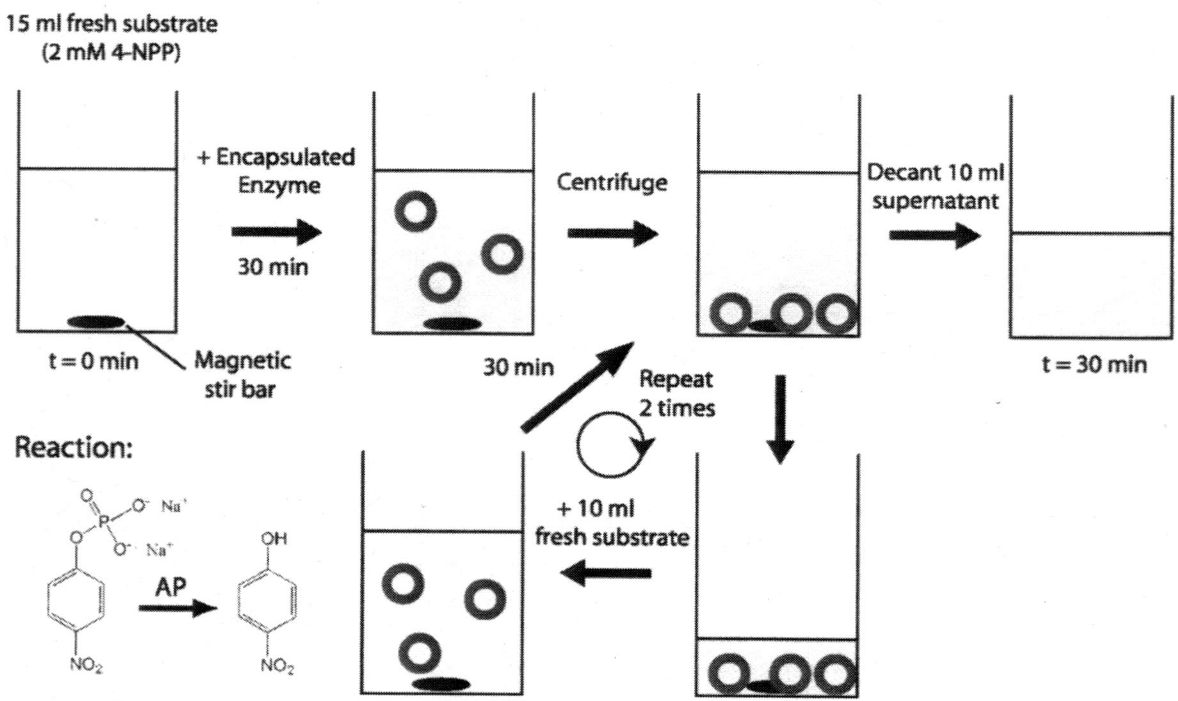

Figure 8. Schematic showing the step-by-step protocol for conducting recoverability studies of AP-NACs.

rpm and two-thirds of the supernatant (containing product and reactant) was replaced with a fresh solution of 4-NPP. This was repeated for two more cycles. The enzymatic activity of the encapsulated AP was measured at the end of each 30-min run by assaying the 10 ml supernatant withdrawn from the reaction mixture. The amount of total enzyme charged to the reactor was set at 0.48 U (0.07 mg), for both encapsulated (16 wt% loading) and free forms. Loss of AP-NACs would have been a significant problem if all the supernatant (instead of 2/3) was removed after centrifugation.

The focus of these data is the comparison of the encapsulated and free forms of AP with regard to recoverability. A brief description of the presented results follows (Figure 9a). Point "0" marks the initial nitrophenol product concentration at the beginning of the runs, and point "1" marks the concentration at the end of the first cycle. Point "2" marks the decreased nitrophenol concentration in the reaction flask after removing two-thirds of the supernatant and replacing with equivalent volume of fresh 4-NPP solution. Point "3" marks the product concentration at the end of the second cycle.

Two conclusions can be gleaned. (1) The free enzyme is more active than the encapsulated form, based on the higher product concentration at the end of the first cycle. Specifically, the molecular activity (which is a reaction rate defined as the number of substrate molecules converted into product per minute per enzyme molecule) of the encapsulated enzyme was measured to be 42% of that of the free enzyme. The lower activity may have been due to transport limitations imposed by the SiO_2/polymer shell wall. Indeed, in a control experiment, we found that AP-PLL-cit aggregates (*i.e.*, without the shell) had the same activity as free enzyme. Since these aggregates were structurally unstable, they were not viable enzyme carriers.

It could be possible that the encapsulated AP molecules are less active due to denaturation. To gain insight into this possibility, we treated AP-NACs with a high-concentration NaCl solution (2 M) to break the capsules. The resultant solution exhibited 92% of the molecular activity of the free enzyme, suggesting that AP denaturation did not occur inside the NACs.

(2) The NACs substantially enhanced enzyme recoverability and re-use. The molecular activities of encapsulated and free AP were normalized to their respective initial molecular activities (from the first cycle; Figure 9c). The free enzyme activity dropped more rapidly than AP-NACs in cycles 2 and 3, due to the unavoidable loss of 2/3 of the enzymes after withdrawal of the supernatant. At the end of the third cycle, only 10% of the initial enzyme amount was left. In contrast, 40% of the encapsulated AP were recovered, supporting the anticipated benefits of NAC encapsulation. The loss of any AP-NACs could be attributed to capsule damage and subsequent enzyme leakage during the centrifugation step. Optical microscopy confirmed that centrifugation of AP-NACs can result in shell damage.

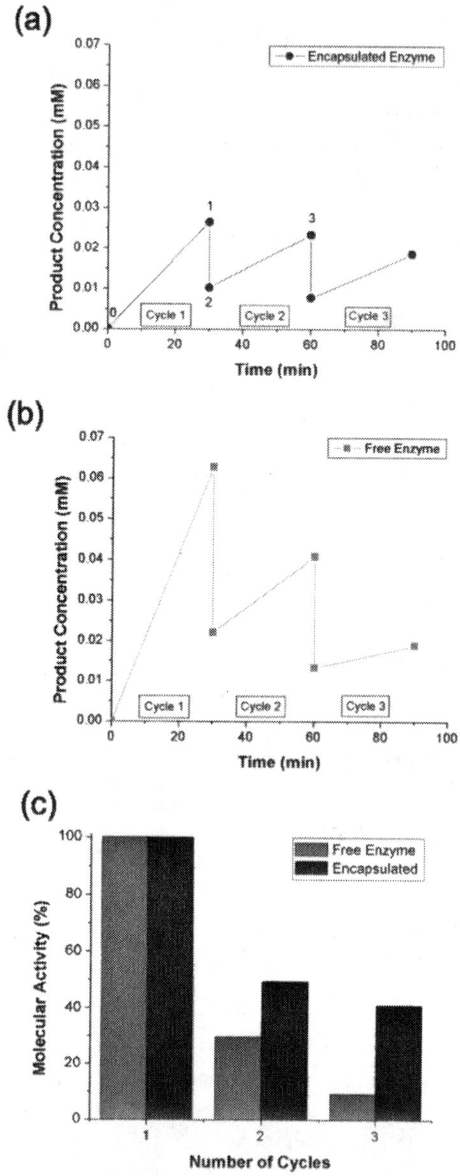

Figure 9. Concentrations of nitrophenol product at the beginning and end of three reaction cycles for (a) encapsulated AP and (b) free AP. The solid lines were drawn to guide the eyes. (c) Relative enzymatic activity of encapsulated and free forms with reaction cycle. Enzyme loading was calculated by measuring the enzyme concentration of the supernatant, after dis-assembly of the AP-NACs. Initial enzyme amount: 0.48 U (0.07 mg)

Conclusions

A facile route for encapsulating sensitive biomolecules such as enzymes in nanoparticle-assembled capsules (NACs) was demonstrated. Combining the enzyme solution with PLL-cit aggregate suspension, and subsequent addition of SiO_2 NPs led to the formation of stable enzyme-containing NACs. Enzyme loadings as high as 18 wt% were achieved under benign synthesis conditions, comparable to the highest enzyme encapsulation reported in literature. Acid phosphatase was located in the NAC interior, with a likely higher concentration at the inner shell wall. Experimentation with BSA-encapsulated NACs showed the protective nature of the SiO_2/polymer shell. The encapsulated AP were active for dephosphorylation, indicative of diffusive transport through the NAC shell wall. Finally, the encapsulated AP could be recovered through centrifugation and re-used. These findings point to NACs as an interesting and new encapsulating material, with opportunities for easily and non-destructively encapsulating high loadings of other enzyme molecules within a robust and permeable shell.

Acknowledgments

We acknowledge Mr. S. Benegal for helpful discussions, Ms. J. Yu for assistance with the TGA measurements, and the financial support of the Smalley/Curl Fund for Innovation and Halliburton Energy Services.

References

1. G. F. Bickerstaff, ed. *Immobilization of Enzymes and Cells*. New Jersey: Humana Press, 1997.
2. P. Gemeiner, ed. *Enzyme Engineering: Immobilized Biosystems*. New York: Ellis Horwood, 1992.
3. R. F. Taylor, ed. *Protein Immobilization: Fundamentals and Applications*. New York: Marcel Dekker, 1991.
4. A. Wiseman, ed. *Handbook of Enzyme Biotechnology*. New York: Ellis Horwood, 1995.
5. H. Uhlig, *Industrial Enzymes and Their Applications*. New York: John Wiley, 1998.
6. R. K. Rana, V. S. Murthy, J. Yu, and M. S. Wong, *Adv. Mater.* 17, 1145-1150 (2005).
7. V. S. Murthy, R. K. Rana, and M. S. Wong, *J. Phys. Chem. B* 110, 25619-25627 (2006).

8. J. Yu, V. S. Murthy, R. K. Rana, and M. S. Wong, *Chem. Commun.*, 1097 (2006).
9. R. B. McComb, G. N. Bowers, and S. Posen, *Alkaline Phosphatase*. New York: Plenum Press, 1979.
10. H. A. Luckarift, J. C. Spain, R. R. Naik, and M. O. Stone, *Nat. Biotechnol.* 22, 211-213 (2004).
11. S. Ge, K. Kojio, A. Takahara, and T. Kajiyama, *J. Biomater. Sci. Polym. Ed.* 9, 131-150 (1998).
12. J. Yu, V. S. Murthy, and M.S. Wong, unpublished results.
13. M. A. M. Abdul-Fadl and E. J. King, *Biophys. J.* 44, 434-435 (1949).

II: Applications of Nanoscale Biocatalysts

Chapter 14

Highly Sensitive and Magnetically Switchable Biosensors Using Ordered Mesoporous Carbons

Dohoon Lee[1], Jinwoo Lee[2,4], Young-Pil Kim[1], Jungbae Kim[3], Taeghwan Hyeon[2], and Hak-Sung Kim[1,*]

[1]Department of Biological Sciences, Korea Advanced Institute of Science and Technology, Daejeon 305–701, South Korea
[2]National Creative Research Initiative Center for Oxide Nanocrystalline Materials and School of Chemical and Biological Engineering, Seoul National University, Seoul 151–744, South Korea
[3]Pacific Northwest National Laboratory, 902 Battelle Boulevard, Richland, WA 99352
[4]Current address: Department of Materials Science and Engineering, Cornell University, Ithaca, NY 14853–1501

Recent progress in the synthesis of nano-structured materials having desirable physico-chemical properties has opened the door for the fabrication of more sensitive and stable biosensors. In this chapter, we will describe the use of ordered mesoporous carbons (OMCs) as matrix materials for enzyme-based biosensors. The unique characteristics of OMCs, including large pore size and surface area, enabled a high loading of biocatalyst, which is a major requirement of sensitive biosensors. Another useful feature, magnetism, was added to produce a high-capacity magnetic enzyme carrier by incorporating magnetic nanoparticles into the OMC. Enzymes immobilized in the magnetic OMC were used to construct a magnetically switchable electrochemical biosensing system in which the catalytic electron flow can be controlled by a magnetic force.

In many biotechnological applications of enzymes including industrial bioprocesses, immobilized enzymes have been widely used owing to their technological and economical advantages over their free enzyme counterparts (*1*). Various immobilization strategies have been developed to enhance the operational and storage stability of biocatalysts, to improve resistance against harsh environmental conditions, such as high temperature and non-aqueous media, and to enable easy separation and repeated use of enzymes. Most of the enzyme-based biosensors also employ the immobilized form of biocatalysts. In this case, performance of the biosensor is highly dependent on two factors: the amount of active enzyme near the electrode and the efficiency of the electron-transfer between enzyme and electrode. Because properties of the enzyme-supporting matrixes have great influence on such factors, the use of appropriate matrix is of great importance in development of highly sensitive biosensors.

Various natural and non-natural polymers are widely used as the matrix for the modification of electrodes (*2, 3*). Enzymes are generally entrapped in the polymeric matrixes or immobilized in layer-by-layer format (*4–6*) in order to deposit a sufficient amount of enzymes near the electrodes. Since few enzymes permit direct electron-transfer from enzyme to the base electrode (*7–9*), mediated transfer relying on diffusion of free electron mediators or relay through static mediators immobilized in matrixes (*7, 10, 11*) is generally adopted. To further improve the biosensor performance in terms of response time and sensitivity, it is necessary to reduce the physical distance between the base electrode and the enzymes to facilitate electrical communication, since not all enzymes are in the proximity of the base electrodes. One method is to use electrically conductive materials, like carbon or metals, as matrixes. However, these materials should be properly shaped to meet some requirements: First, they should have large empty space to accommodate enzymes, which also should be easily accessible by macromolecules. In addition, they should be well equipped with sub-structures, like channels or pores, to facilitate the transport of various chemicals necessary for the enzyme and the electrochemical reactions. Surely, some advanced synthesis techniques enabling percise control over sizes and spatial arrangement of sub-structures are necessary to fabricate such materials.

Ordered Mesoporous Carbon Materials

Despite the recent progress in nano-scale manufacturing technologies, bottom-up synthesis of nano-structured materials with desirable properties still remains a challenge. There exist a few alternative and more practical ways. The first attempt was to use porous carbon materials as matrixes for biosensor application (*12*). The pores can provide space for enzyme loading and also can act as channels for substrate transport. However, the conventional methods for introducing pores into solid materials, such as physical and chemical activation processes, generally result in irregular pore structures with broad pore size distribution, which can cause poor and irreproducible loading of biocatalysts. Furthermore, micropores (< 2 nm) are dominantly generated in such processes in general, which poses a limitation on the size of biomolecules to be immobilized.

Synthesis and Structural Characteristics

In recent years, a novel approach has been developed for the synthesis of carbon materials with more defined structures, which are designated as ordered mesoporous carbon (OMC) materials (*13–15*). They are produced through template-based synthesis procedures, and mesoporous silicas with various structures are generally used as the templates. If precise information on the structure of the templates is available, it is possible to predict or finely tune the dimensions and the organization of the sub-structures by modification of the synthesis conditions. As a result, OMCs are characterized by highly ordered and controllable pore structures with narrow pore size distribution. Such unique structural properties ensure homogeneous and reproducible immobilization of enzymes. In addition, the mesopores (with diameters > 2 nm, < 50 nm) and large surface area and pore volume of the OMCs enable a high loading of biocatalyst (*16, 17*). In this regard, OMCs synthesized using appropriate templates are expected to provide well-characterized matrixes for the development of enzyme-based biosensors.

Fabrication of biosensors using enzymes immobilized in OMCs was reported (*18*). For the immobilization of glucose oxidase (GOx) with a molecular weight of ca. 160 kDa and a size of 5.2 nm × 6.0 nm × 7.7 nm (*19*), we synthesized an OMC containing interconnected hollow cells with a diameter of ca. 30 nm (Figure 1). A mesoporous silica having similar pore structure, called MSU-F (*20*), was used as the template, and controlled incorporation and polymerization of carbon precursors were used to obtain the OMC with large cellular pores. The resulting OMC, designated as MSU-F-C, was revealed to possess three kinds of sub-structures (see Figure 1): (a) interconnected hollow spherical cells (diameter ca. 30 nm). (b) channels with the size of ca. 6 nm surrounding the cells, which were left empty after removal of the silica template. (c) micropores with the size of ca. 0.6 nm occurring throughout the carbon framework. From the size of GOx, it is expected that most of the enzymes would be immobilized within the cellular pores. The smaller mesopores and micropores can provide the channels for transport of small chemicals (e.g. glucose).

Immobilization of Enzymes in MSU-F-C

The particulate form of MSU-F-C made the enzyme immobilization procedure easy. Enzymes were readily immobilized in the pores of MSU-F-C

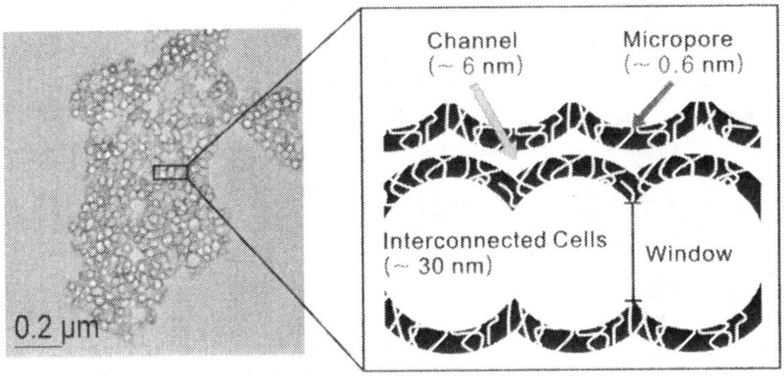

Figure 1. TEM image (left) and schematic illustration (right) of MSU-F-C showing the morphology and the sizes of sub-structures.

during mixing and shaking of enzyme solution with carbon matrixes. Kinetic study using GOx revealed that the maximum loading of enzyme reached a level of 30~40 wt. %, and equilibrium was obtained in less than 1 hr regardless of initial enzyme concentration. The enzyme immobilization isotherm was different from the Langmuir type (*18*), implying that the immobilization cannot totally be explained by adsorption on the exterior carbon surface only. Reduction in gas adsorption capacity of MSU-F-C after enzyme immobilization indicates that the cellular pores are filled with enzymes (*21, 22*). A high correlation between enzyme loading level and pore size was found (*18*), showing that large pore size is a prerequisite for high enzyme loading. Leaching of enzyme was negligible even under high-speed shaking conditions, indicating that there might be a strong interaction between enzyme and carbon surface, like hydrophobic interactions. The connecting windows (Figure 1) are also believed to make a contribution to the stability of the immobilized enzymes by providing a bottleneck-like structure, which can prevent the leaching of enzymes from MSU-F-C (*23–25*).

Enzyme immobilization in another mesoporous media, mesoporous silica, has been extensively studied (*21, 22, 26–32*). Despite several advantages of silica matrixes, including an easily modifiable surface for covalent attachment of enzymes, comparable enzyme loadings were observed only for small (≤ 3 nm) and basic (pI ≥ 10) enzymes, probably owing to the negatively charged silica surface (pI $\approx 2\sim3$) (*22, 31*). Furthermore, to prevent the leaching of enzymes, a common problem of silica matrixes (*22, 26, 32*), additional treatments, such as covalent attachment of enzymes (*33*) or formation of a protection layer around the matrix (*22*), are necessary. By contrast, OMCs generally exhibit high loading and stability for enzymes regardless of sizes and overall charge states, which, along with the inherent conductivity of carbon, makes them a better choice for the matrix for enzyme-based biosensors.

Fabrication of Glucose Biosensor

A glucose biosensor was fabricated by immobilizing MSU-F-C particles containing GOx on glassy carbon electrodes (GCEs) in Nafion film (*34, 35*). Dissolved oxygen molecules were employed as electron mediators, which are reduced to hydrogen peroxides by accepting electrons when GOx oxidizes glucose, and then the resulting hydrogen peroxides are reoxidized on electrode surface, generating an anodic current. Though a bare GCE is not a good hydrogen peroxide detector (*12*), modification with MSU-F-C significantly increased its sensitivity to hydrogen peroxide by increasing the effective electrode surface area (Table I).

Table I. Sensitivity to Hydrogen Peroxide and Measured Surface Area of MSU-F-C-modified Glassy Carbon Electrode

Amount of MSU-F-C Deposited (μg)	*Sensitivity to H_2O_2 (nA/mM)*	*Electrode Surface Area (cm^2)*
0.0	7.8	0.07
0.7	61.5	0.05
3.5	392.8	0.28
7.0	1405.4	1.53
21.0	3931.7	15.01
42.0	4768.7	23.43

Note: The electrode surface area was measured by chronocoulometry.

This result can be explained as follows: On GCE, most of the MSU-F-C particles are electrically connected through physical contact between them. For this reason, not only the GCE, but the MSU-F-C particles themselves can act as electrodes due to their inherent electrical conductivity. Because each MSU-F-C particle has large surface area (ca. 7 cm^2/μg), the total effective electrode surface area increases significantly. In addition, as mentioned earlier, the conductive matrixes can reduce the length of the diffusion path through which the electron mediators must traverse. In other words, the produced hydrogen peroxide can be oxidized immediately on the carbon wall of MSU-F-C near the enzymes. Consequently, the constructed biosensor showed a high sensitivity even without additional hydrogen peroxide transducers like metal compounds which are often employed by carbon-based enzyme electrodes to overcome their poor sensitivity to hydrogen peroxide (*35*).

Magnetically Switchable Bioelectrocatalytic System

As described in the previous section, electrically interconnected OMC/enzyme particles can be used as a sensitive biosensing electrode. The connection can be made reversible if the movement of the particles is controllable. An example is a swichable electrochemical biosensing system using external magnetic force (Figure 2) (*36, 37*). In this system, by changing the location of enzymes immobilized in magnetic carriers, an electrical connection between redox enzymes and electrode can be achieved or disrupted, which, in turn, can switch on and off the current signal.

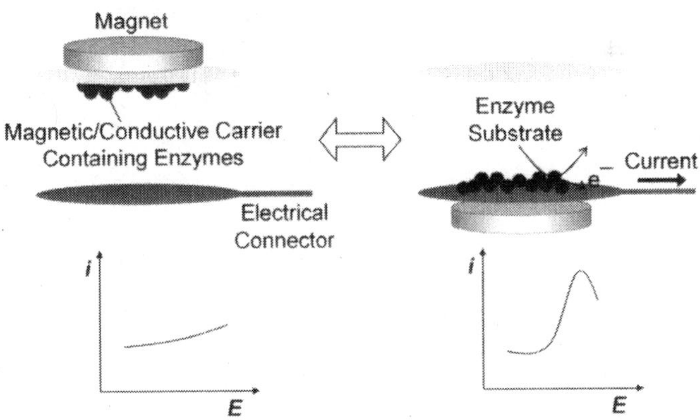

Figure 2. Principle of the magnetically switchable electrochemical biosensing. Expected voltammetric results are depicted schematically for switched off (left) and on (right) states.

As demonstrated in the previous section, OMCs can be excellent enzyme carriers in terms of loading capacity, retention of biocatalysts, and electrical properties. In order to make a magnetically switchable bioelectrocatalytic system using OMC as an enzyme carrier, we attempted to synthesize a magnetic OMC by incorporating magnetic nanoparticles into the OMC.

Synthesis of a Magnetic OMC, Mag-MCF-C

A simple synthesis route to magnetic OMC was developed (36), in which the carbonization and the generation/incorporation of magnetic nanoparticles proceeds simultaneously through the solid phase conversion of iron salt to nanoparticles (*38*). The resulting material, designated as Mag-MCF-C, was found to have an overall carbon structure similar to MSU-F-C, characterized by the co-existence of the three types of sub-structures mentioned earlier. Electron microscopy and magnetic property analysis revealed that ferromagnetic nanoparticles were generated and well distributed mainly on the exterior surface of Mag-MCF-C (*36*).

Enzyme Immobilization and Construction of a Magnetically Switchable Biosensing System

Owing to the structural similarity, the Mag-MCF-C showed enzyme loading capacity, immobilization kinetics, and stability for GOx similar to MSU-F-C's (*36*). A switchable glucose sensing system was constructed using GOx immobilized in the Mag-MCF-C. A ferrocene derivative was employed as a soluble electron mediator. By changing the location of Mag-MCF-C using an external magnetic field, the current signal could be switched repeatedly without loss of signal (*36*).

Conclusion and Perspectives

The unique structural properties and inherent electrical conductivity of OMC enabled the development of highly sensitive and stable biosensors. Further improvement can be made in some aspects. For example, electron mediators co-immobilized with enzymes can improve the electron-transfer efficiency. However, direct incorporation of small-size redox mediators is difficult because they can be easily leached out owing to the relatively large pore size of OMCs and the chemical modification of carbon surface for covalent attachment of mediators is not a simple task. Metallic nanoparticles (*9, 39*) or polymers modified with mediators (*6*) might be used as alternatives. OMCs can have applications other than enzyme-based biosensors. In immunoassay, antibodies labeled with enzymes are widely used for sensitive detection. By increasing the number of enzyme labels, the detection signal can be further amplified (*40, 41*). Due to the high protein loading capacity of OMCs, it is possible to immobilize probe proteins, such as antibodies or avidin, together with a large number of enzyme labels that transform the recognition events into highly amplified optical or electrochemical signals. It is expected that through the combination of these possibilities, OMCs can find wide applications in the field of enzyme-based biosensing system.

Acknowledgment

We acknowledge financial supports from the Nano/Bio Science & Technology Program (M1053609000205N3609-00210) and the Nano Science & Technology Program (M10503000218-05M0300-21810) of MOST, the Korea Health 21C R&D Project (0405-MN01-0604-0007) of MHW, and the Next-Generation New-Technology Development Program of MOCIE. We appreciate Michael B. Fischback for his help in the preparation of manuscript.

References

1. Bornscheuer, U. T. *Angew. Chem. Int. Ed.* **2003**, *42*, 3336-3337.
2. Adhikari, B.; Majumdar, S. *Prog. Polym. Sci.* **2004**, *29*, 699-766.
3. Zhang, M.; Smith, A.; Gorski, W. *Anal. Chem.* **2004**, *76*, 5045-5050.
4. Beissenhirtz, M. K.; Scheller, F. W.; Lisdat, F. *Anal. Chem.* **2004**, *76*, 4665-4671.
5. Yoon, H. C.; Kim, H. S. *Anal. Chem.* **2000**, *72*, 922-926.
6. Yoon, H. C.; Hong, M. Y.; Kim, H. S. *Anal. Chem.* **2000**, *72*, 4420-4427.
7. Habermüller, K.; Mosbach, M.; Schuhmann, W. *Fresenius J. Anal. Chem.* **2000**, *366*, 560–568.
8. Heller, A. *Nat. Biotechnol.* **2003**, *21*, 631–632.
9. Xiao, Y.; Patolsky, F.; Katz, E.; Hainfeld, J. F.; Willner, I. *Science* **2003**, *299*, 1877-1881.
10. Joshi, P. P.; Merchant, S. A.; Wang, Y.; Schmidtke, D. W. *Anal. Chem.* **2005**, *77*, 3183-3188.
11. Zhang, M.; Gorski, W. *J. Am. Chem. Soc.* **2005**, *127*, 2058-2059.
12. Sotiropoulou, S.; Gavalas, V.; Vamvakaki, V.; Chaniotakis, N. A. *Biosens. Bioelectron.* **2003**, *18*, 211-215.
13. Lee, J.; Han, S.; Hyeon, T. *J. Mater. Chem.* **2004**, *14*, 478-486.
14. Ryoo, R.; Joo, S. H.; Kruk, M.; Jaroniec, M. *Adv. Mater.* **2001**, *13*, 677-681.
15. Vinu, A.; Streb, C.; Murugesan, V.; Hartmann, M. *J. Phys. Chem. B* **2003**, *107*, 8297-8299.
16. Dai, Z.; Xu, X.; Ju, H. *Anal. Biochem.* **2004**, *332*, 23-31.
17. Xu, X.; Tian, B.; Kong, J.; Zhang, S.; Liu, B.; Zhao, D. *Adv. Mater.* **2003**, *15*, 1932-1936.
18. Lee D.; Lee, J.; Kim, J.; Kim, J.; Na, H. B.; Kim, B.; Shin, C. –H.; Kwak, J. H.; Dohnalkova A.; Grate, J. W.; Hyeon, T.; Kim, H. –S. *Adv. Mater.* **2005**, *17*, 2828-2833.

19. Hecht, H. J.; Kalisz, H. M.; Hendle, J.; Schmid, R. D.; Schomburg, D. *J. Mol. Biol.* **1993**, *229*, 153-172.
20. Kim, S. –S.; Pauly, T. R.; Pinnavaia, T. J. *Chem. Commun.* **2000**, 1661-1662.
21. Takahashi, H.; Li, B.; Sasaki, T.; Miyazaki, C.; Kajino, T.; Inagaki, S. *Microporous Mesoporous Mater.* **2001**, *44-45*, 755-762.
22. Wang, Y.; Caruso, F. *Chem. Mater.* **2005**, *17*, 953-961.
23. Corma, A.; Garcia, H. *Eur. J. Inorg. Chem.* **2004**, 1143-1164.
24. Weber, W. A.; Phillips, B. L.; Gates, B. C. *Chem. Eur. J.* **1999**, *5*, 2899-2913.
25. Yamamoto, T.; Shido, T.; Inagaki, S.; Fukushima, Y.; Ichikawa, M. *J. Am. Chem. Soc.* **1996**, *118*, 5810-5811.
26. Lee, J.; Kim, J.; Kim, J.; Jia, H.; Kim, M. I.; Kwak, J. H.; Jin, S.; Dohnalkova, A.; Park, H. G.; Chang, H. N.; Wang, P.; Grate, J. W.; Hyeon, T. *Small* **2005**, *1*, 744-753.
27. Vinu, A.; Murugesan, V.; Tangermann, O.; Hartmann, M. *Chem. Mater.* **2004**, *16*, 3056-3065.
28. Fan, J.; Yu, C.; Gao, F.; Lei, J.; Tian, B.; Wang, L.; Luo, Q.; Tu, B.; Zhou, W.; Zhao, D. *Angew. Chem. Int. Ed.* **2003**, *42*, 3146-3150.
29. Fan, J.; Lei, J.; Wang, L.; Yu, C.; Tu, B.; Zhao, D. *Chem. Commun.* **2003**, 2140-2141.
30. Lei, C.; Shin, Y.; Liu, J.; Ackerman, E. J. *J. Am. Chem. Soc.* **2002**, *124*, 11242-11243.
31. Han, Y. –J.; Watson, J. T.; Stucky, G. D.; Butler, A. *J. Mol. Catal. B: Enzym.* **2002**, *17*, 1-8.
32. Días, J. F.; Balkus Jr., K. J. *J. Mol. Catal. B: Enzym.* **1996**, *2*, 115-126.
33. Nomura, A.; Shin, S.; Mehdi, O. O.; Kauffmann, J. –M. *Anal. Chem.* **2004**, *76*, 5498-5502.
34. Wang, J.; Musameh, M.; Lin, Y. *J. Am. Chem. Soc.* **2003**, *125*, 2408-2409.
35. Karyakin, A. A.; Kotel´nikova, E. A.; Lukachova, L. V.; Karyakina, E. E. *Anal. Chem.* **2002**, *74*, 1597-1603.
36. Lee, J.; Lee, D.; Oh, E.; Kim, J.; Kim, Y. –P.; Jin, S.; Kim, H. –S.; Hwang, Y.; Kwak, J. H.; Park, J. –G.; Shin, C. –H.; Kim, J.; Hyeon, T. *Angew. Chem. Int. Ed.* **2005**, *44*, 7427-7432.
37. Willner, I.; Katz, E. *Angew. Chem. Int. Ed.* **2003**, *42*, 4576-4588.
38. Geng, J.; Jefferson, D. A.; Johnson, B. F. G. *Chem. Commun.* **2004**, 2442-2443.
39. Gan, X.; Liu, T.; Zhong, J.; Liu, X.; Li, G. *ChemBioChem* **2004**, *5*, 1686-1691.
40. Wang, J.; Liu, G.; Jan, M. R. *J. Am. Chem. Soc.* **2004**, *126*, 3010-3011.
41. Munge, B.; Liu, G.; Collins, G.; Wang, J. *Anal. Chem.* **2004**, *77*, 4662-4666.

Chapter 15

Continuous-Flow Applications of Silica-Encapsulated Enzymes

Heather R. Luckarift[1] and Jim C. Spain[2,*]

[1]Air Force Research Laboratory, 139 Barnes Drive, Suite # 2, Tyndall Air Force Base, FL 32403
[2]School of Civil and Environmental Engineering, Georgia Institute of Technology, 311 Ferst Drive, Atlanta, GA 30332

Recent studies have demonstrated the applicability of biomineralization reactions to create an inorganic support matrix suitable to enzyme immobilization. The enzyme/ inorganic nanocomposites exhibit excellent mechanical stability and provide an effective method for developing immobilized enzyme reactors, applicable to biocatalysis, biosensors and drug discovery.

Introduction

Enzymes are remarkably versatile catalysts, but in their native soluble form are often labile in the absence of stabilizing agents and are difficult to recover from reaction mixtures. Immobilization of enzymes is therefore frequently employed in an attempt to stabilize enzyme activity and allow reuse of the catalyst. Enzyme immobilization methods primarily involve adsorption, attachment or encapsulation of biomolecules onto or into a solid phase *(1-7)*. A range of silicates have been investigated for enzyme immobilization, either by attachment to functionalized mesoporous silica or encapsulation within sol-gel composites, but processing limitations have restricted widespread applicability *(2,4,6,7)*.

Silicification for enzyme immobilization

Silaffin polypeptides in diatoms catalyze the biomineralization of silica to form the exoskeleton *(8,9)*. The biosilicification reaction can be mimicked *in vitro* by utilizing synthetic peptides (e.g. R5) based on the native silaffin sequence or from silica-binding peptides identified from combinatorial peptide libraries *(8-13)*. Silica formation is also observed in the presence of simple cationic polymers such as polyethyleneimine and by proteins such as lysozyme and silicateins, producing silica nanospheres with a range of morphologies (Figure 1) *(14-18)*.

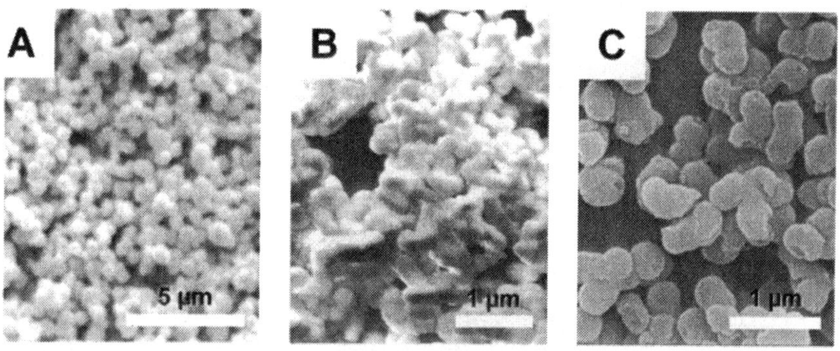

Figure 1. Synthesis of silica nanospheres catalyzed by tetramethylorthosilicate and (a) polyethyleneimine, (b) R5 peptide and (c) lysozyme , as viewed by SEM analysis. (From references 15, 18, 19).

The silica-precipitating species becomes entrapped during the generation of the silica matrix suggesting the potential of the silicification reaction to also encapsulate additional enzymes inside the silica matrix. In practice, the mild encapsulation chemistry and high biocompatibility of the reaction provide a rapid and highly efficient method for immobilizing a wide range of biomolecules (Table I). The biomimetic silicification reaction yields a network of fused silica nanospheres, providing a high surface area for encapsulation and permitting high enzyme loading capacities of up to 20% w/w *(19)*.

The exceptional stability of the silica-immobilized enzymes under operational conditions, dramatically increases the versatility of the biocatalysts. Silica-immobilized butyrylcholinesterase (BuChE), for example, could be stored in aqueous solution at room temperature with no loss of initial enzyme activity, whereas free enzyme under identical conditions lost activity rapidly. The thermostability of the immobilized enzyme was also significantly enhanced. Silica-immobilized BuChE for example retained enzyme activity after heat-

Table I. Immobilization efficiency of a range of enzymes in silica nanoparticles

Enzyme	*Immobilization efficiency (expressed enzyme activity)*	*Data from Reference*
Butyrylcholinesterase	> 90%	*(19)*
Catalase	100%	*(20)*
Soybean peroxidase	65 - 85%	
Horseradish peroxidase	> 90%	*(20)*
Bromoperoxidase	35 - 48%	
Hydroxylaminobenzene mutase	44 - 67%	*(21)*
Organophosphate hydrolase	25 - 35%	
Nitrobenzene nitroreductase	~ 80% [a]	*(15)*

[a] Immobilized in silica formed from polyethyleneimine and tetramethylorthosilicate (TMOS). All other enzymes are immobilized in silica formed from R5 peptide and TMOS.

treatment of up to 65°C; conditions which caused rapid denaturation of soluble-BuChE. The enhanced enzyme stability can be attributed to the stabilizing effect of the silica support matrix, which prevents the conformational changes typical of enzyme denaturation *(19)*.

Biotechnological application to continuous flow systems

A stable immobilized-enzyme preparation is attractive for a wide range of applications, particularly facilitating application to continuous flow-systems. Enzymes catalyze a wide variety of processes that can be exploited for example, in biocatalysis; for the production of novel synthons or drug intermediates. Enzymes also possess a wide range of pharmacological activities and are often investigated for therapeutic effects in drug discovery. Inhibitors of cholinesterase enzymes for example, can be used for the treatment of disorders such as Alzheimer's disease (*22,23*) and nitroreductase enzymes are key activators of prodrugs for cancer therapy *(24,25)*. The applicability of silica-encapsulated enzymes was, therefore, further evaluated with respect to the specific systems described above to provide insight into the versatility of the method.

Immobilized Enzyme Reactors for Cholinesterase Inhibition Studies

Immobilization of enzymes in packed columns specifically designed for continuous flow systems are often referred to as immobilized enzyme reactors (IMERs) *(26-30)*. IMERs consisting of immobilized cholinesterase for example have been investigated in drug screening to identify inhibitors for treatment of disorders of the central nervous system, such as Alzheimer's disease. Current IMER configurations however, often exhibit specific drawbacks such as low loading capacity and long preparation times *(31-34)*.

An IMER consisting of silica-immobilized BuChE was investigated in an attempt to circumvent some of the current limitations of IMER preparations. Silica-immobilized BuChE was prepared in two alternate column configurations; 1) a fluidized bed and 2) a packed-bed. For the fluidized bed system, substrate conversion was complete for over 12 hours of continuous flow with no loss in enzyme activity or conversion efficiency. The fluidized-bed system could also be operated at higher flow rates with no loss in activity, but with comparably lower conversion efficiency due to a reduced contact time within the column. In the packed-bed system, however, the conversion rate decreased with time. The silica particles became packed under continuous flow conditions, leading to compression and eventual channeling of the silica particles *(19)*. Thus, the mechanical stability of the silica-immobilized enzyme was well suited to flow-through systems but the configuration of the column packing required optimization.

In order to avoid the above mechanical limitations, the silica-immobilization technique was modified such that the silica particles form and attach simultaneously to a commercial pre-packed column via affinity binding of a histidine-tag on the silica-precipitating peptide *(35)* (Figure 2). The modified method was used to prepare a butyrylcholinesterase immobilized enzyme reactor (BuChE-IMER). A metal ion affinity chromatography column charged with cobalt ions selectively binds histidine residues on proteins or peptides. A $(His)_6$-homologue of the R5 peptide therefore selectively binds to the cobalt coated surface. The silicification reaction occurs and integrates with the peptide bound to the column, resulting in formation of silica nanospheres attached to the surface of the agarose beads and the concurrent immobilization of the enzyme (Figure 2). The location of the histidine-tag on the silica-nucleating peptide rather than on the protein eliminates any need for recombinant modification of the protein of interest.

The addition of histidine residues at either the carboxyl-terminus or amino-terminus of the silica-precipitating peptide did not affect the silicification reaction. The resulting BuChE-IMER exhibited high loading capacities and an immobilization efficiency approaching 100%. When connected to a liquid chromatography system, the columns could be operated at a wide range of flow rates (up to 3 ml/min) with low back pressure (Figure 3). Multiple substrate injections by means of an auto sampler provided rapid and reproducible analysis, with no significant loss in enzyme activity or conversion efficiency during continuous flow.

The hydrolysis of substrate by cholinesterases is decreased by the presence of inhibitors from which inhibition constant (IC_{50}) values can be derived. The

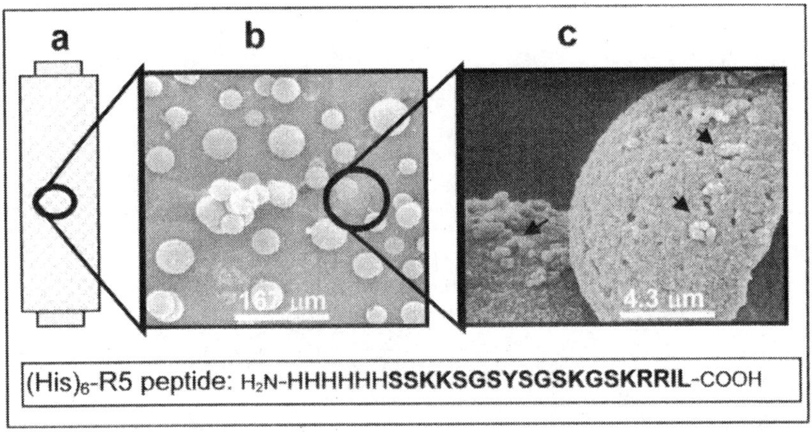

Figure 2. Immobilization of silica beads to agarose beads via affinity binding. A pre-packed affinity chromatography column (a) containing agarose beads (b) is coated with silica (c) by incubation of a histidine tagged-synthetic R5 peptide and TMOS (as viewed by SEM analysis). The $(His)_6$-R5 peptide sequence is shown. Data from reference 35.

BuChE-IMER can thus be utilized for rapid analysis of inhibition characteristics. A range of cholinesterase inhibitors were investigated and exhibited a concentration-dependent response, from which inhibition constants could be determined (shown in Figure 3b for the inhibitor, Tacrine). The IMER was stable for more than 50 hours of continuous use. In addition, the reusability of the IMER significantly reduces the amount of enzyme required for analysis.

The bioencapsulation strategy described above provides a rapid route for synthesizing IMER systems and provides a model system applicable to a range of formats. The method is also scalable to applications ranging from a microfluidic format for biosensors to large-scale for biocatalysis. Two further examples of silica-immobilized enzymes in packed columns for 1) drug discovery and 2) biocatalysis for drug synthesis will be discussed in more detail below.

Microfluidic Immobilized Enzyme Reactors for Drug Discovery

Nitroreductase enzymes are used to activate prodrugs to a cytotoxic derivative specific to tumor cells (*24,25*). Nitroreductase enzymes catalyze the reduction of a nitro group (strongly electron-withdrawing) to the corresponding hydroxylamine (electron-donating), which results in a large electronic change and provides an effective enzyme-mediated electronic 'trigger'. Nitroreductase

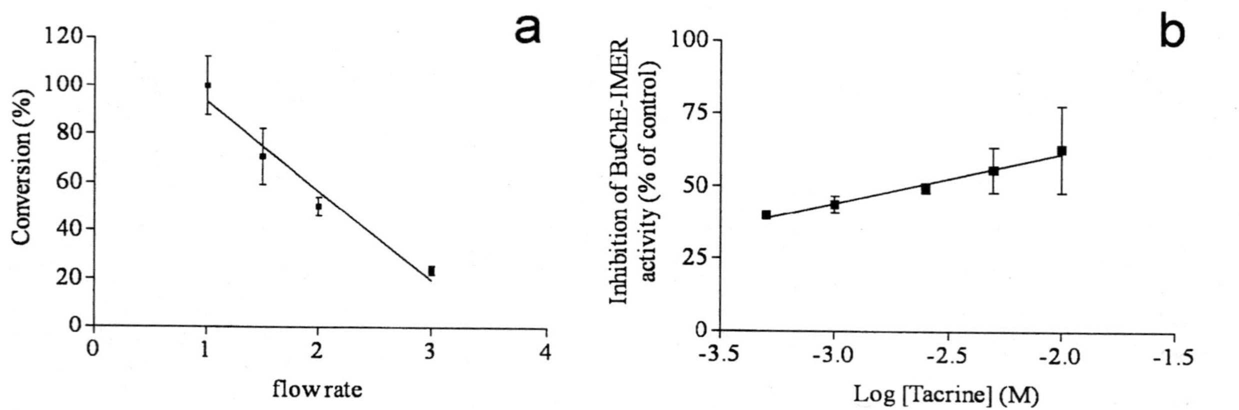

Figure 3. BuChE-IMER activities during continuous flow. Panel (a): Effect of flow rate on the conversion efficiency; (b) Determination of IC_{50} for tacrine using the BuChE-IMER. Data from reference 35

enzymes are also known to activate nitrofuran antibiotics by reduction to the corresponding hydroxylamine intermediate, which causes the fragmentation of DNA *(36,37)*. Despite the pharmacological relevance of nitroreductases, there are few reports documenting immobilization of such enzymes for drug discovery.

Nitroreductase enzymes were encapsulated in silica formed by a simple cationic polymer; polyethyleneimine (PEI). PEI precipitates silica in a reaction homologous to biogenic systems but with a significant reduction in cost *(15)*. The resulting silica particles proved suitable for encapsulation of nitrobenzene nitroreductase (NbzA) from *P. pseudoalcaligenes* JS45 with immobilization yields of greater than 80% and high retention of enzyme activity (45 – 55 %) *(15)*.

The main selection criteria for prodrug formulations are a high affinity for substrate and a differential toxicity between the active species and the prodrug. The affinity of the immobilized-NbzA was therefore determined for nitrobenzene, an anticancer prodrug (CB1954) and a proantibiotic (nitrofurazone). The kinetic properties of the immobilized NbzA were comparable to those of the soluble enzyme, indicating that immobilization was not detrimental to enzyme activity. Nitrofurazone was a poor substrate for NbzA (high K_m value) whereas the K_m value for CB1954 was very low, indicating a high affinity for substrate activation of CB1954 in comparison to other bacterial nitroreductases (Table II).

Table II. Kinetic characteristics of nitroreductase enzymes

	E. coli (NTR)	*P. pseudoalcaligenes JS45 (NbzA)*	
		Soluble	Silica-Immobilized
Nitrobenzene	ND	2.3 (± 0.35)	2.0 (±0.23)
CB1954	862	11.7 (± 1.0)	33.7 (± 4.5)
Nitrofurazone	64	1763 (± 572)	5123 (± 687)

Data from reference 15. All data represent K_m values in μM

As demonstrated previously for encapsulation of BuChE, the silica-immobilization method conferred enhanced stability to NbzA. Immobilized NbzA retained enzyme activity when stored at 4°C for several weeks and exhibited dramatically higher thermostability than the soluble enzyme *(15)*. The enhanced stability in this system is thought to be a consequence not only of the physical support provided by the silica matrix but also the protective nature of PEI itself *(38-40)*.

The small size (< 1 μm diameter) of the silica-encapsulated NbzA particles provides a high surface to volume ratio considered suitable for microfluidic flow-through systems *(41)*. Silica-encapsulated NbzA was therefore packed into a microfluidic device and demonstrated high conversion efficiencies under continuous flow conditions and at a range of flow rates. At 1 $\mu l/min^{-1}$, for example nitrobenzene, CB1954 and nitrofurazone were all converted stoichiometrically and conversion of nitrobenzene (>90%) could be maintained for more than 3 days of continuous operation.

Immobilized Enzyme Reactors in Biocatalysis

In whole cells of *Pseudomonas pseudoalcaligenes* JS45, the NbzA described above reduces nitrobenzene to hydroxylaminobenzene (HAB), which undergoes further transformation by HAB mutase to form *ortho*-aminophenol (Figure 4). The activity of NbzA and HAB mutase in concert catalyze the conversion of a range of nitroarenes to yield novel *ortho*-aminophenols *(42,43)*, but the use of NbzA for biocatalysis is limited by its requirement for NADPH. The NADPH-dependent reduction of nitroarenes can be replaced however, by a zinc-catalyzed chemical reduction. HAB can then be enzymatically rearranged to *ortho*-aminophenol by HAB mutase, an enzyme with no cofactor requirements.

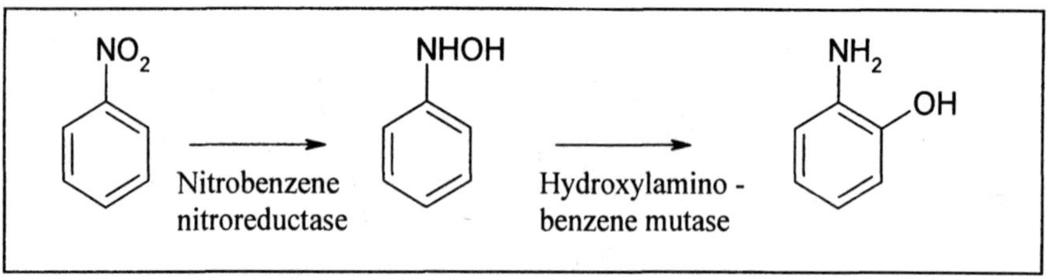

Figure 4. Enzymatic synthesis of ortho-aminophenol

To demonstrate the applicability of the approach, HAB mutase was immobilized and packed into a column and connected in series to a second column containing zinc. Nitrobenzene (1 mM) was pumped through the two columns and *ortho*-aminophenol was produced continuously for over 5 hours with a conversion efficiency approaching 90%. The flow-through system could be operated at higher flow rates and substrate concentration (5 mM at 0.5 ml/min) with a conversion efficiency of approximately 70%, which could be maintained for over 8 hours *(21)*.

The use of the zinc/mutase flow-through columns was applied to the formation of a novel antibiotic. The biosynthesis of antibiotics using bacterial cells is limited due to the biocidal properties of the product, for which the immobilized enzyme system provides an attractive alternative. Chloramphenicol contains an active nitro substituent which was converted stoichiometrically to the corresponding aminophenol analog by passage through the zinc and immobilized mutase column in series. At a flow rate of 0.25 ml/min, continuous synthesis of the novel product was maintained for a period of 24 hours *(21)*.

Conclusion

Enzyme immobilization methods have been widely investigated for many years, but recent developments in stabilizing enzymes within biomimetic inorganic matrices substantially extends the range of operational stabilities. Nano-sized materials offer a number of intrinsic advantages such as high surface areas which lead to high loading capacities. Silica-encapsulation has proven to be a versatile method for immobilizing biocatalytic activity and is applicable to a wide range of biomolecules. In addition, the morphology of the particles can be controlled by modifying the reaction conditions during silica formation *(13, 44-46)*. The primary limitation to broad application is the cost of synthesizing peptides required for silica formation. The use of synthetic polyamines for the silica precipitation reaction however, provides a significant reduction in cost and a realistic opportunity to develop the method for large-scale synthesis of immobilized enzyme preparations. The use of lysozyme to catalyze the silica precipitation *(18)* also imparts the additional benefit of antimicrobial activity to the resulting silica nanoparticles. Lysozyme/silica composites therefore provide antifouling properties to the encapsulated catalysts, for potential use as antibacterial coatings. We anticipate that the resulting silica-encapsulated catalysts will find significant and widespread application in the design of biosensors and for biocatalysis and drug discovery.

References

1. Cao. L. *Curr. Opin. Chem. Biol.* **2005**, *9*, 217-226.
2. Gill, I.; Ballesteros, A. *Trends. Biotech.* **2000**, *18*, 282-296.
3. Gill, I.; Ballesteros, A. *Trends. Biotech.* **2000**, *18*, 469-479.
4. Kim, Y.D.; Dordick, J.S.; Clark, D.S. *Biotechnol. Bioeng.* **2001**, *72*, 475-482.
5. Lei, C.; Shin, Y.; Liu, J.; Ackerman, E.J. *J. Am. Chem. Soc.* **2002**, *124*, 11242-11243.
6. Avnir, D.; Braun, S.; Lev, O.; Ottolenghi, M. *Chem. Mater.* **1994**, *6*, 1605-1614.
7. Gill, I.; Ballesteros, A. *J. Am. Chem. Soc.* **1998**, *120*, 8587-8598.
8. Kroger, N.; Duetzmann, R.; Sumper, M. *Science* **1999**, *286*, 1129-1132.

9. Kroger, N.; Lorenz, S.; Brunner, E.; Sumper, M. *Science* **2002**, *298*, 584-586.
10. Naik, R.R.; Brott, L.L.; Clarson, S.J.; Stone, M.O. *J. Nanosci. Nanotech.* **2002**, *2* (1), 95-100.
11. Cha, J.N.; Stucky, G.D.; Morse, D.E.; Deming, T.J. *Nature* **2000**, *403*, 289-292.
12. Cha, J.N.; Shimizu K.; Zhou Y.; Christiansen S.C.; Chmelka B.F.; Stucky G.D.; Morse D.E. *Proc. Natl. Acad. Sci. U S A.* **1999**, *96*(2), 361-365.
13. Tomczak, M.M.; Stone, M.O.; Naik, R.R. In *Nano-scale science and Technology in Biomolecular Catalysis*. Wang, P.; Kim, J. ACS symposium series; American Chemical Society, Washington DC.
14. Belton, D.J.; Patwardhan, S.V.; Perry, C.C. *J. Mater. Chem.* **2005**, *15*, 4629-4638.
15. Berne, C., Betancor, L., Luckarift, H.R., Spain, J.C. *Biomacromol.* **2006**, *In Press.*
16. Roth, K.M.; Zhou, Y.; Yang, W.; Morse, D.E. *J. Am. Chem. Soc.* **2005**, *127* (1), 325-330.
17. Shimizu, K.; Cha, J.; Stucky, G.D.; Morse, D.E. *Proc. Nat. Acad. Sci. USA* **1998**, *95*, 6234-6238.
18. Luckarift, H.R.; Dickerson, M.B.; Sandhage, K.H.; Spain, J.C. *Small* **2006**, *2* (5), 640-643.
19. Luckarift, H.R.; Spain, J.C.; Naik, R.R.; Stone, M.O. *Nat. Biotech.* **2004**, *22* (2), 211-213.
20. Naik, R.R.; Tomczak, M.M.; Luckarift, H.R.; Spain, J.C.; Stone, M.O. *Chem. Commun. (Cambridge)* **2004**, 1684 -1685.
21. Luckarift, H.R.; Nadeau, L.J.; Spain, J.C. *Chem. Commun. (Cambridge)* **2005**, 383-384.
22. Holden, M.; Kelly, C. *Adv. Psychiatr. Treat.* **2002**, *8*, 89-96.
23. Liston, D.R.; Nielsen, J.A.; Villalobos, A.; Chapin, D.; Jones, S.B.; Hubbard, S.T.; Shalaby, I.A.; Ramirez, A.; Nason, D.; Frost White, W. *Eur. J. Pharmacol.* **2004**, *486*, 9-17.
24. Denny, W.A. *Curr. Pharm. Des.* **2002**, *8*, 1349-1361.
25. Anlezark, G.M.; Melton, R.G.; Sherwood, R.F.; Coles, B.; Friedlos, F.; Knox, R.J. *Biochem. Pharmacol.* **1992**, *44*, 2289-2295.
26. Urban, P.L.; Goodall, D.M.; Bruce, N.C. *Biotech. Adv.* **2006**, *24* (1), 42-57.
27. Krenkova, J.; Foret, F. *Electrophoresis.* **2004**, *25*, 3550-3563.
28. Calleri, E.; Temporini, C.; Furlanetto, S.; Loidice, F.; Fracchiolla, G. ; Massolini, G. *J. Pharma. Biomed. Analysis* **2003**, *32*, 715-724.
29. Girelli, A.M.; Mattei, E.J. *J. Chromatogr. B* **2005**, *819*, 3.
30. Markoglou, N.; Wainer, I.W. *J Chromatogr A.* **2002**, *948*(1-2), 249-56.
31. Bartolini, M.; Cavrini, V.; Andrisano, V. *J. Chromatogr. A.* **2004**, *1031*, 27-34.

32. Bartolini, M.; Cavrini, V.; Andrisano, V. *J. Chromatogr. A.* **2005**, *1065*, 135-144.
33. Andrisano, V.; Bartolini, M.; Gotti, R.; Cavrini, V.; Felix, G. *J. Chromatogr. B.* **2001**, *753*, 375-383
34. Dong, Y.; Wang, L.; Shangguan, D.; Zhao, R.; Liu, G. *J. Chromatogr. B.* **2003**, *788*, 193-198.
35. Luckarift, H.R.; Johnson, G.R.; Spain, J.C. *J. Chromatogr. B.* **2006**, *In Press.* doi: 10.1016/j.jchromb.2006.06.036
36. Whiteway, J.; Koziarz, P.; Veall, J.; Sandhu, N.; Kumar, P.; Hoecher, B.; Lambert, I.B. *J. Bacteriol.* **1998**, *180*, 5529-5539.
37. Jenks, P.J.; Ferrero, R.L.; Tankovic, J.; Thiberge, J.M.; Labigne, A. *Antimicrob. Agents. Chemother.* **2000**, *44*, 2623-2629.
38. Jin, R-H.; Yuan, J-J. *Macromol. Chem. Phys.* **2005**, *206*, 2160-2170.
39. Andersson, M.M.; Hatti-Kaul, R. *J. Biotechnol.* **1999**, *72*, 21-25.
40. Lopez-Gallego F.; Betancor L.; Hidalgo A.; Alonso N,; Fernandez-Lafuente R,; Guisan J.M. *Biomacromol.* **2005**, *6*, 1839-1842.
41. Srinivasan, A.; Wu, X.; Lee, M-Y.; Dordick, J.S. *Biotech. Bioeng.* **2003**, *81* (5), 563-569.0020
42. Nadeau, L.J.; He, Z.; Spain, J.C. *J. Ind. Micro. Biotech.* **2000**, *24*, 301-305.
43. Kadiyala, V.; Nadeau, L.J.; Spain, J.C. *Appl. Environ. Micro.* **2003**, *69*, 6520-6526.
44. Naik, R.R.; Whitlock, P.W.; Rodriguez, F.; Brott, L.L.; Glawe, D.D.; Clarson, S.J.; Stone, M.O. *Chem. Commun. (Cambridge)* **2003**, 238-239.
45. Tomczak, M.M.; Glawe, D.D.; Drummy, L.F.; Lawrence, C.G.; Stone, M.O.; Perry, C.C.; Pochan, D.J. *J. Am. Chem. Soc.* **2005**, *127* (36), 12577-12582
46. Dickerson, M.B.; Naik, R.R.; Sarosi, P.M.; Agarwal, G.; Stone, M.O.; Sandhage, K.H. J. *Nanosci. Nanotechnol.* **2005**, *5*(1), 63-67.

Chapter 16

Enzyme–Nanofiber Composites for Biocatalysis Applications

Byoung Chan Kim[1,6], Sujith Nair[2], Seong H. Kim[2], Byoung In Sang[3], Jungbae Kim[4,*], and Man Bok Gu[5,*]

[1]Advanced Environmental Monitoring Research Center (ADEMRC), Gwangju Institute of Science and Technology (GIST), Gwangju, Republic of Korea
[2]Department of Chemical Engineering, Pennsylvania State University, University Park, PA 16802
[3]Hazardous Substances Research Center, Korea Institute of Science and Technology, Seoul 136–791, Korea
[4]Pacific Northwest National Laboratory, 902 Battelle Boulevard, Richland, WA 99352
[5]School of Life Sciences and Biotechnology, Korea University, Anam-dong, Sungbuk-gu, Seoul 136–701, Republic of Korea
[6]Current address: Diagnostics Group, Institut Pasteur Korea, Hawolgok-dong, Sungbuk-gu, Seoul 136–791, Republic of Korea

Nanofibers fabricated via electrospinning have drawn much attention nowadays due to its large surface area, and easy production of nanoscaled structure for interaction with biomolecules. This chapter presents an overview for the recent applications of nanofibers as scaffolds for enzyme immobilization. This chapter contains a) brief introduction to conventional nanofiber production via electrospinning, b) recent application of nanofibers as scaffolds for enzyme immobilization and c) its application to biocatalytic system.

Nanometer-sized materials are of considerable interest in analytical chemistry and biotechnology due to their potential applications in bioassay, medical diagnostics, and catalysis. The recent advances in the creation of new nanomaterials and nanoscaled devices trigger off the new direction of biological research (*1*). The nanofibers for the application in many biotechnology areas have drawn much attention nowadays. Depending on polymer materials used, various surface functionalized nanofibers could be produced. In this chapter, an overview on the recent application of nanofibers for biological applications, especially as scaffolds for enzyme immobilization, is presented.

Electrospinning for the production of nanofibers

The electrospinning, melt-blow, phase separation, self-assembly and template synthesis methods have been used to produce polymer nanofibers for different usage for a long time (*2-6*). Among these methods, electrospinning method uses very simple apparatus and units for operation, and very flexible in the range of polymeric materials selections. Typically, electrospinning apparatus consists of capillary connected with a metallic needle, a high voltage supplier, and conductive collectors. When high voltage is applied, polymer solution in the capillary under electrostatic repulsion and Columbic force excluded on the surface of collectors from the needle. During the elongation of polymer solution on the collectors, the solvent is evaporated and nano-sized fiber is deposited in non-woven mat forms randomly (Figure 1.). The property of nanofibers generated depends on viscosity of polymer solution, the distance between the collector and needle, the strength of voltage applied, and the diameter of the capillary used. Depending on the electrospinning apparatus used, nanofibers could be also fabricated in a hollowed fiber form, porous form, or core-sheath form, respectively (*7-10*) .

Many polymeric materials can be used for producing nanofibers via electrospinning process. Early work on nanofiber fabrication with the electrospinning method dealt with organic polymers mostly. Recently, materials used for electrospinning process expanded to inorganic/polymer or nanomaterials to confer new physicochemical characters to nanofibers. Furthermore, nanofibers could be functionally modified with a variety of different materials having surface functional groups, which were derived from polymer itself after electrospinning process (*8*). For many applications, nanofibers could be controlled in spatial orientation of 1D structure based on the collector format. Rotating drum as collectors allowed that nanofibers were oriented to each other in parallel (*11*). Li and Xia demonstrated that the production of geometrical alignment of nanofibers using pair of split electrode. By using spatial electrode, nanofibers could be aligned axially between electrodes (*8,12*).

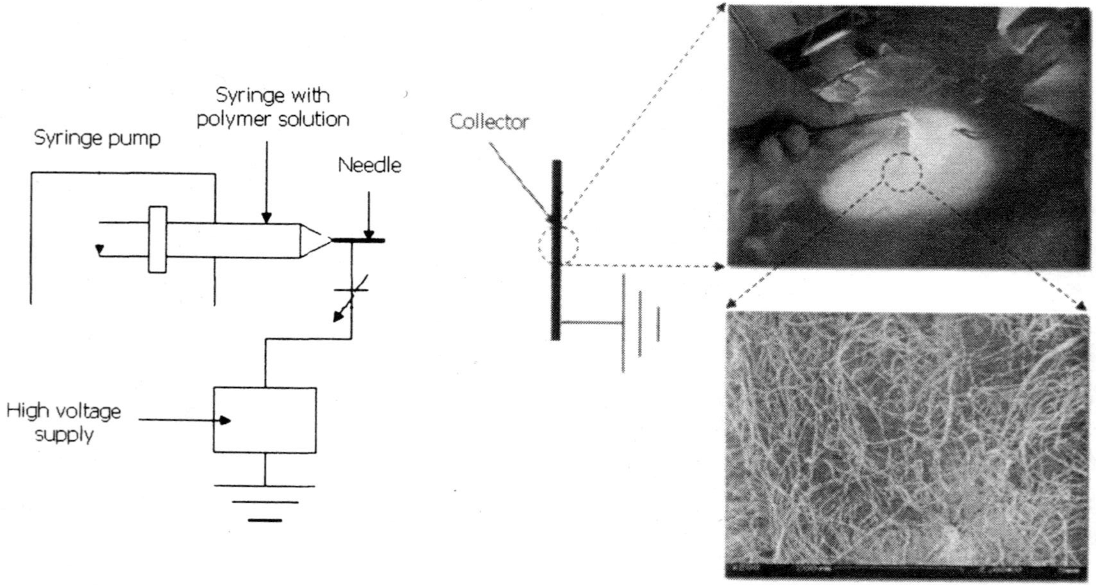

Figure 1. Typical process of nanofiber processing via electrospinning methods. Adapted with permission from Reference [17]. Copyright 2005 IOP Publishing Ltd.

Strategies on enzyme immobilization with nanofibers

Nanofibers supply large surface areas and small size-strings for immobilization of biomolecules. Concerning the practical approach of enzyme immobilization, the bigger surface to volume ratio from supporting scaffolds is desirable to achieve high catalytic efficiency. Therefore, it is quite hopeful to use nanofibers as scaffolds to confer high enzyme loading in immobilization. Also, the porous nanofibers can reduce diffusional resistance of the substrate in the reaction. In terms of polymeric materials selected, various nanofibers could be produced easily (*13,14*), depending on the demanding characteristic between enzymes and nanofibers. As mixing different polymer solutions, nanofibers could be produced directly via electrospinning and they could be used as scaffolds for biomolecules immobilization without further functionalization process. Because of these attractive features, electrospun nanofibers have drawn much attention as scaffolds for enzyme immobilization. Stability is also important for the application of biocatalytic system. Generally speaking, immobilized enzyme may have good stability in process than free enzyme, but their reusability still is in question. The scheme of enzyme immobilization on nanofibers followed conventional enzyme immobilization methods, which are covalent attachment, enzyme aggregate coating or entrapment in solid supports (Figure 2). There are several strategies studied for efficient enzyme immobilization with nanofibers.

Jia and co-workers pioneered work of enzyme immobilization on the surface of nanofibers and resulted in high loading capacity of enzymes because of large surface area of nanofibers (*15*). They achieved functionalization of polystyrene nanofibers after electrospinning and α-chymotrypsin was immobilized with 27 % coverage of enzyme molecules on the external surface of nanofibers. The activity of immobilized enzymes on the nanofibers was improved in both aqueous and organic solvents such as hexane and isooctane. The half-life of the covalently attached enzyme on the surface of nanofibers in organic solvents was 18-fold higher than that of free enzymes, suggesting that the covalent bonding enhanced the enzyme stability (Figure 2 (a)). Other approach was also introduced the immobilization of enzyme on the surface of nanofibers. Poly (arculonitrile-co-maleci acid) (PANCMA) was used to fabricate nanofibers for enzyme immobilization by Ye and co-workers (*16*). In their work, lipase from *Candida rugosa* was immobilized via activating carboxylic groups of PANCMA nanofibers with 1–ethyl–3–((dimethylamono) propyl) carodiimide hydrochloride/N-hydroxylsuccinimide (EDC/NHS) coupling reaction. The diameter of PANCMA nanofiber was controlled from 100 nm to 600 nm by changing concentration of PANCMA in solution. Depending on fiber diameters, amounts of enzyme loading could be controlled. However, this needs further functionalization after nanofiber fabricated for enzyme immobilization.

Recently, Kim and co-workers reported successful development of enzyme-nanofiber composite for highly active and stable biocatalysis application by

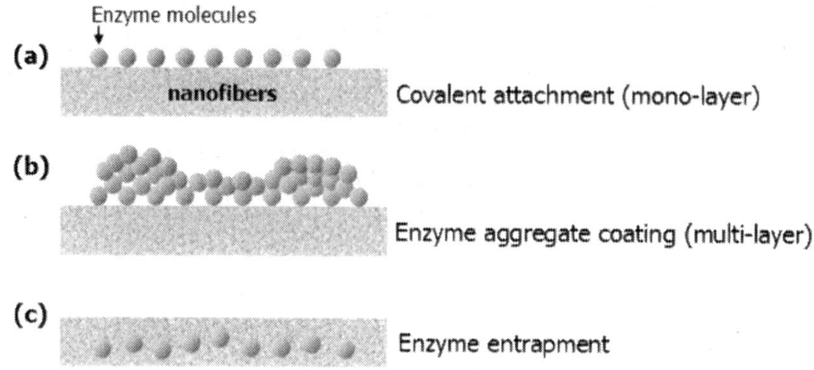

Figure 2. Strategies of enzyme immobilization using nanofibers (a) covalent attachment, (b) enzyme aggregation coating and (c) enzyme entrapment.

aggregating enzyme coating on the surface of nanofibers (*17*). Nanofibers consisting of a mixture of polystylene (PS) and poly(styrene-co-maleic anhydride) (PSMA) were fabricated via electrospinning. In this study, maleic anhydride group of PSMA acts as the linker for covalent attachment of enzyme molecules (α-chymotrypsin). It leads to direct immobilization of enzymes on the surface of nanofibers without further surface functionalization of nanofibers after fabricating. Covalently attached enzyme molecules on the nanofibers act as seed and cross-linker was treated to achieve enzyme-aggregate coatings around covalently attached enzymes (Figure 2 (b)). These enzyme-nanofiber composites could be recovered easily and has reusable characteristic for long a while more than a week without loss of initial activity in aqueous media (Figure 3).

Herricks and co-workers prepared polymer nanofibers carrying enzymes by electrospinning a solution mixed with enzyme and polymer in toluene directly (Figure 2 (c)) (*18*). This method used internal structure of nanofibers to entrap enzymes while covalent attachment or enzyme aggregating approach uses only external surface of nanofibers. For this, an enzyme (α-chymotrypsin) was transferred and stabilized in organic solvent as treating ionic surfactant. Ionic surfactant changed the surface charge of enzyme molecules in the organic solvents and it makes enzyme and polymer solution exist homogenously. Enzyme-polymer composites have been electrospinned to produce enzyme entrapped nanofibers directly. Additional cross-linker (glutaraldehyde) treatment could stabilize the enzyme activity more than two weeks. Also, they showed improved activity of enzyme entrapped nanofibers over bulk films due to the increased mass-transfer rate of substrate. This approach produces highly durable nanofiber-based biocatalytic mats but still needs to solve the mass transfer limitation because outer layer of nanofibers around enzyme entrapped still be act as a barrier to contact with substrates. Xie and Hsieh studied enzyme

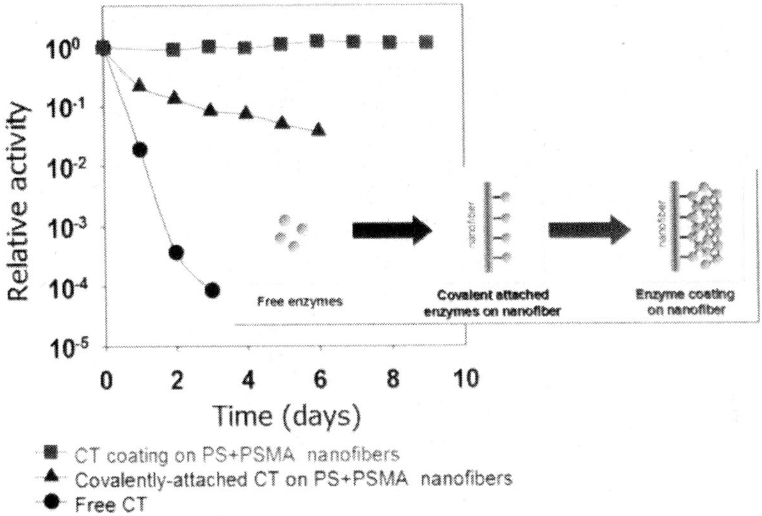

Figure 3. Comparison of stability of α-chymotrypsin (CT) coating nanofibers, covalently attached chymotrypsin on the nanofibers and free enzyme. Adapted with permission from Ref [17]. Copyright 2005 IOP Publishing Ltd.

encapsulation in nanofibers more systematically (*19*). In this study, spinning status was observed by controlling ratio of polymer and casein. They showed an optimum ratio existed between polymer and enzyme in order to make a continuous fiber form via electrospinning process.

Application of enzyme immobilized nanofibers

Covalent attachment, aggregate coating or entrapment of enzymes with nanofibers yields highly active and stable systems for the applications in the various fields such as bioconversion, biomedicals, and biosensors. There are several recent applications in these fields using enzyme-nanofiber composites. Wu and co-workers have demonstrated that porous membranes made of poly (vinyl alcohol) (PVA) are good scaffolds for cellulase. This application showed that cellulase entrapped in nanofibers could be applied to enzymatic biotransformation process (*20*). In their application, cellulase and PVA mixture was directly electrospunned and formed porous membrane. As comparison with PVA casting films, nanofiberous membrane enhanced not only an enzyme loading capacity but also an enzyme activity per gram of scaffolds. They reused it up to 6 cycles with 40 % initial activity. But the reusability of enzyme entrapped nanofibers was less hopeful in their application.

On the concerns about drug delivery systems, nanofibers could be also used as carriers for protein drugs which should be released at position of target organ. Zeng and co-workers investigated protein release properties from the nanofiber scaffolds (*21*). Although they used non-drug enzyme in their study, they showed the possibility using enzyme-nanofiber composites in this application. BSA or luciferase loaded poly (vinyl alcohol) (PVA) nanofibers were produced via electrospinning and these composites were coated with ploy (p-xylene) (PPX) by chemical vapor deposition. Based on the coating thickness of PPX, release of BSA was retarded and controllable. Although systematic studies on structural and morphological characteristics of nanofibers used as drug carrier-scaffolds is still needed, it could be possible to use nanofibers as drug delivery scaffolds.

Because of flexibility in the use of various polymeric solutions, nanofibers could be fabricated easily with inorganic or organic nanomaterials during electrospinning for the biosensor application. By incorporating nanomaterials such as carbon nanotubes (CNTs) to nanofibers, new characters could be conferred. The CNTs inside the nanofibers behave either as electron acceptors and donors for the immobilized redox enzymes. It could lead high activity of composed enzyme-nanomaterials. In addition, nanofibers in a formed membrane form are good scaffolds to hold nanomaterials, which make composite materials easily applicable to various application areas. Wang and co-workers produced multi-wall carbon nanotubes (MWCNTs) containing poly(acrylonitrile-*co*-acrylic acid) (PANCCAA) nanofibers (*22*). Redox enzyme (catalase) was immobilized by EDC/NHS coupling to produce nanofibers-MWCNTs-enzyme composites. The PANCC/MWCNTs nanofibers could hold more enzymes than that without MWCNTs because the porosity was increased. Also, enzyme activity with MWCNTs was enhanced by 42% because MWCNTs was attributed to promote electron. This result showed nanofibers could be used as scaffolds for enzyme mediated biosensors and enhanced its sensitivity by blending with other appropriate nanofibers.

Non-woven mat form of nanofibers could be used to hand held biosensors. Patel and co-workers produced a non-woven mat formed nanoporous silica containing nanofibers as a matrix for encapsulation of horseradish peroxidase (*23*). Porous silica nanofiber mats were fabricated with higher surface area than that of only using a polymer solution and showed four fold enhancements in the enzyme activity. By introducing several viscous solutions with enzymes and nanoporous silica, diameters and pore size could be controlled. The nanofiber mats containing enzymes showed its activity by converting hydrogen peroxide to water in the conversion process of 4-aminoantipyrine to quinoneimine. This reaction turned color of nanofiber mats from white to red, and so the presence of target compounds could be identified with naked eyes.

Not only enzymes but also tissues have been incorporated with nanofibers. Because nanofibers could mimic the fibrous organ in biological body, their application will be amended gradually and more biological application will be developed with various types of nanofibers in the future. Especially, biodegradable scaffolds must be important in tissue engineering. Polymeric nanofibers are considered as scaffolds for tissues such as bones, blood vessels, nerve and so on because of its ability to mimic natural extracelluar matrix and versatility with other materials such as inorganic components, growth factors, other cell regulatory biomolecules (*2,24,25*). Aligned poly(L-lactide-co-caprolactone) (PLLA-CA) nanofibers was studied for blood vessel application (*26*). Recently, Yang and co-workers produced poly (L-lactic acid) (PLLA) nanofibers as scaffolds for nerve stem cells (*25,27*). Not only these applications, nanofibers scaffolds for other biomolecules are of interest in various biotechnology fields.

Concluding remarks

The recent advances in nanofiber fabrications with various materials have drawn considerable attentions in many fields of biotechnology. The fabricating methods of nanofibers with electrospinning are suitable for various applications with various nanomaterials and biomolecules. Because of its easy fabrication and flexibility to choose various polymeric materials, nanofibers have drawn attentions for application in many fields of biotechnology. Especially, nanofibers as scaffolds of enzyme immobilization were briefly reviewed in this article. There is growing interest on enzyme-nanofiber composites as sources of stable and active biocatalytic systems. They are attractive for bioremediation, biosensors and bioconversion, which needs stable enzyme systems. However, controlling process for uniform diameters, defect-free, or alignment of nanofibers scaffolds still need to be solved. Still remained much improvement, it is believed that nanofibers will have more biotechnological applications in the very near future.

Acknowledgement

This work was supported by the Water Remediation & Environmental Research Center of Korea Institute of Science and Technology (KIST).

References

1. Whitesides, G. M. *Small* **2005**, *1*, 172-179.
2. Zhang, Y. Z.; Lim, C. T.; Ramakrishna, S.; Huang, Z. M. *J. Mater. Sci.-Mater. Med.* **2005**, *16*, 933-946.

3. Chun, I.; Reneker, D. H.; Fong, H.; Fang, X. Y.; Dietzel, J.; Tan, N. B.; Kearns, K. *J. Adv. Mater.* **1999**, *31*, 36-41.
4. Fong, H.; Chun, I.; Reneker, D. H. *Polymer* **1999**, *40*, 4585-4592.
5. Reneker, D. H.; Yarin, A. L.; Fong, H.; Koombhongse, S. *J. Appl. Phys.* **2000**, *87*, 4531-4547.
6. Reneker, D. H.; Kataphinan, W.; Theron, A.; Zussman, E.; Yarin, A. L. *Polymer* **2002**, *43*, 6785-6794.
7. Li, D.; Xia, Y. N. *Nano Lett.* **2004**, *4*, 933-938.
8. Li, D.; Xia, Y. N. *Adv. Mater.* **2004**, *16*, 1151-1170.
9. Li, D.; Babel, A.; Jenekhe, S. A.; Xia, Y. N. *Adv. Mater.* **2004**, *16*, 2062-+.
10. McCann, J. T.; Li, D.; Xia, Y. N. *J. Mater. Chem.* **2005**, *15*, 735-738.
11. Kim, J. S.; Reneker, D. H. *Polym. Eng. Sci.* **1999**, *39*, 849-854.
12. Li, D.; Wang, Y. L.; Xia, Y. N. *Nano Lett.* **2003**, *3*, 1167-1171.
13. Huang, Z. M.; Zhang, Y. Z.; Kotaki, M.; Ramakrishna, S. *Composites Science And Technology* **2003**, *63*, 2223-2253.
14. Jayaraman, K.; Kotaki, M.; Zhang, Y. Z.; Mo, X. M.; Ramakrishna, S. *J. Nanosci. Nanotechnol.* **2004**, *4*, 52-65.
15. Jia, H. F.; Zhu, G. Y.; Vugrinovich, B.; Kataphinan, W.; Reneker, D. H.; Wang, P. *Biotechnol. Prog.* **2002**, *18*, 1027-1032.
16. Ye, P.; Xu, Z. K.; Wu, J.; Innocent, C.; Seta, P. *Macromolecules* **2006**, *39*, 1041-1045.
17. Kim, B. C.; Nair, S.; Kim, J.; Kwak, J. H.; Grate, J. W.; Kim, S. H.; Gu, M. B. *Nanotechnology* **2005**, *16*, S382-S388.
18. Herricks, T. E.; Kim, S. H.; Kim, J.; Li, D.; Kwak, J. H.; Grate, J. W.; Kim, S. H.; Xia, Y. N. *J. Mater. Chem.* **2005**, *15*, 3241-3245.
19. Xie, J. B.; Hsieh, Y. L. *J. Mater. Sci.* **2003**, *38*, 2125-2133.
20. Wu, L. L.; Yuan, X. Y.; Sheng, J. *J. Membr. Sci.* **2005**, *250*, 167-173.
21. Zeng, J.; Aigner, A.; Czubayko, F.; Kissel, T.; Wendorff, J. H.; Greiner, A. *Biomacromolecules* **2005**, *6*, 1484-1488.
22. Wang, Z. G.; Xu, Z. K.; Wan, L. S.; Wu, J.; Innocent, C.; Seta, P. *Macromol. Rapid Commun.* **2006**, *27*, 516-521.
23. Patel, A. C.; Li, S. X.; Yuan, J. M.; Wei, Y. *Nano Lett.* **2006**, *6*, 1042-1046.
24. Yoshimoto, H.; Shin, Y. M.; Terai, H.; Vacanti, J. P. *Biomaterials* **2003**, *24*, 2077.
25. Yang, F.; Murugan, R.; Wang, S.; Ramakrishna, S. *Biomaterials* **2005**, *26*, 2603-2610.
26. Xu, C. Y.; Inai, R.; Kotaki, M.; Ramakrishna, S. *Biomaterials* **2004**, *25*, 877.
27. Yang, F.; Murugan, R.; Ramakrishna, S.; Wang, X.; Ma, Y. X.; Wang, S. *Biomaterials* **2004**, *25*, 1891-1900.

Chapter 17

Biomimetic Synthesis of an Active H_2 Catalyst Using the Ferritin Protein Cage Architecture

Zachary Varpness[1,3], Chad Shoopman[1,3], John W. Peters[1,3], Mark Young[2,3], and Trevor Douglas[1,3,*]

Departments of [1]Chemistry and Biochemistry, [2]Plant Sciences, and [3]Center for Bio-Inspired Nanomaterials, Montana State University, Bozeman, MT 59717

[*]Corresponding author: tdouglas@chemistry.montana.edu

The possibility of developing hydrogen gas as an environmentally friendly alternative fuel has received considerable attention[1-6]. However, the majority of hydrogen produced for energy yielding applications is generated by the process of reforming methane or fossil fuels[7, 8]. Thus, we are not in a position to develop a hydrogen energy economy that is independent of nonrenewable fossils fuels. A renewable source for the production of hydrogen is needed for the development of hydrogen as an alternative fuel.

We have taken a biomimetic synthetic approach to the synthesis of efficient catalyst systems for H_2 formation. We have combined the advantages of biological control and the strengths of synthetic chemistry to create novel materials with tailored catalytic properties. A major goal of biomimetic chemistry is to couple biological molecules with synthetic capacity to create new functional materials showing dramatic property enhancements. By interfacing hard inorganic materials (metals) with soft biological materials (proteins), we can create new composite materials with controlled multifunctionality.

With the high cost and limited supply of platinum, it is necessary to explore ways of maximizing the catalytic efficiency of Pt on a per atom basis in order to develop economically feasible catalysts. In a particle based approach toward developing a Pt catalyst, it is necessary to minimize the diameter of the particle and thus increase the surface area (i.e. the number of exposed Pt atoms per particle). A number of different synthetic approaches have been used to synthesize platinum nanoparticles using different passivating layers[9-18].

However, the passivating layers generally interfere with the exposed Pt atoms and reduce catalytic efficiency[9]. In this study, we have employed the protein cage of ferritin as a synthetic platform, which, unlike a passivating layer, does not coat the entire surface of the nanoparticle but still isolates the nanoparticle in solution and prevents aggregation.

We have previously demonstrated that protein cage architectures can be used as bio-templates for the synthesis of encapsulated nanomaterials[19-25]. Protein cage architectures are structurally-defined, container-like assemblies. They are self-assembled from a limited number of protein subunits in which the interior and exterior surfaces are chemically distinct. Through genetic and chemical modifications, reaction sites can be inserted into specific locations within the cage architecture to create designed functionality[22]. Protein cages have been used to not only to define the size and shape of nanomaterials but also the crystal phase[23].

We have adopted this biomimetic approach to create a ferritin-based catalyst for the production of H_2. Ferritin is a remarkably stable protein architecture that assembles from 24 subunits into a 12 nm cage defining an 8 nm interior cavity (Figure 1). Ferritin is an iron storage protein found a nearly all organisms and

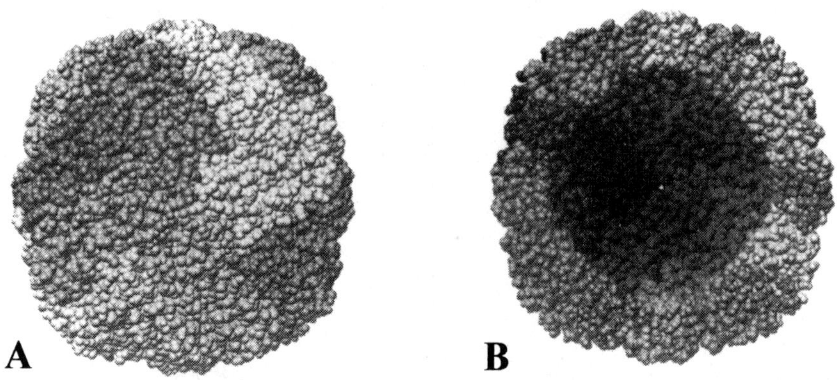

Figure 1. (A) Space filling representation of the horse spleen ferritin cage (pdb: 1AEW). (B) Cut-away view of ferritin showing the interior cavity of the cage.

has been used as a platform for the synthesis of a range of nanoparticles, including catalysts such as palladium[26-28]. We have previously shown that catalytically active platinum nanoparticles can be synthesized inside another protein cage, the small heat shock protein from *Methanococcus jannaschii*[25]. In the current study, we have demonstrated synthesis and activity of Pt nanoparticles within the ferritin protein cage architecture.

The synthesis of ferritin encapsulated Pt nanoparticles was undertaken using synthetic methodology similar to previously reported[25]. Briefly outlined, demineralized and purified apo-ferritin (5 mg, 0.01 μmoles) was incubated with either 250 or 1000 $PtCl_4^{2-}$ per protein cage at 65°C for 15 minutes. Reduction with dimethylamine borane complex ($(CH_3)_2NBH_3$) resulted in the formation of a brown colored solution. Characterization of the reaction product by size exclusion chromatography revealed retention volumes identical to untreated ferritin and showed co-elution of protein (280 nm) and Pt (350 nm) components (Figure 2). Dynamic light scattering indicated no change in the particle

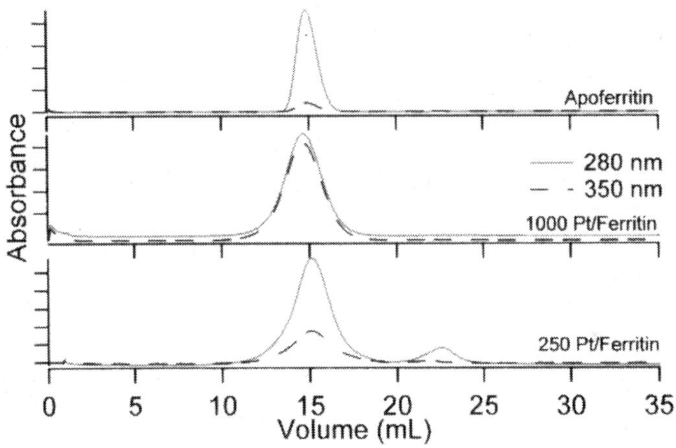

Figure 2. Size exclusion chromatography of ferritin. (A) Unmineralized ferritin, (B) Ferritin mineralized with 1000 Pt/ferritin showing coelution of protein (280 nm) and mineral (350 nm), and (C) Ferritin mineralized with 250 Pt/ferritin showing coelution of protein (280 nm) and mineral (350 nm).

diameter after the reaction. Transmission electron microscopy (TEM) of the Pt-treated ferritin (Fn-Pt) revealed homogeneously sized electron dense cores 1.9 ± 0.8 nm in diameter (Figure 3A). These particles were identified as Pt metal by electron diffraction as a powder pattern was observed (Figure 3A, inset) with the d-spacings shown in Table I. Negatively stained TEM samples displayed 12 nm protein cages with Pt particles clearly localized within the cage structure (Figure 3B). For theoretical loadings of 1000 Pt/cage, nanoparticles of 1.9 ± 0.8 nm were observed (Figure 3C) and for loadings of 250 Pt/cage, particles of 1.8 ± 1.2 nm were observed (Figure 4). Ferritin-free control reactions resulted in the formation of aggregated Pt colloids, which rapidly precipitated from solution. Control reactions using bovine serum albumin (BSA), at the same total protein concentration, also resulted in bulk precipitation and only a fraction of the Pt

Table I. Measured d-spacings

Miller Index	d-spacing	Measured d-spacing
111	2.265	2.304
200	1.961	1.965
220	1.387	1.140
113	1.183	1.195

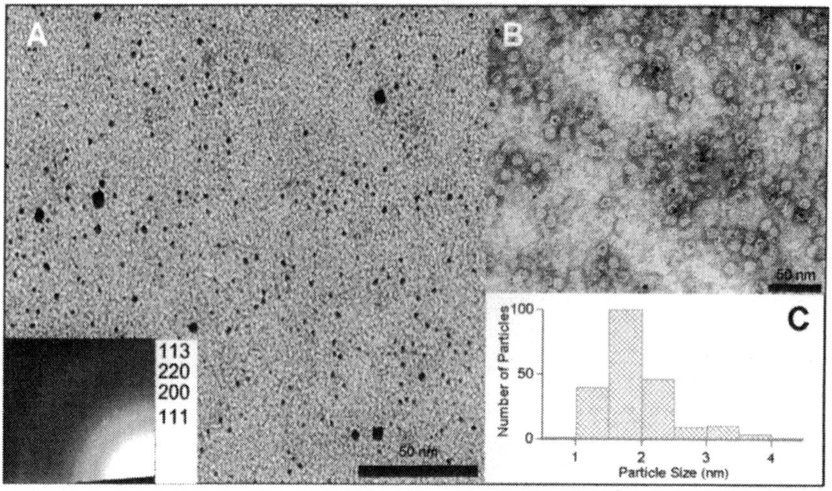

Figure 3. (A) TEM of ferritin 1000 Pt unstained. The inset shows electron diffraction of Pt^0 from ferritin 1000 Pt. (B) TEM of ferritin 1000 Pt stained with 2% uranyl acetate. (C) Histogram of Pt particle diameters in ferritin 1000 Pt. Average 1.9 ± 0.8 nm from 215 particles. Scale bar = 50 nm.

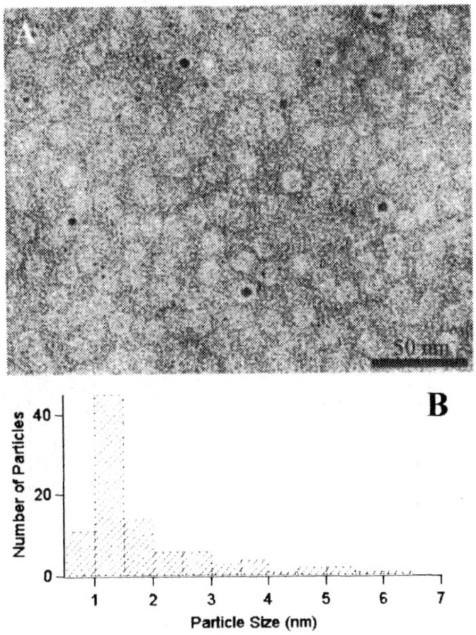

Figure 4. (A) TEM of ferritin 250 Pt stained with 2 % uranyl acetate. (B) Histogram of Pt particle diameters in ferritin 250 Pt. Average 1.8 ± 1.2 nm from 99 particles . Scale bar = 50 nm.

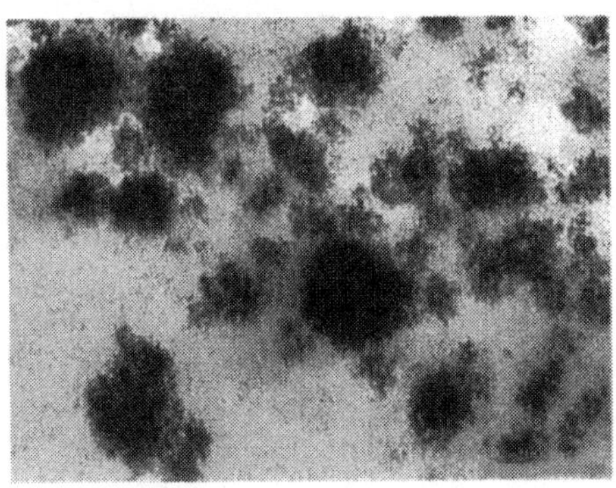

Figure 5. BSA mineralization control reaction. Scale Bar = 200 nm.

remained in solution with a wide distribution of particle sizes (3 nm to 120 nm) when observed by TEM (Figure 5).

The Pt-Fn protein cage composites are highly active catalysts able to reduce H^+ to form H_2 when supplied with an appropriate reducing agent. We have used visible light and a co-catalyst ($Ru(bpy)_3^{2+}$) to generate MV^+ through oxidation of simple organics such as EDTA (Figure 6) to assay H_2 production from the Fn-Pt catalyst. In this assay, the solution was illuminated at 25°C with a 150W Xe-arc lamp equipped with an IR filter and a UV cut-off filter (360 nm). The Pt-Fn (0.5 μg, 1.1 x10^{-12} moles) was illuminated in the presence of MV^{2+} (0.5 mM), $Ru(bpy)_3^{2+}$ (0.2 mM) and EDTA (200 mM) at pH 5.0 and the resulting H_2 was quantified by gas chromatography (Shimadzu GC-8A).

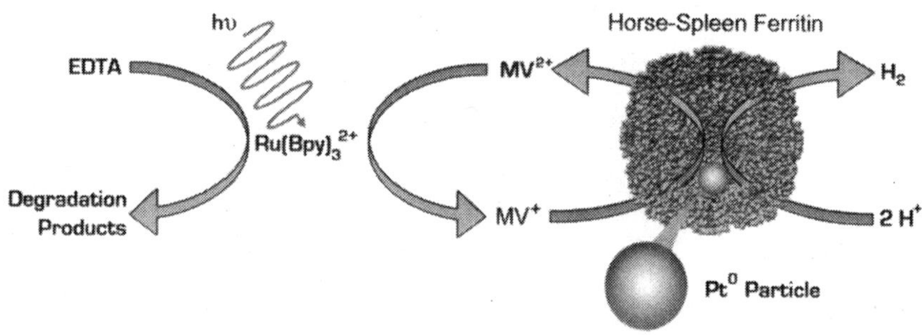

Figure 6. Schematic of the light mediated H_2 production from Pt-Fn. Methyl viologen (MV^{2+}) is used as an electron transfer mediator between the $Ru(bpy)_3^{2+}$ photocatalyst and the Pt-Fn responsible for H_2 production.

The initial rates of H_2 formation, when calculated on a per cage basis, were 14.9 x 10^3 H_2/sec/cage (± 2.77 x 10^3 H_2/sec) for a loading factor of 1000 Pt per ferritin (Figure 7) and 1.82 x 10^3 H_2/sec/cage (± 8.47 x 10^2 H_2/sec) for a loading factor of 250 (Figure 8). When the H_2 production rates are calculated on a per Pt basis, they compare very favorably with other reported Pt nanoparticles. The initial rates for Pt-Fn with 1000 Pt/cage are 893 H_2/Pt/min, as compared with the reported literature values (20 H_2/Pt/min[9], 16 H_2/Pt/min[15], and 6.5 H_2/Pt/min[29]). In addition, initial H_2 production rates for Pt-Fn are more active than the previous reports using Pt nanoparticles synthesized inside of small heat shock protein (Pt-Hsp). The initial rates of H_2 production from Pt-Hsp with 1000 Pt/cage are 268 H_2/Pt/min[25]. The Pt-Fn synthesis reaction produced Pt particles with the size of 1.9 ± 0.8 nm as compared to particles of 2.6 ± 0.2 nm for previous synthesis in Hsp under the same conditions. The particular protein cage platform employed in the synthesis therefore has a significant effect on the size and activity of the catalyst nanoparticle. After 20 minutes of illumination, the $Ru(bpy)_3^{2+}$ begins to photobleach and the methyl viologen undergoes platinum catalysed hydrogenation, which significantly decreases the hydrogen production. The smaller size of the particles in Pt-Fn compared to Pt-Hsp can,

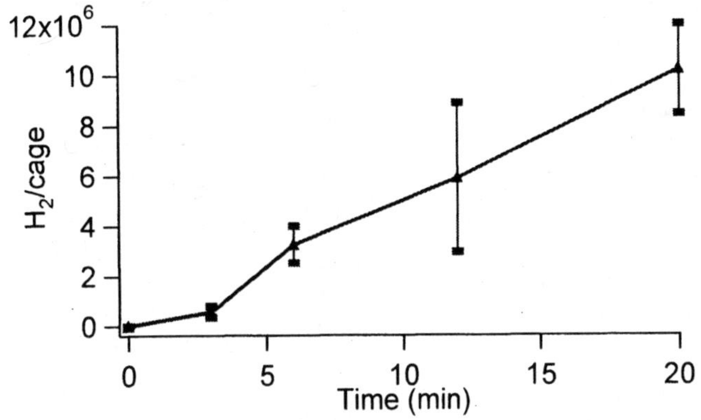

Figure 7. H_2 production from 1000 Pt/Ferritin in 0.2 mM $Ru(bpy)_3^{2+}$, 0.5 mM methyl viologen, 200 mM EDTA, and 500 mM acetic acid pH 5.0. ■ 1000 Pt/ferritin 1.1 x 10^{-9} moles Pt.

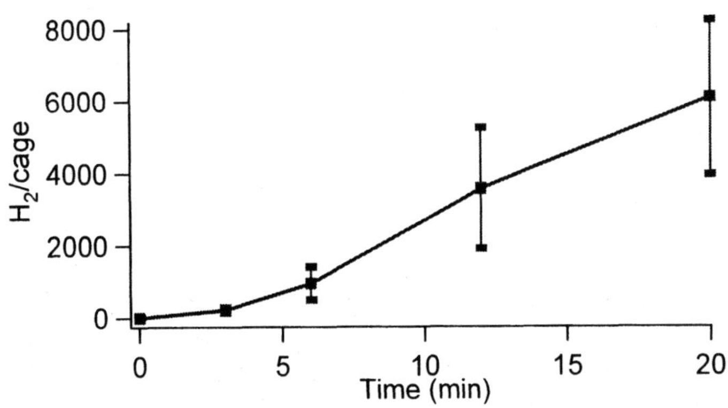

Figure 8. H_2 production from 250 Pt/Ferritin in 0.2 mM $Ru(bpy)_3^{2+}$, 0.5 mM methyl viologen, 200 mM EDTA, and 500 mM acetic acid pH 5.0. ■ 250 Pt/ferritin 3.3 x 10^{-10} moles Pt.

in part, explain the increased H_2 production. Based on the particle size histograms from Pt-Fn and Pt-Hsp, it appears that Pt-Fn has a narrower size distribution compared to Pt-Hsp. This distribution suggests that the Pt-Hsp will have more particles that are in the range too small to be detected by TEM, which are inactive catalysts. Pt-Fn has fewer particles that are below the threshold of detection giving rise to an overall more active preparation. Pt-Fn could more efficiently mineralize catalytically active platinum particles by making more particles over the catalytically active threshold of 50 atoms of platinum[30] while Pt-Hsp has more platinum in particles that are inactive lowering the per Pt efficency.

We have used a well-defined thermally stable protein cage architecture to synthesize an encapsulated catalyst for producing H_2. We have introduced nano-scale metal clusters, in a spatially selective manner, to the interior of the ferritin protein cage architecture. The encapsulated Pt nanoparticles act as active sites for the reduction of H^+ to form H_2, and the activity of these bio-composite materials are significantly better than previously described Pt nanoparticles. The protein cage architecture of ferritin prevents the agglomeration of the nanoparticles while minimizing effects of passivating the surface of the catalyst. The Pt-Fn composite illustrates the utility of using protein architectures for the formation of functional nanomaterials with applications in energy production and utilization.

Acknowledgements

This work was funded in part by grants from the Office of Naval Research (19-00-R0006). Molecular graphics images were produced using the UCSF Chimera package from the Resource for Biocomputing, Visualization, and Informatics at the University of California, San Francisco (supported by NIH P41 RR-01081).

References

1. Chum, H. L.; Overend, R. P.; *Fuel Processing Technology* **2001**, *71*, (1-3), 187-195.
2. Nandi, R.; Sengupta, S.; *Critical Reviews in Microbiology* **1998**, *24*, (1), 61-84.
3. Ssen, Z.; *Progress in Energy and Combustion Science* **2004**, *30*, (4), 367-416.
4. Ting, C. H.; Lin, K. R.; Lee, D. J.; Tay, J. H.; *Water Science and Technology* **2004**, *50*, (9), 223-228.

5. Tseng, P.; Lee, J.; Friley, P.; *Energy* **2005**, *30*, (14), 2703-2720.
6. Winter, C. J.; *International Journal of Hydrogen Energy* **2005**, *30*, (7), 681-685.
7. Armor, J. N.; *Applied Catalysis a-General* **1999**, *176*, (2), 159-176.
8. Armor, J. N.; *Catalysis Letters* **2005**, *101*, (3-4), 131-135.
9. Brugger, P. A.; Cuendet, P.; Gratzel, M.; *Journal of the American Chemical Society* **1981**, *103*, (11), 2923-2927.
10. Chen, C. W.; Tano, D.; Akashi, M.; *Journal of Colloid and Interface Science* **2000**, *225*, (2), 349-358.
11. Chen, D. H.; Yeh, J. J.; Huang, T. C.; *Journal of Colloid and Interface Science* **1999**, *215*, (1), 159-166.
12. Eklund, S. E.; Cliffel, D. E.; *Langmuir* **2004**, *20*, (14), 6012-6018.
13. Gomez, S.; Erades, L.; Philippot, K.; Chaudret, B.; Colliere, V.; Balmes, O.; Bovin, J. O.; *Chemical Communications* **2001**, (16), 1474-1475.
14. Jiang, D. L.; Choi, C. K.; Honda, K.; Li, W. S.; Yuzawa, T.; Aida, T.; *Journal of the American Chemical Society* **2004**, *126*, (38), 12084-12089.
15. Keller, P.; Moradpour, A.; *Journal of the American Chemical Society* **1980**, *102*, (24), 7193-7196.
16. Teranishi, T.; Kurita, R.; Miyake, M.; *Journal of Inorganic and Organometallic Polymers* **2000**, *10*, (3), 145-156.
17. Zhao, S. Y.; Chen, S. H.; Wang, S. Y.; Li, D. G.; Ma, H. Y.; *Langmuir* **2002**, *18*, (8), 3315-3318.
18. Narayanan, R.; El-Sayed, M. A.; *Nano Letters* **2004**, *4*, (7), 1343-1348.
19. Allen, M.; Willits, D.; Mosolf, J.; Young, M.; Douglas, T.; *Advanced Materials* **2002**, *14*, (21), 1562-1565.
20. Allen, M.; Willits, D.; Young, M.; Douglas, T.; *Inorganic Chemistry* **2003**, 42, (20), 6300-6305.
21. Douglas, T.; Young, M.; *Nature* **1998**, *393*, (6681), 152-155.
22. Flenniken, M. L.; Willits, D. A.; Brumfield, S.; Young, M. J.; Douglas, T.; *Nano Letters* **2003**, *3*, (11), 1573-1576.
23. Klem, M.; Willits, D.; Solis, D.; Belcher, A.; Young, M.; Douglas, T.; *Advanced Functional Materials* **2005**, *15*, 1489-1494.
24. Usselman, R. J.; Klem, M.; Allen, M.; Walter, E. D.; Gilmore, K.; Douglas, T.; Young, M.; Idzerda, Y.; Singel, D. J.; *Journal of Applied Physics* **2005**, *97*, (10).
25. Varpness, Z.; Peters, J. W.; Young, M.; Douglas, T.; *Nano Letters* **2005**, *5*, (11), 2306-2309.
26. Kramer, R. M.; Li, C.; Carter, D. C.; Stone, M. O.; Naik, R. R.; *Journal of the American Chemical Society* **2004**, *126*, (41), 13282-13286.
27. Ueno, T.; Suzuki, M.; Goto, T.; Matsumoto, T.; Nagayama, K.; Watanabe, Y.; *Angewandte Chemie-International Edition* **2004**, *43*, (19), 2527-2530.

28. Iwahori, K.; Yoshizawa, K.; Muraoka, M.; Yamashita, I.; *Inorganic Chemistry* **2005**, *44*, (18), 6393-6400.
29. Song, Y. J.; Yang, Y.; Medforth, C. J.; Pereira, E.; Singh, A. K.; Xu, H. F.; Jiang, Y. B.; Brinker, C. J.; van Swol, F.; Shelnutt, J. A.; *Journal of the American Chemical Society* **2004**, *126*, (2), 635-645.
30. Greenbaum, E.; *Journal of Physical Chemistry* **1988**, *92*, (16), 4571-4574.

Chapter 18

Carbon Nanotube Composite Electrodes for Biofuel Cells

Hongfei Jia[1,4], Xueyan Zhao[2], Jungbae Kim[3], and Ping Wang[2,5,*]

[1]Department of Chemical and Biomolecular Engineering, The University of Akron, Akron, OH 44325
[2]Department of Bioproducts and Biosystems Engineering, University of Minnesota, St. Paul, MN 55108
[3]Pacific Northwest National Laboratory, 902 Battelle Boulevard, Richland, WA 99352
[4]Current address: Toyota Technical Center, 2350 Green Road, Ann Arbor, MI 48105
[5]Current address: Department of Bio-Based Products, University of Minnesota, St. Paul, MN 55108

Applications of biofuel cells are largely restricted by their low power output density, to which the performance of the biocatalysts in terms of activity, stability and electrical conductivity is the limiting factor in most cases. To address those issues, researchers have demonstrated substantial interests in recent years in nanostructured biocatalysts. By tailoring the surrounding environments of enzymes, people can develop 'nanobiocatalysts' with much improved stability and efficiency. In this chapter, the construction and use of composite electrodes based on enzyme-carbon nanotubes-Nafion® for glucose-O_2 biofuel cell will be examined.

Introduction

Clean and efficient power generation is actively sought worldwide, as the demands in energy rapidly grow while concerns over environment pollution and petroleum resources remain potent. Fuel cells, as an alternative power generation technology to thermo-mechanical processes, directly convert chemical energy of non-nuclear fuels into electricity with a theoretical efficiency of 100% and almost zero pollution emission. Even though the concept of the technology has been demonstrated for decades, there are several challenges to overcome to make fuel cells economically competitive for general-purpose applications. For example, conventional fuel cells typically use noble metals such as platinum for catalysts on electrodes. But the precious Pt costs about $38,000 per kg, which makes the fuel cells too expensive for many large-scale applications. Metals are not renewable and resource-limited. In addition, metallic catalysts usually have to be operated at high temperatures; they are subject to inactivation by trace amount of impurities such as CO and sulfur in fuels; and may cause pollution to the environment upon discarded (1). Biofuel cells refer to a class of fuel cells that utilize microbial or enzymatic biocatalysts, which can effectively catalyze redox reactions under ambient conditions and neutral pH values. In contrast to noble metals, biocatalysts are renewable and the cost of production can be very low, as microorganism and enzymes such as detergent enzymes can be economically produced from large-scale fermentation processes once the market is developed. In addition, because of the unique catalytic activity of biocatalysts, biofuel cells can use a variety of special fuels such as carbohydrates and organic pollutants in wastewater (2-4).

One major challenge in the development of biofuel cells is their low power output density, typically 2-3 orders of magnitude lower than that of chemical fuel cells. Many factors may contribute to that. Among others, the inefficiency of electron conduction between biocatalysts and electrodes is believed to be critical. In chemical fuel cells, metallic catalysts are deposited on electrodes and, because both catalysts and electrode materials are conductive, electrons can transfer between the catalytic sites and the surface of electrodes. Such direct electron transfer (DET) is available only for a few biocatalysts. Bacteria including *Rhodoferax ferrireducens* (5), *Geobacter sulfurreducens* (6) and *Shewanella putrefaciens* (7), along with several small enzymes such as laccase and hydrogenase, were found capable of conducting electrocatalysis with DET. Enzymes that are capable of DET have been mostly examined for biosensors (8-13), although biofuel cells using physically adsorbed laccase and hydrogenase have been also reported (14). However, these biofuel cells generally produced low power and current output (15).

Electron transfer between two sites separated by nonconductive medium is possible should the distance between them is short enough (16, 17). For most enzymes, except the smaller enzymes such as laccase and hydrogenase, the protein shells surrounding the active sites are often too thick for DET. One strategy to overcome this barrier is to transform the protein shell to be conductive by chemical modification (18-21). Another popular approach is the use of redox mediators, which carry the electrons and shuttle between the enzyme active sites and the surface of electrodes. Many of such mediated catalytic systems have been reported for both microbial and enzymatic biofuel cells (2-4, 22). Even though the mediators introduce an additional step in the redox reaction chains from fuel to electron generation, much higher performance of biofuel cells was usually observed. One shortcoming of such systems lies in the fact that redox mediators are usually small molecules and are difficult to retain during continuous operations. To address this issue, substantial efforts have been made during the last decade to co-immobilization of enzyme and mediators (23-29).

The past decade also witnessed the explosive growth of nanotechnology. Progresses in synthetic and fabrication methods have greatly extended the application of nano-structured materials for many purposes such as the development of electronic, optoelectronic, biomedical, and sensing devices. It has been demonstrated that catalysts supported by nano-structures well performed for nitrooxide reduction (30), hydrogenation (31), hydroformylation (32), and electrochemical catalysis in sensors (33) and fuel cells (34). Recently various configurations of nanomaterials such as nanofibers (35-37), nanotubes (38-40), nano-particles (41, 42) and nanoporous materials (43-46) have been reported for use as the supports for enzymes. Many of this type of "nanobiocatalysts" displayed enhanced performance in terms of catalytic efficiency and durability in comparison with their native parent enzymes (47). The emergence of nanobiocatalysts may also bring up chances for the fabrication of high-performance bio-electrodes for biofuel cells. Particularly, electrodes based on conductive nanomaterials may offer promising opportunities to improve the power output of biofuel cells. An example has been reported by Xiao et al (48). In that work, glucose oxidase (GOx) was "wired" to electrode by gold nanoparticles and showed higher activities than native enzymes. Although the electrode was not suited for biofuel cell application due to the high over potential, it did show the potentials in developing novel biocatalysts using nanoscale manipulation approaches.

Carbon nanotube (CNT) attracted a lot of attention for its unique physical, chemical and mechanical properties. It also represents a class of ideal materials for electrochemical applications due to the good conductivity and extraordinary stability. Previously, CNT has been extensively studied for biosensing and many CNT-based biosensors showed better selectivity and/or improved sensitivity (4, 19, 39, 49-55). In this chapter, we explore the application CNT in biofuel cells. Nano-composite electrodes based on GOx, Nafion® and multiwall carbon nanotubes (MWCNTs) were prepared and examined as anodes in a model glucose-O_2 biofuel cell.

Materials and Methods

MWCNTs (NanoLab Inc.) with an outer diameter of 30±15 nm and a length of 1~5 μm were pretreated with a mixture of 98% H_2SO_4 and 65% HNO_3 (3:1, v/v) at room temperature for 8 h to increase the hydrophilicity and generate carboxylic groups on the surface (56). The MWCNTs were then functionalized via carbodiimide coupling. Typically, 2.5 mg of acid-treated MWCNTs, 10 mg of N-Ethyl-N'-(3-dimethylaminopropyl)-carbodiimide (EDC) and 1.7 mg of N-hydroxysuccinimide were added into a 2-ml centrifuge tube containing 1.5 ml of pH 7.4 phosphate buffer (0.05 M). The suspension was first sonicated for 15 min and then incubated at room temperature with vigorous shaking for 4 h. The activated MWCNTs were rinsed with buffer solution and stored at 4°C.

Composite electrodes were prepared by depositing a casting solution onto electrode backing materials such as glassy carbon (CH Instruments, Inc.), carbon paper (AcCarb™ P50, purchased from FuelCellStore.com), carbon cloth (purchased from ElectroChem, Inc.) and carbon felt (purchased from Alfa Aesar). The casting solution consisted of 2 mg/ml of MWCNTs, 5 mg/ml of GOx (Sigma, Type X-S from *Aspergilus niger*) and Nafion® (0.5% by wt) in a phosphate buffer (pH 7.4, 0.05 M). Before casting onto the electrode backing materials, the suspension was incubated at 4°C with gentle shaking overnight to immobilize the enzyme onto MWCNTs. Typically 20 μl of the resulted suspension was cast on the backing electrodes and allowed to dry at room temperature for 2 h, followed by washing with phosphate buffer (pH 7.0, 0.1 M).

The electrochemical properties of the MWCNTs-Nafion® composites were studied using cyclic voltammetry (CV) on glassy carbon electrodes. The three-electrode testing cell used glassy carbon electrode as the working electrode, Ag/AgCl as the reference and a Pt wire as the counter electrode. Typically, the electrolyte contains 0.5 mM of hydroquinone (HQ) in pH 7.0 phosphate buffer. All the CV tests were performed on a DLK-60 Electrochemical Analyzer (Analytical Instrument Systems, Inc.) at scanning rates ranging from 2.5 to 25 mV/s.

The enzyme loading in the composite electrodes was estimated by indirect mass balance method. The leached enzyme during washing was collected and determined spectrophotmetrically (at 280 nm) and enzyme loading in the electrode was calculated from the difference of the amount of enzyme added to the casting solution and that detected in the washing solutions. The specific catalytic activities of prepared electrodes towards glucose oxidation were measured by UV-visible spectrophotometry with oxygen and BQ as the electron acceptors respectively. When oxygen was used as the oxidant, the activity was determined by monitoring the generation of hydrogen peroxide during the reaction. Typically, a working solution was first prepared by dissolving *o*-dianisidine (10 mg), horseradish peroxidase (1.1 mg) and glucose (200 mM) in 150 ml of phosphate buffer (0.1 M, pH 7.0). The electrode with immobilized enzyme was then incubated in 15 ml of working solution and the generation of oxidation product of *o*-dianisidine was measured spectrophotometrically at a wavelength of 460 nm for 30 min. In case of BQ as the electron acceptor, same protocol was followed except that a different working solution was used, which contained 200 mM glucose and 5 mM BQ in the same buffer. The formation of HQ was monitored by monitoring the absorbance at 290 nm.

The enzyme-MWCNTs-Nafion® composite electrodes were examined as the anode in glucose/O_2 biofuel cells. Electrodes with glassy carbon backing (nonporous) were placed ~2 mm above the Nafion® membrane, which separated anode and cathode electrodes during the biofuel cell operation. In case of electrodes with porous backing materials, they were pressed against the membrane with a graphite rod (1.6 mm in diameter). The anode compartment of the cell typically contains 30 ml of 0.1 M phosphate buffer (pH 7.0) containing glucose and BQ. Nitrogen was bubbled in the electrolyte reservoir during the operation. An Ag/AgCl electrode was used as the reference electrode to monitor the polarization of electrodes. For all tests, a membrane gas diffusion electrode (obtained from Fuelcellstore.com) was used as the cathode for oxygen reduction (4.7 cm in diameter, with a Pt loading equals to 4.0 mg/cm^2). The biofuel cell was connected to an adjustable external load with a resistance range from 0 to 500 kΩ. The output voltage and current were measured by multimeters (EXTECH Multipro 510). The potentials of the anode and cathode vs. the reference electrode were measured by the DLK-60 Electrochemical analyzer. Impedance measurement was performed on an Impedance/Gain-Phase Analyzer (Solartron, 1260) with frequencies ranging from 0.1 to 10^6 Hz.

Results and Discussions

The composite electrodes were typically prepared by casting solution containing MWCNT-immobilized GOx and Nafio® onto backing materials such as glassy carbon, carbon paper, carbon cloth and carbon felt. The effectiveness of MWCNT was first examined by CV in the presence of benzoquinone (BQ). As shown in Figure 1, composite-coated electrode showed higher peak height and narrower peak separation than the bare glassy carbon and that coated with Nafion®, indication an apparent improvement of reaction kinetics. The $KD^{1/2}$ (parameter reflecting electrochemical flux) and apparent heterogeneous reaction rate constants were estimated and listed in Table 1. The enhanced electrochemical flux was most likely due to the larger surface area as a result of the use of MWCNT. HQ/BQ redox couple typically displays quasi-reversible to irreversible behaviors on electrodes such as Pt, Au and glassy carbon. However, a reversible reaction cycle was observed previously with pyrolytic graphite electrodes (57). Similar to graphite, MWCNT surface also has a highly organized structure. That might contribute to the kinetic improvement.

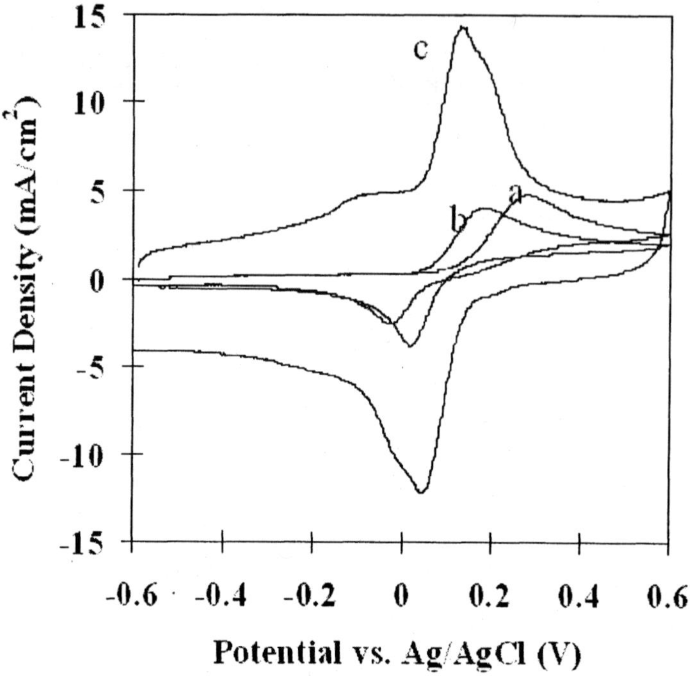

Figure 1. Representative cyclic voltammograms of HQ/BQ redox couple on base glassy carbon (a), Nafion®-coated glassy carbon (b) and MWCNT-Nafion® coated glassy carbon(c) electrodes. Testing conditions: 0.5 mM HQ in 0.1 M pH 7.0 phosphate buffer, scanning rate: 10 mV/s.

Table 1. Reaction Kinetics Parameters for Various Electrodes

Electrodes	$KD^{1/2}$ [a] $(10^{-3}\ cm^2/s)$	*Apparent heterogeneous reaction rate constant* (k_s) [b] 10^{-5} *cm/s*
Glassy carbon	1.6	1.8
Nafion® coated glassy carbon	1.2	20
CNT-Nafion® coated glass carbon	3.7	130

[a] *K*: extraction coefficient; *D*: diffusion coefficient; In diffusion-limited CV system, electrochemical flux is proportional to $KD^{1/2}$ (58, 59). Apparent electrode surface area was used for the calculation: 0.071 cm^2 for bare and Nafion®-coated glassy carbon electrodes, 0.332 cm^2 for MWCNTs-Nafion® modified electrodes.

[b] A diffusion coefficient of 8.4×10^{-6} cm^2/s was used for the estimation of k_s (60).

GOx was immobilized onto the acid-treated MWCNTs via NHS-enhanced carbodiimide coupling (Figure 2). In the activation step, carboxyl groups first reacted with EDC to form unstable intermediates, which then reacted with NHS and generated active succinimide ester groups on the surface of the MWCNTs. The active ester groups then reacted directly with amino groups of the enzyme and form peptide bonds between enzyme molecules and the MWCNTs.

The resulting electrodes were examined in the biofuel cell as illustrated in Figure 3. In the anode, BQ was used as the mediator to facilitate the electron transfer between the enzyme and the electrode. At the active sites of the enzyme, BQ accepts two H from the cofactor $FADH_2$ and is reduced to HQ. HQ then diffuses to the surface of MWCNTs where it is oxidized back to BQ. Two protons and two electrons are generated from this process. Protons are released to the buffer solution and eventually reach cathode through the electrolyte and the proton exchange membrane. Electrons pass through the MWCNTs network, first to the backing electrode, then to the anode, and finally reach cathode via external loop. At the catalyst layer of the cathode, oxygen reacts with protons and electrons and water is generated. Because this study was focused on the enzymatic anode, the influence of cathode polarization (potential drop with current density) on the biofuel cell performance was minimized by using a small anode (6.4 mm in diameter, projective area of 0.32 cm^2) and a much larger cathode (47 mm in diameter, projective area of 17.4 cm^2). Typical potentials of the cathode were observed in the range of 630~660 mV vs. Ag/AgCl with polarization usually less than 30 mV during the operation of the biofuel cells.

A maximum power density of 44 $\mu W/cm^2$ was achieved using composite coated glassy carbon as the anode (Figure 4). It appeared that the voltage of the biofuel cells decreased with current mostly due to the anodic polarization (Figure 5).

Generally speaking, fuel cell performance is determined by electrochemical activation (electrode reaction kinetics), internal resistance and mass transfer (diffusional) limitation. Polarization from electrochemical activation refers to the potential loss because of the slow kinetics of the heterogeneous electrode reaction. Because reaction rate constant increases exponentially with the working potential of anode, this type of polarization usually causes voltage drop in the low current density region, where working potential is close to reversible electrode potential. For the biofuel cell examined in this work, this seemed not to be a problem because sharp increase of anodic potentials was not observed. With glassy carbon backing, the internal resistance of biofuel cell was about 225 Ω, as estimated by impedance analysis. Considering that the short circuit current obtained from this type of anodes was less than 60 μA, the contribution of this internal resistance to the anodic polarization should be less than 15 mV. Therefore, it appeared that the ohmic internal resistance should not limit the biofuel cell performance in this study.

Enzyme loading is also a critical factor for high output current density. For example, when GOx is randomly packed as monolayer on a flat surface, a

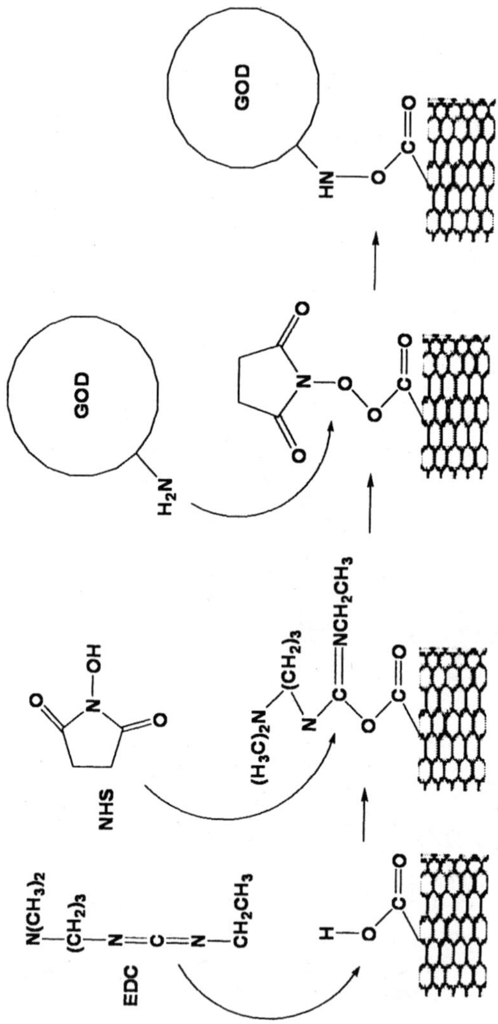

Figure 2. Scheme of covalent attachment of GOx to MWCNT

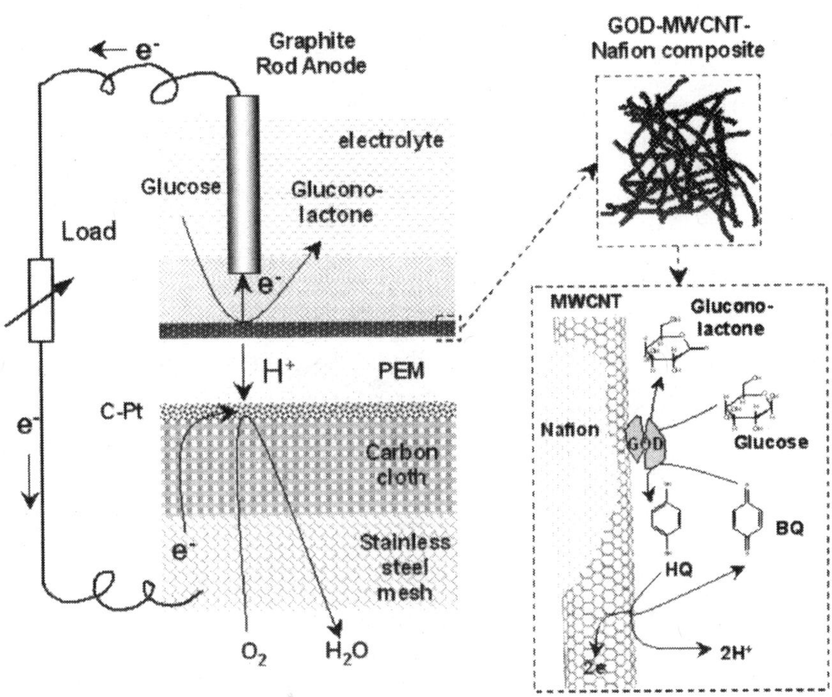

Figure 3. Configuration of GOx-O_2 biofuel cell. PEM: proton exchange membrane (Nafion® 117); C-Pt: carbon-supported platinum, cathode catalyst; BQ: benzoquinone; HQ: hydroquinone.

loading of 1.7×10^{-12} mol/cm^2 (0.27 μg/cm^2) may be achieved. If assuming the enzyme molecules are 100% active with a turnover number of 600 s^{-1}, the upper limit of the current density is only about 0.2 mA/cm^2 (18-21). The enzyme loading on the composite-coated glassy carbon electrode was estimated to be 43±3 μg (~130 μg/cm^2), much higher than the theoretical monolayer coverage loading. The activity of the resulting electrodes towards glucose oxidation (non-electrochemical) was measured using oxygen and BQ as the electron acceptor respectively Eq. (1-3):

$$\text{Glucose} + FAD \rightarrow \text{Gluconolactone} + FADH_2 \quad (1)$$

$$FADH_2 + O_2 \rightarrow FAD + H_2O_2 \quad (O_2 \text{ as the electron acceptor}) \quad (2)$$

$$\text{or } FADH_2 + BQ \rightarrow FAD + HQ \; (\text{BQ as the electron acceptor}) \quad (3)$$

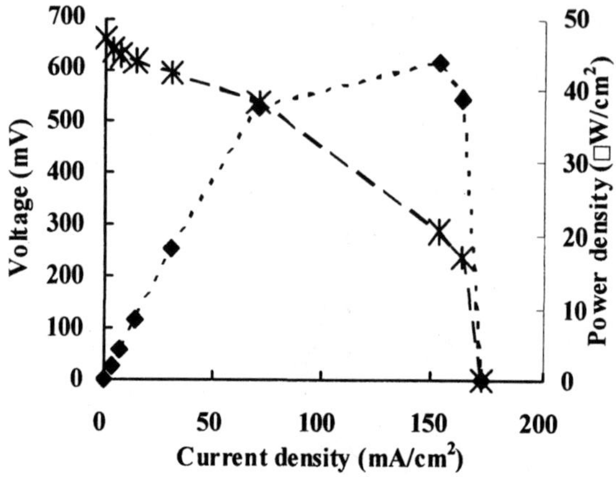

Figure 4. Performance of biofuel cell using composite coated glassy carbon anode. Dashed line: voltage; Dotted line: power density.

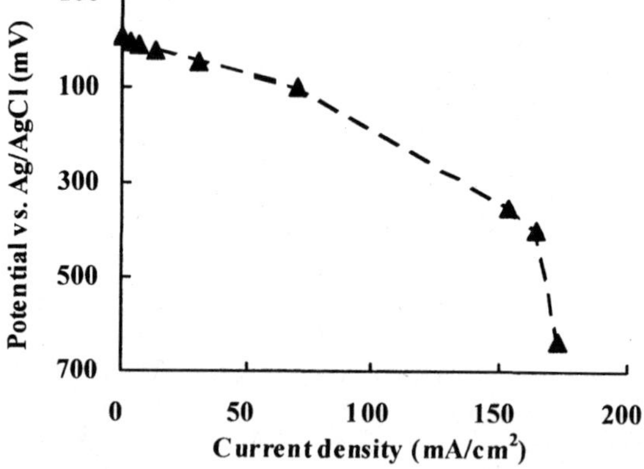

Figure 5. Polarization curve of the anode.

where *FAD* and $FADH_2$ (cofactors) represent the oxidized and reduced enzymes, respectively. With BQ as the electron acceptor (Eq. 3), an initial glucose oxidation rate of 0.15 μmol/min was observed under the specified conditions. Based on the estimated enzyme loading (43 μg), the specific activity of GOx in the composite electrode was 0.0035 μmol/(min.μg enzyme), only accounting for ~1% of the activity of free enzyme in homogeneous reaction system. However, if converting this activity into current density, it corresponded to a current density of 1.4 mA/cm^2, 8 times higher than the observed shirt circuit current density (SCCD). In another activity test with GOx-CNT (before casted into the composite electrode) suspended in reaction solution and activity of 0.34 μmol/min was observed, equivalent to a current density of 3.3 mA/cm^2. These tests imply that there should be enough enzyme on the electrode for much higher reaction rate than the observed current.

Several diffusion processes are involved in the anode, including that of glucose from bulk solution to enzyme active sites, gluconolactone from electrode to bulk solution, BQ from anode to enzyme active sites and HQ from enzyme to anode. For a rough estimation, the diffusion coefficient of HQ and BQ can be taken as 8.4×10^{-6} cm^2/s (60), while assuming glucose has a similar diffusion coefficient. That will transfer to maximum diffusion rates of the redox mediator and glucose as 8.5×10^{-10} and 3.4×10^{-8} $mol/(s\cdot cm^2)$, respectively, by accounting the concentration of each species available in the reaction medium. These maximum mass transfer rates are equivalent to currents of 0.16 and 6.6 mA/cm^2 by assuming 100% reaction efficiency. Compare to the observed current density of ~0.17 mA/cm^2, it indicates that the diffusion of redox mediator might be the limiting step.

To improve the mass transfer inside the composite electrodes, porous backing materials such as carbon cloth (ElectroChem, Inc.; 0.5 mm in thickness), carbon felt (Alfa Aesar, 3.2 mm in thickness) and carbon paper (AcCarb™ P50, Fuelcellstore.com; 0.17 mm in thickness) were examined. Carbon paper appeared to be more porous than carbon cloth, but less porous than carbon felt (Figure 6). The key performance parameters of biofuel cells using different porous backing materials were summarized in Table 2, including open circuit voltage (OCV), maximum power density (PD_{max}), working voltage and SCCD. Although it was expected that the porous structure of the backing materials might promote the diffusion within the composite, significant improvement of power density was only observed using carbon paper. A PD_{max} of 152 $\mu W/cm^2$ was observed, ~3 folds of the PD_{max} using glassy carbon backing.

Conclusions

In this work we explored the use of nano-composite electrodes for biofuel cell application. Enzymatic electrodes were prepared by casting GOx-MWCNT-

Figure 6. Porous electrode backing materials. ***a, b****, and* ***c:*** *SEM images of carbon cloth, carbon felt and carbon paper, respectively. Thickness: 0.35, 3.18 and 0.175 mm, respectively.*

Table 2. Key performance parameters of biofuel cells using different porous backing materials

Anode Backing	*OCV (mV)*	PD_{max} *(μW/cm²)*	*SCCD (μA/cm²)*
Carbon Cloth	534	12.0 @ 459 mV	56
Carbon Felt	510	58 @ 353 mV	215
Carbon Paper	585	152 @ 356 mV	522

NOTE: Power and current densities were calculated based on the projective area of the anodes (0.32 cm^2).

Nafion® composites on carbon-based backing materials. It has been found that MWCNT-Nafion® composite electrodes afforded much-enhanced electrochemical flux and reaction kinetics. In addition, the properties of electrode backing materials showed strong impacts on the performance of the electrodes and biofuel cells. That observation was related to the diffusion and reaction rates within the composite electrode, and it showed a possible limitation from the mass transfer of the redox mediators. When carbon paper was used to replace glassy carbon as the backing material, an improved power density of 152 $\mu W/cm^2$ was achieved. Overall, our results showed that the use of nanostructures was able to provide an enzyme loading high enough to generate about 8 times higher current density than what currently observed; yet mass and electron transfer processes within the composite electrode become the limiting factors.

References

1. N. Kim, Y. Choi, S. Jung and S. Kim, *Biotechnology and Bioengineering*, **2000**, *70*, 109-114.
2. G. T. R. Palmore and G. M. Whitesides, *ACS Symposium Series*, **1994**, *566*, 271-290.
3. K. Lewis, *Bacteriological Reviews*, **1966**, *30*, 101-113.
4. E. Katz, N. A. Shipway and I. Willne, in *Handbook of fuel cells - Fundamentals technology and applications*, eds. W. Vielstich, A. Lamm and H. A. Gasteiger, John Wiley and Sons Ltd, Chichester, Editon edn., 2003, vol. 1, pp. 355-381.
5. S. K. Chaudhuri and D. R. Lovley, *Nature Biotechnology*, **2003**, *21*, 1229-1232.

6. D. R. Bond and D. R. Lovley, *Applied and Environmental Microbiology*, **2003**, *69*, 1548-1555.
7. H. J. Kim, H. S. Park, M. S. Hyun, I. S. Chang, M. Kim and B. H. Kim, *Enzyme Microb. Technol.*, **2002**, *30*, 145-152.
8. W. H. Scouten, J. H. T. Luong and R. S. Brown, *Trends Biotechnol.*, **1995**, *13*, 178-185.
9. K. Habermuller, M. Mosbach and W. Schuhmann, *Fresenius' Journal of Analytical Chemistry*, **2000**, *366*, 560-568.
10. R. S. Freire, C. A. Pessoa, L. D. Mello and L. T. Kubota, *Journal of the Brazilian Chemical Society*, **2003**, *14*, 230-243.
11. S. D. Varfolomeev, I. N. Kurochkin and A. I. Yaropolov, *Biosensors & Bioelectronics*, **1996**, *11*, 863-871.
12. A. L. Ghindilis, P. Atanasov and E. Wilkins, *Electroanalysis*, **1997**, *9*, 661-674.
13. W. Schuhmann, *Reviews in Molecular Biotechnology*, **2002**, *82*, 425-441.
14. M. R. Tarasevich, V. A. Bogdanovskaya, N. M. Zagudaeva and A. V. Kapustin, *Russian Journal of Electrochemistry*, **2002**, *38*, 335.
15. R. A. Bullen, T. C. Arnot, J. B. Lakeman and F. C. Walsh, *Biosensors & Bioelectronics*, **2006**, *21*, 2015-2045.
16. R. A. Marcus, *Reviews of Modern Physics*, **1993**, *65*, 599-610.
17. R. A. Marcus and N. Sutin, *Biochimica et Biophysica Acta, Reviews on Bioenergetics*, **1985**, *811*, 265-322.
18. I. Willner, V. Heleg-Shabtai, R. Blonder, E. Katz, G. Tao, A. F. Bueckmann and A. Heller, *J. Am. Chem. Soc.*, **1996**, *118*, 10321-10322.
19. A. Guiseppi-Elie, C. Lei and R. H. Baughman, *Nanotechnology*, **2002**, *13*, 559-564.
20. Y.-D. Zhao, W.-D. Zhang, H. Chen and Q.-M. Luo, *Anal. Sci.*, **2002**, *18*, 939-941.
21. C. Cai and J. Chen, *Analytical Biochemistry*, **2004**, *332*, 75-83.
22. G. Govil and A. Saran, *Journal of the Indian Chemical Society*, **1982**, *59*, 1226-1228.
23. A. Heller, *Phys. Chem. Chem. Phys.*, **2004**, *6*, 209-216.
24. T. Chen, S. C. Barton, G. Binyamin, Z. Gao, Y. Zhang, H.-H. Kim and A. Heller, *J. Am. Chem. Soc.*, **2001**, *123*, 8630-8631.
25. N. Mano, F. Mao, W. Shin, T. Chen and A. Heller, *Chem. Commun.*, **2003**, 518-519.
26. N. Mano, F. Mao and A. Heller, *J. Am. Chem. Soc.*, **2002**, *124*, 12962-12963.
27. N. Mano, F. Mao and A. Heller, *J. Am. Chem. Soc.*, **2003**, *125*, 6588-6594.
28. V. Soukharev, N. Mano and A. Heller, *J. Am. Chem. Soc.*, **2004**, *126*, 8368 -8369.
29. *US Pat. Application*, 2004101741, 2004.

30. H. Ogihara, S. Takenaka, I. Yamanaka and K. Otsuka, *Carbon*, **2004**, *42*, 1609-1617.
31. C. Park and R. T. K. Baker, *J. Phys. Chem. B*, **1998**, *102*, 5168-5177.
32. C. Gao, M. J. Whitcombe and E. N. Vulfson, *Enzyme Microb. Technol.*, **1999**, *25*, 264-270.
33. H. Klefenz, *Engineering in Life Sciences*, **2004**, *4*, 211-218.
34. R. P. Raffaelle, B. J. Landi, J. D. Harris, S. G. Bailey and A. F. Hepp, *Materials Science & Engineering, B: Solid-State Materials for Advanced Technology*, **2005**, *B116*, 233-243.
35. H. Jia, G. Zhu, B. Vugrinovich, W. Kataphinan, D. H. Reneker and P. Wang, *Biotechnology Progress*, **2002**, *18*, 1027-1032.
36. Z.-G. Wang, Z.-K. Xu, L.-S. Wan, J. Wu, C. Innocent and P. Seta, *Macromolecular Rapid Communications*, **2006**, *27*, 516-521.
37. K. Sawicka, P. Gouma and S. Simon, *Sensors and Actuators, B: Chemical*, **2005**, *B108*, 585-588.
38. K. Rege, N. R. Raravikar, D.-Y. Kim, L. S. Schadler, P. M. Ajayan and J. S. Dordick, *Nano Letters*, **2003**, *3*, 829-832.
39. S. G. Wang, Q. Zhang, R. Wang, S. F. Yoon, J. Ahn, D. J. Yang, J. Z. Tian, J. Q. Li and Q. Zhou, *Electrochem. Commun.*, **2003**, *5*, 800-803.
40. J. J. Davis, M. L. H. Green, H. A. O. Hill, Y. C. Leung, P. J. Sadler, J. Sloan, A. V. Xavier and S. C. Tsang, *Inorganica Chimica Acta*, **1998**, *272*, 261-266.
41. H. Jia, G. Zhu and P. Wang, *Biotechnology and Bioengineering*, **2003**, *84*, 406-414.
42. M. Qhobosheane, S. Santra, P. Zhang and W. Tan, *Analyst (Cambridge, U. K.)*, **2001**, *126*, 1274-1278.
43. A. Heilmann, N. Teuscher, A. Kiesow, D. Janasek and U. Spohn, *Journal of Nanoscience and Nanotechnology*, **2003**, *3*, 375-379.
44. B. El-Zahab, H. Jia and P. Wang, *Biotechnology and Bioengineering*, **2004**, *87*, 178-183.
45. J. K. Kim, J. K. Park and H. K. Kim, *Colloid Surf. A-Physicochem. Eng. Asp.*, **2004**, *241*, 113-117.
46. P. Wang, S. Dai, S. D. Waezsada, A. Y. Tsao and B. H. Davison, *Biotechnology and Bioengineering*, **2001**, *74*, 249-255.
47. J. Kim, H. Jia and P. Wang, *Biotechnology Advances*, **2006**, *24*, 296-308.
48. Y. Xiao, F. Patolsky, E. Katz, J. F. Hainfeld and I. Willner, *Science*, **2003**, *299*, 1877-1881.
49. L. G. Bachas, S. A. Law, V. Gavalas, J. C. Ball and R. Andrews, in *Abstracts of Papers, 223rd ACS National Meeting, Orlando, FL, United States, April 7-11, 2002*, Editon edn., 2002, pp. ANYL-064.
50. C. V. Nguyen, L. Delzeit, A. M. Cassell, J. Li, J. Han and M. Meyyappan, *Nano Letters*, **2002**, *2*, 1079-1081.
51. S. Sotiropoulou and N. A. Chaniotakis, *Analytical and Bioanalytical Chemistry*, **2003**, *375*, 103-105.

52. J. Wang, M. Musameh and Y. Lin, *J. Am. Chem. Soc.*, **2003**, *125*, 2408-2409.
53. H. Xue, W. Sun, B. He and Z. Shen, *Synthetic Metals*, **2003**, *135-136*, 831-832.
54. X. Yu, D. Chattopadhyay, I. Galeska, F. Papadimitrakopoulos and J. F. Rusling, *Electrochem. Commun.*, **2003**, *5*, 408-411.
55. J. Chen, J. Bao, C. Cai and T. Lu, *Analytica Chimica Acta*, **2004**, *516*, 29-34.
56. Z. Grabarek and J. Gergely, *Analytical Biochemistry*, **1990**, *185*, 131-135.
57. T.-W. Hui, K.-Y. Wong and K.-K. Shiu, *Electroanalysis*, **1996**, *8*, 597-601.
58. R. S. Nicholson, *Anal. Chem.*, **1965**, *37*, 1351-1355.
59. C. M. A. Brett and A. M. O. Brett., *Electrochemistry : Principles, Methods, and Applications*, Oxford University Press, New York, 1993.
60. I. M. Kolthoff and J. I. Lingane, *Polarography*, Interscience Publishers, New York, 1952.

Chapter 19

In Situ Characterization Techniques for Design and Evaluation of Micro- and Nano-Enzyme-Catalyzed Power Sources

Michael J. Cooney and Bor Yann Liaw

Hawaii Natural Energy Institute, School of Ocean and Earth Science and Technology, University of Hawaii at Manoa, Honolulu, HI 96822

Although the literature is full of presentations on various aspects of bio-fuel cell performance, whether they are focused on components, such as anode or cathode; or whole cell, there is a need for an overriding view of how the bio-fuel cell should be characterized in terms of both defining performance of a component on a consistent basis that can be compared across the discipline, but also in defining the relative impacts on cell performance. Here, we suggest a view of the bio-fuel cell in the context of how performance can be judged by researchers spanning disciplines and laboratories. We then review many characterization techniques available and provide guidelines on how their use relates to characterizing overall component and cell performance.

An enzyme-catalyzed fuel cell is a fuel cell in which the fuel is oxidized at the anode, and the oxidant is reduced at the cathode, by use of enzyme as the catalyst [1]. A typical bio-fuel cell configuration for laboratory investigation is depicted in Figure 1. The anode captures the electrons freed from the enzyme-catalyzed oxidation of the fuel while an equal number of protons are liberated and migrated to the cathode via electrochemical diffusion (which is a combined effect of chemical diffusion and migration) in the electrolyte. Simultaneously, a catalyzed reduction of an oxidant takes place at the cathode, such as a platinum-catalyzed reduction of oxygen, where the migrated protons and electrons can react with the oxygen to form water. The energy (exhibited as power) that can be harnessed by this transformation of chemical energy to electricity is harvested as external work conducted by the electrons to a load in an external circuit. The electrolyte plays an important role in this process in that it only allows transport of ionic species, thereby forcing the electrons to migrate through the external circuit. As such energy is harnessed via the electrical work conducted on the external load. To facilitate this process, sometimes a proton permeable membrane is used to separate the two half-cells as well as the reactants. The use of proton permeable membranes may be eliminated in applications that, with careful selection of enzymes (and their mutant derivatives) to exclusively and effectively catalyze the desired reaction at the respective electrode, the cell can operate with high efficiency and without interference by the crossing of the reactants [2,3].

Performance of an Enzyme-Catalyzed Power Source

The flow of electrons through the external circuit yields the current I which flows in direction opposite to the electrons and through a resistive load (R_{load}) where the work is done. The potential measured across this load gives the cell voltage (V_{cell}). To conduct work, the cell's electromotive force has to overcome its own internal resistance (R_{cell}), which is a lumped parameter that consists of a number of serial and parallel resistances, including resistance to electron transfer and/or proton transport via electrolyte bulk diffusion and migration across the double layer. Therefore,

$$V_{cell} = I \cdot R_{load} = V^{\circ}_{cell} - I \cdot R_{cell} \quad (1)$$

The power (P) is given by the product of the cell voltage by the current:

$$P = V_{cell} \cdot I \quad (2)$$

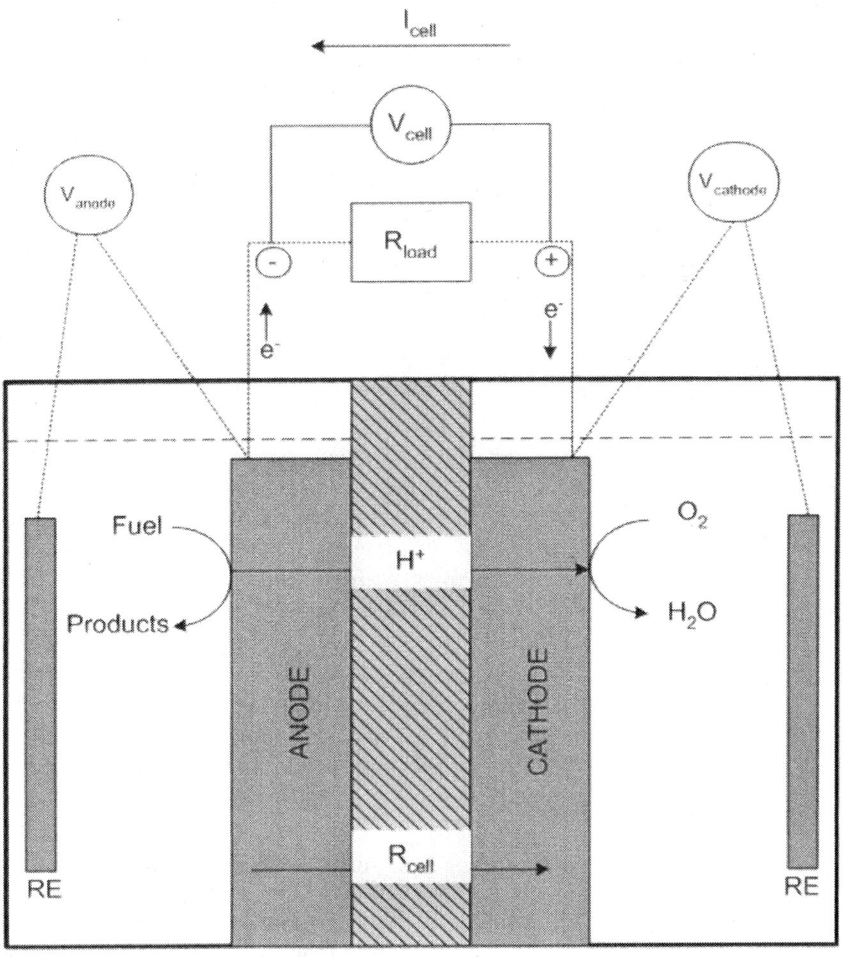

Figure 1. Schematic of an enzyme catalyzed bio-fuel cell.

There is a common plot with two curves, presented in Figure 2, to depict cell performance. In such a plot, the cell voltage (V_{cell}) is measured and plotted along with the power density against the current density. Traditionally, the current and power densities are calculated with the cross section area (CSA) of the electrode, under the presumption of a pseudo-2-D cell configuration. By contrast, in the field of catalysis, only adsorption surface area (often measured by the Brunauer-Emmett-Teller or BET technique) is used [4]. The problem with CSA-based representations is that they make no reference to the accessible surface area (ASA) in a unit volume (cm^2/cm^3), where the electrode is actually a 3-D structure; e.g., [5-16]. For example, the BET technique uses, more or less, a static mode (in equilibrium) to measure both flow-through pores and blind pores within the porous medium [13], some of which might not provide catalytically active surface area. Therefore, increase in current or power density (based on CSA) cannot distinguish between better 3-D electrode geometric engineering and improved catalysis. This can be corrected by using the ASA instead of the CSA. The ASA is the accessible area defined by the flow-through pores in a porous electrode medium, and its value might be dependent upon the flow rate. Although the ASA is difficult to determine, it should be pursued because it represents the true surface area that is accessible by the reactants [12]. Ideally, with increasing capability in characterizing electrode surface, and the utilization of high surface area materials in catalytic electrode fabrication [6-11], the ASA of the electrode should be used to provide a more comprehensible and informative presentation of the current or power density.

An increasing number of pertinent discussions of these issues have been reported in the literature. For example, Kim et al. [14] showed that with careful manipulation of co-polymer blends, one can synthesize controlled mesopore structure in silica. Comparing mesopore structures with other porous media, Thommes et al. [15] showed that hysteresis observed in the adsorption-desorption critically related to the pore structure in the porous networks, suggesting that accessibility to different size of pores and structures depends on the adsorption-desorption mechanism and the pore size and shape in the network structure. Jena and Gupta [16] recently demonstrated that, using envelope surface area analysis, one can distinguish the accessible area in flow-through pore channels, which is also dependent on the flow rate itself. There are also a number of emerging research activities [17,18] that have addressed the macro and mesoporous pore structure of chitosan scaffolds (CS) in terms of pore structure and its influence on fluid flow and diffusion coefficients. While these studies have focused on scaffolds for tissue implants and drug delivery, there is now an increasing surge in literature reports that are proposing chitosan as an electrode matrix for biosensors and bio-fuel cells, in part because of their attractive pore structure [18-21]. Some typical scaffolds fabricated in the lab of the authors are presented in Figure 3.

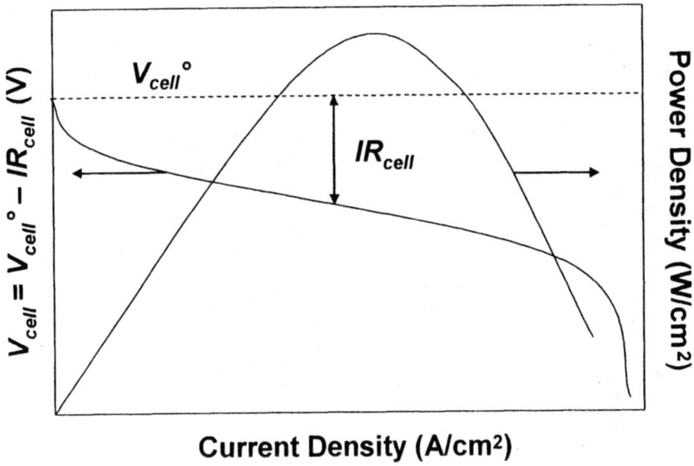

Figure 2. A schematic diagram that depicts an enzyme bio-fuel cell performance, in which cell voltage and power density are plotted against current density.

It should be noted that Figure 2 actually plots the difference between V°_{cell}, the equilibrium cell potential as defined by thermodynamics, and the polarization potential, $I \cdot R_{cell}$. This relation depicts the reality that, with increasing current, the cell voltage drop reflects that portion of the potential that must be used to move the electrons against the cell resistance [22]. This trade-off between potential and current is analogous to fluid systems whereby the height a fluid possessing at a given pressure head (e.g., height of head) that can be pumped decreases with the velocity at which the fluid is pumped through the pipe. The faster a fluid is pumped the greater the resistance to its flow and hence the more of the potential head that must be used to overcome the resistance to flow.

One should also realize that the cell voltage (V_{cell}) represents the difference between the cathode and anode potential:

$$V_{cell} = V_{cathode} - V_{anode} \tag{3}$$

The potential of the cathode ($V_{cathode}$) and anode (V_{anode}) can be measured relative to a (standard) reference electrode to yield the formal potential for the corresponding half-cell reaction at equilibrium, thus $V^{\circ}_{cathode}$ and V°_{anode}; and under polarization at different current densities, thus $V_{cathode} = V^{\circ}_{cathode} - I \cdot R_{cathode}$ and $V_{anode} = V^{\circ}_{anode} - I \cdot R_{anode}$. This representation is shown in Figure 4.

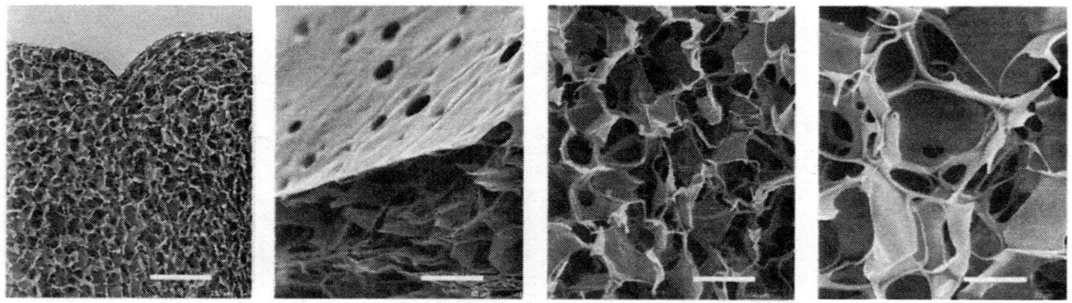

Figure 3. Chitosan scaffolds. Scale bar dimension (L to R): 225 μm, 56.25 μm, 56.25 μm, and 22.5 μm.

Any attempt to improve bio-fuel cell power density, one can begin with an equation equivalent to (2) as a guide:

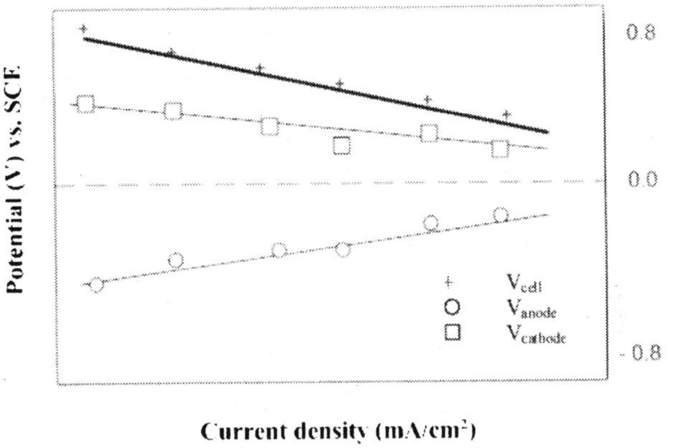

Figure 4. A schematic showing the change of cell voltage and electrode potentials as a function of current density during a fuel cell operation. The polarization on each electrode can be monitored against a common reference. (Reproduced with permission from the author. Copyright held by author.)

$$\overline{P} = V_{cell} \cdot \overline{I} \qquad (4)$$

Here the overhang bar represents the notation for density. Substituting equation (1) with density notation into (4) yields

$$\overline{P} = \left(V^{\circ}_{cell} - \overline{I} \cdot R_{cell}\right) \cdot \overline{I} = V^{\circ}_{cell} \cdot \overline{I} - \overline{I}^{2} \cdot R_{cell} \qquad (5)$$

Equation (5) suggests that for any given rate of operation in the fuel cell, there are two clear paths to achieve improved power density. The first one is the thermodynamic consideration as represented in the first term of equation (5), where the equilibrium potential of the cell V°_{cell} is governed by the cell chemistry through a judicious choice of the oxidation and reduction reaction at the anode and cathode, respectively. In a battery or conventional fuel cell, this would be defined by the electrochemical potential of the respective anodic oxidation and cathodic reduction reactions. In an enzyme bio-fuel cell, however, this potential is dictated by the formal potential of the last mediator to transfer the charge to the electrode. For example, an ethanol enzyme bio-fuel cell concept similar to that proposed by Palmore et al. [23] is shown in Figure 5 for illustration.

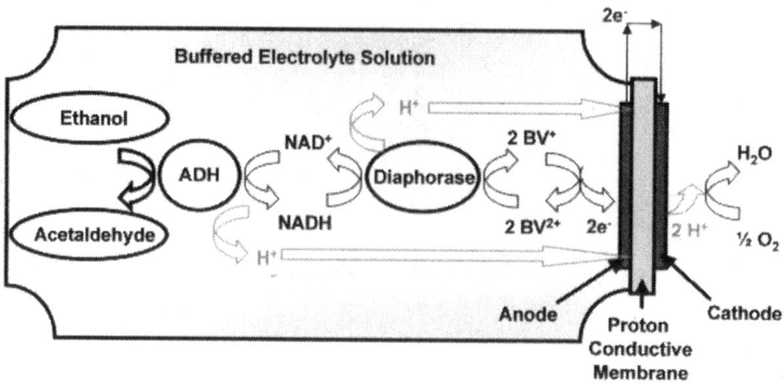

Figure 5. A schematic showing an ethanol enzymatic bio-fuel cell concept based on a similar system proposed by Palmore et al. [23]

In this scheme ethanol is oxidized by the alcohol dehydrogenase (ADH) in a proposed two-electron charge transfer process that reduces the required cofactor NAD^+ to NADH. NADH needs to further transfer the charge to the electrode in order to release the electrons to conduct work, while recycling NADH to NAD^+ to perpetuate the anodic reaction cycle. Direct oxidation of NADH at the electrode surface has been found rather difficult [24], a process that requires relatively high activation energy or overpotentials. Palmore et al. [23] proposed to use benzyl viologen (BV) as a mediator, due to the favorable formal potential for the oxidation of BV^+ to BV^{2+}, which is reasonably close to, but less negative than the formal potential for the oxidation of NADH to NAD^+ on the reduction potential scale. In conjunction with the use of diaphorase, which has been suggested capable of catalyzing the regeneration of NAD^+ from NADH, this ethanol bio-fuel cell scheme represents an optimal thermodynamic consideration, as shown in Figure 6.

From this perspective, Palmore and coworkers paid attention to V°_{cell} in equation (5) by establishing a thermodynamically favorable potential at the anode for ethanol oxidation. In practice, however, BV is difficult to use, requiring strict anaerobic conditions. This is not a surprise, due to BV's high formal potential for oxidation, as observed in Figure 5. Therefore, unless the BV is bound directly to the electrode, and in a manner that supports direct electron transfer, it will pass its electrons to any species with a greater affinity for electrons than the electrode – in this case, oxygen in the solution. This competition introduces an interference and internal leakage, via the side reaction, that terminates the charge transfer pathway in the anode and resulting in no useful energy that can be harnessed. To overcome this, one solution is to develop an efficient mechanism by which the BV can preferentially transfer its charge to the electrode surface. This can be achieved, presumably, through the direct immobilization of BV to the electrode surface without inhibiting the reaction kinetics. On the other hand, we could protect the immobilized BV layer with an oxygen barrier to retard the diffusion of oxygen to the charge transfer sites.

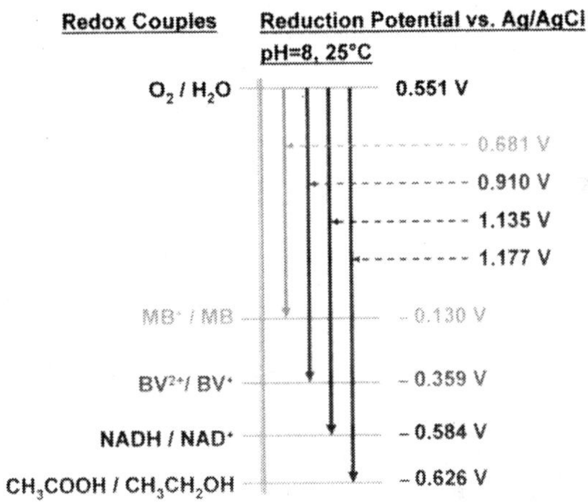

Figure 6. Redox couples in an enzymatic ethanol fuel cell and their reduction potential versus the Ag/AgCl reference electrode at pH 8 and 25°C. (Reproduced with permission from the author. Copyright by held by author.)

If immobilization of the thermodynamically appropriate mediator (e.g., BV) is not an option, another solution would be to choose an alternative mediator. Meldola's blue (MB), which is also on the chart, has been proposed as one such mediator because of its ability to bind to the enzyme and is less reactive with oxygen [25-30]. Similar studies [23] have reported the use of redox mediators such as phenazine or methosulfate, despite their relatively unattractive formal potentials (relative to NADH/NAD^+ oxidation) to shuttle electrons from the enzyme to the electrode. Like MB these mediators were presumably less reactive with oxygen but at a cost in terms of formal potential relative to the NADH/NAD^+ oxidation. Another favorable approach has been to covalently attach or electrochemically polymerize the mediator to the electrode surface. This has been reported for MB on glassy carbon [31], but proved unsuccessful on platinum (data not shown). It has also been shown for poly-(methylene green) (poly-(MG)) on glassy carbon or platinum [32-34]. Despite the reports of these immobilized mediators exhibiting improved stability, favorable reaction rate for NADH oxidation, and decreased sensitivity to oxygen, the formal potentials of these polymerized mediators usually require a tradeoff between thermodynamic consideration and kinetic stability. This tradeoff has inspired recent reports on direct electron transfer mechanisms to give the maximum potential difference between the anode and cathode [8].

While the above discussion has emphasized the importance of getting the thermodynamics right, any attempt to improve power density must also eliminate kinetic limitations [8,20,35]. In other words, to improve power density one should optimize both the thermodynamics and kinetics, for a given chemistry. The importance of kinetics is emphasized by the second term of equation (5), where the cell internal resistance R_{cell} is a crucial parameter in determining the power density in a cell reaction. As we discussed earlier in equation (1), R_{cell} is a lump sum of various attributes, including charge transfer resistance of the electrode and the electrolyte conductance. From this perspective, improving kinetics is not solely an issue of finding a better enzyme, or improving its activity within the immobilized state. Taking the ethanol bio-fuel cell scheme in Figure 4 as an example, the kinetics, and thus power density, can also be improved by the avoidance or minimization of side reactions such as the oxygen interference in the charge transfer pathway.

To better emphasize this point, we can consider the use of immobilization strategies to enhance enzyme loading as a means to increase the density of active catalytic sites [8,20,35-38]. In general these most often consist of entrapment [39,40] or covalent bonding techniques to condense enzyme population inside a polymer backbone [41]. In a simple smooth 2-D electrode surface, where ASA is equivalent to the CSA, an increase in enzyme loading simply reflects the improvement of enzyme packing density. In this case the relationship between the feedstock and the reaction sites can be discussed with a simple 1-D (diffusion + migration) model, without consideration of the flow field. When the concept is extended, however, to a 3-D electrode, where the equivalency between ASA and CSA departs, the effect of enzyme loading on the electrode performance becomes less trivial when the effects of porosity and electrode thickness on kinetics need to be considered. To begin with, the porosity and tortuosity of the porous electrodes under the influence of specific electric field and mass transport condition will influence the fluid flux and flow patterns [9]. As such, this expansion in geometry from a simple 1-D to a much more complicated 2-D or even 3-D framework immediately increases in the number and complexity of terms that influence the kinetics [10]. In any case the improvement of enzyme loading per unit ASA resembles an increase of interfacial conductance for the charge transfer; thus, leading to a lower charge transfer resistance in the R_{cell}.

Besides the enzyme loading improvement, one should also pay attention to the supply of feedstock and removal of the residual unusable products. As the increase in enzyme loading provides higher capacity to handle current flux per unit ASA, sufficient fuel supply is needed to sustain the power generation. In the context of 3-D porous electrodes, the handling of the fuel logistics can be similarly complex. There is concomitant need for large internal void volume of interconnected pores to support mass transport of fuel. As such, it is important to realize that the ASA, porosity, the average pore size and its distribution are parameters that need to be all optimized [42]. Furthermore, it becomes critical that the pore shape and its distribution are both multidimensional and multidirectional [43]. A well-connected pore channel network, similar to a cave with serpentine formation or that found in bone, are good examples that show co-existence of large surface area with channeled pore network, which will reduce the drag in fuel transport through the porous electrode and contribute as much to power density as the enhancement of the enzyme loading [9]. In this regard, the reduction in the diffusional and migration resistance will also be reflected in the reduction of R_{cell}.

The above discussion presents an important perspective on the relationship depicted in equation (5) for enzyme bio-fuel cell performance optimization. Before beginning the characterization of any aspect of the enzyme bio-fuel cell performance; whether it be the use of a new mediator, enzyme immobilization technique, electrode material, or fuel delivery system; we suggest that one should look at the basic governing equations of the bio-fuel cell operation and to define precisely upon which term(s) in equation (5) the proposed approach will improve and estimate if those gains are significant relative to the magnitude of the other terms. For example, if it is proposed to replace the platinum catalyzed cathode with an enzyme system, this should be justified in terms of improving either the thermodynamic potential difference across the cell ($V_{cathode} - V_{anode}$) or the reduction of the resistance within the cell (i.e., R_{cell}), and maybe both.

Characterization Techniques and Cell Performance Criteria

Understanding the thermodynamic and kinetic constraints in enzyme bio-fuel cell performance in power generation, we need tools and techniques to help us characterize these constraints so we can begin to identify challenges, alleviate harsh requirements, overcome technical barriers, and quantify improvements. For example, enzyme bio-fuel cells require operation typically in (chemically and electrochemically) near-neutral conditions and in liquid environments. Such environments pose unique requirements on the characterization and analysis methodology applied, since most of the *ex situ* characterization techniques often require post operation manipulation of the sample such as using drying and/or application of vacuum. The requirements of these *ex situ* characterization methodologies can introduce artifacts that may not represent the actual characteristics of the system, materials, physical structure, or cell performance in the liquid phase environments. Besides the unique need for *in situ* characterization, it is also important to investigate the system across dimensions, from macroscopic properties to microscopic behavior for both temporal and spatial correlation. For instance, an enzyme's catalytic behavior is sensitive to the immediate environment surrounding the enzyme, in the dimension of nanometers [37]. By contrast, power generation is the collective measure of catalytic behavior over the macroscopic dimension of at least several hundred microns, if not more. This spatial correlation demands innovative integration of unique instrumentation and analytical capabilities that span over several orders of magnitude in dimensions. In the following sections, we outline a range of analytic techniques that we consider useful and pertinent to the characterizations of the bio-fuel cell performance. We begin with some well-known techniques to understand what they can be utilized for characterizing the performance of bio-fuel cell and its components in terms of thermodynamic and kinetic properties, and then describe how they can be integrated with other physical characterization techniques to derive spatial and temporal correlations. In particular, for the fabrication of 3-D porous electrode structure and architecture, we will discuss tools to monitor the active material behavior and the surrounding morphological dependence.

Potentiometry

Potentiometry is an established technique, primarily designed to determine the activity of the electrochemical active species in a solution with respect to a known reference state imposed by a suitable reference electrode and its condition [22]. The technique is typically used to measure an electrode potential under open circuit condition. With a suitable choice of reference electrode, the equilibrium open circuit potential of the working electrode (which is of interest) is measured against the reference [22]. The potential difference measured by the potentiometry is related to the difference between the activity a_i of the electroactive species i present on the surface of the working electrode and the $a_{i,ref}$ on the reference electrode, as depicted by the Nernst equation:

$$V = V_{ref} + \frac{RT}{zF}\ln\left(\frac{a_i}{a_{i,ref}}\right) \tag{6}$$

where V_{ref} is the potential of the reference electrode in its referent state (not necessarily standard), R the gas constant, T temperature in Kelvin, z the valence of species i, and F the Faraday constant. Hence, the potential difference measured between the electrode and the reference is

$$\Delta V = V - V_{ref} = \frac{RT}{zF}\ln\left(\frac{a_i}{a_{i,ref}}\right) \tag{7}$$

Equation (7) is the simplest form defining the relation between redox potential of an electrode (electrode potential) versus a reference. It can be used to verify correct functionality of an electrode. In a liquid system, for simplicity, one can usually substitute the activity with the concentration of the species.

In the context of enzyme bio-fuel cells, potentiometry can be used to measure the open circuit voltage (OCV) of the cell:

$$V_{cell} = V^{\circ}_{cell} + \frac{RT}{zF}\ln(K_{eq}) \tag{8}$$

where V°_{cell} is the OCV at the standard state and K_{eq} the equilibrium constant of the cell reaction. Similar to equation (7), equation (8) allows us to measure the OCV of the cell, but it also takes into account the effect of pH on V_{cell} so that we can correct for the effect of pH on the formal potential of the working electrode against a known reference. The effect of pH is a particularly important term to consider in bio-fuel cells since the use of an enzyme-catalyzed electrode can be a non-neutral pH condition, such as seen with the use of laccases from yeasts [39,44]. Considering the oxidation of ethanol to acetaldehyde as catalyzed by the NADH-dependent ADH, as we discussed earlier, here we will present the case of a bio-fuel cell that uses polymerized methylene green on the anode to avoid the use of diaphorase [34]. This is a two-proton, two-electron process whereby the ethanol, in the presence of its cofactor NAD^+, is oxidized to acetaldehyde while NAD^+ is reduced to NADH with one proton released to the solution (when facilitated)

$$\text{Ethanol} + NAD^+ \rightarrow \text{Acetaldehyde} + \text{NADH} + H^+ \tag{9}$$

NAD^+ is then regenerated by the mediator releasing the second proton and two electrons

$$\text{NADH} \rightarrow NAD^+ + H^+ + 2e^- \tag{10}$$

The overall half-cell reaction on the anode is therefore

$$\text{Ethanol} \rightarrow \text{Acetaldehyde} + 2H^+ + 2e^- \tag{11}$$

If we apply equation (8) to reaction (11), we get

$$V = V^{\circ} + \frac{RT}{2F}\ln\frac{[\text{Acet}][\text{H}^{+}]^{2}}{[\text{Eth}]} \qquad (12)$$

The correct redox potential for reaction (11) as represented in equation (12) is however not attainable in the bio-fuel cell with a buffered solution, since the H^+ concentration will be controlled by the buffer solution. To mediate this issue, we can transfer the potential variation from the tempering of the H^+ concentration to a mediated standard potential term, as shown in equations (13) to (16):

$$V = V^{\circ} + \frac{RT}{2F}\ln[\text{H}^{+}]^{2} + \frac{RT}{2F}\ln\frac{[\text{Acet}]}{[\text{Eth}]} \qquad (13)$$

$$V = V^{\circ} - 2.303\frac{RT}{F}\text{pH} + \frac{RT}{2F}\ln\frac{[\text{Acet}]}{[\text{Eth}]} \qquad (14)$$

$$V = V'^{\circ} + \frac{RT}{2F}\ln\frac{[\text{Acet}]}{[\text{Eth}]} \qquad (15)$$

$$V'^{\circ} = V^{\circ} - 2.303\frac{RT}{F}\text{pH} \qquad (16)$$

This mediated potential term allows the reaction formal potential to be defined in the specific buffered solution used. In this manner the potentiometry technique can be used to direct improvement of a specific anode and cathode couple from a thermodynamic point of view.

Transient Potentiometry

In contrast to static potentiometry applied to equilibrium, transient potentiometry is a time-dependent technique that measures the electrode resting potential as a function of time [45]. This technique can be useful when the transient behavior of an electrode is of interest. For example, in the fabrication of enzyme bio-electrodes with physical adsorption, we can monitor the transient potential change to observe the kinetics of adsorption under different preparation conditions, thus offer a rapid electrode screening tool. An example is given in Figure 7, which presents the open circuit potential of two electrodes made by different techniques and reveals their behavior in the presence of hydrogen gas

[45]. In the first instance the electrodes are left blank to investigate their reactions with the gas stream, as a probe to their reactivity toward the gas, and, in the second, they are placed in the presence of the hydrogenase enzyme to screen for the rate and effectiveness of immobilization. The change in resting potential of the electrode with time is taken as a proxy for the adsorption of the enzyme.

The blank pyrolytic carbon paper (PCP) was shown to be inert to hydrogen [45] (e.g., the resting potential remained around +0.2 V vs. Ag/AgCl in either gas phase). However, when placed in the presence of enzyme in solution the PCP electrode resting potential began to drop, indicative of enzyme reactivity, after about 20 minutes, eventually stabilizing at around +0.10 V vs. Ag/AgCl. The phenomenon observed above suggests that the enzyme had directly absorbed to the electrode surface and catalyzed the oxidation of hydrogen. The electrode, however, did not reach the full reversible hydrogen potential (RHP), suggesting that the hydrogen activity is low on the electrode surface. It also reflects that the enzyme reactivity was high, but that the charge transfer was limited, and was thus unable to release proton with a prominent rate to dominate the cell potential.

The packed graphite column (PGC) electrodes, which possess higher ASA from the graphitic matrix embedded in nickel foams (Ni/PGC) or stainless steel meshes (SS/PGC), exhibited a much better immobilization and charge transfer efficiency [45]. Both PGC electrodes, post enzyme immobilization, and in hydrogen atmosphere, exhibited electrode potential close to the RHP, indicating that near unity hydrogen activity was achieved by the immobilized hydrogenase enzyme, and with high charge transfer efficiency. This result suggested that the adsorption of hydrogenase enzyme to the PGC support was much better than to the PCP. In this manner, transient potentiometry is a useful *in-situ* characterization tool for screening electrode materials, probing enzyme immobilization kinetics and efficiency, gauging enzyme reactivity toward fuels of interest, and detecting charge transfer efficiency.

Chronopotentiometry

In addition to the open circuit measurements, potentiometry can be carried out also with current passage (i.e., chronopotentiometry) [22]. Under a constant current mode, for instance, we can monitor the electrode or cell potential change with time to derive some useful information. The Sand equation, which is similar to that of the Cottrell diffusion model, is used for potentiometric analysis:

$$I = nF(CSA)\cdot D\left(\frac{\partial C}{\partial x}\right)_{x=0} \tag{17}$$

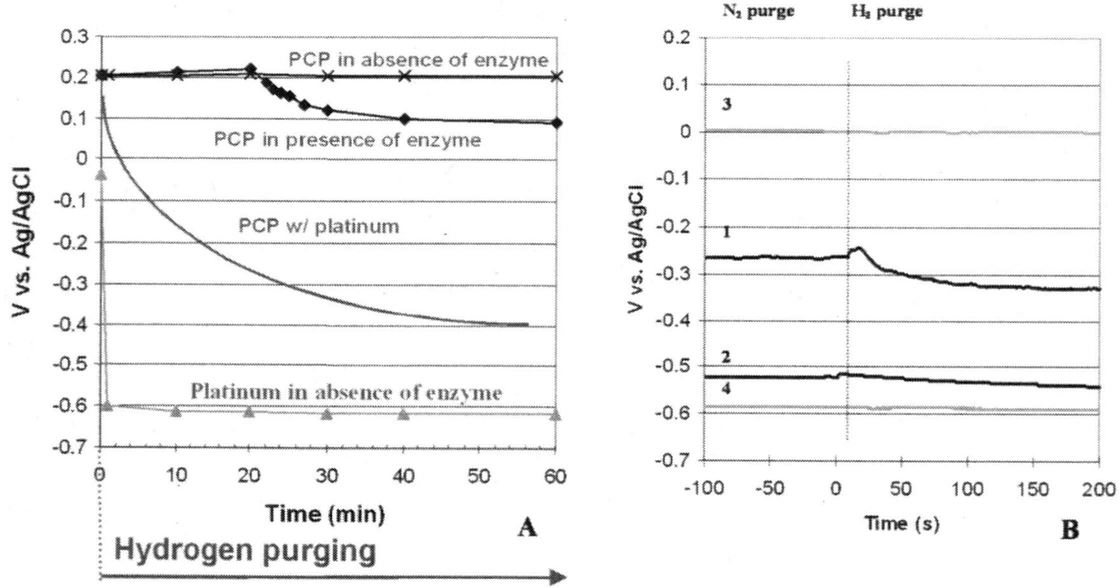

Figure 7. Transient potentiometry applied to adsorption of hydrogenase to (A) pyrolytic carbon paper (PCP) and (B) Ni/packed graphite column (PGC) and SS/PGC electrodes. Symbols: 1, Ni/PGC without enzyme; 2, Ni/PGC with enzyme; 3, SS/PGC without enzyme; 4, SS/PGC with enzyme. Time zero was set to represent change of gas sparging from nitrogen to hydrogen (at 75 ℃) [45].

Here the last term in equation (17) is a Fick's diffusion flux based on chemical concentration gradient [9].

Cyclic Voltammetry

Cyclic voltammetry (CV) is a well-established technique, in which we sweep the potential of a working electrode against a reference within a potential window at a specific or a range of sweep rate [22]. The current-potential (*I-V*) correspondence from the cyclic sweep (voltammogram) can be used to analyze the behavior of the working electrode in the presence of a redox reaction or a series of redox reactions.

CV has been used extensively in bio-fuel cell investigations. Depending upon the purpose of the investigation, the cell configuration and evaluation protocols are adjusted accordingly. A common use of CV, for example, is to determine the (formal) redox potential for a given chemistry and the nature of the chemistry will reflect on the *I-V* curves at a given sweep rate (e.g., [33]). Any redox couple will appear as oxidation and/or reduction *I-V* peaks on the curve, depending on the nature of the kinetics (e.g., [46]). The peak position permits the determination of the redox potential of the reaction. For an electrode that exhibits a Nernstian behavior in a redox reaction written as

$$\mathrm{Re} = \mathrm{Ox} + ne^- \tag{18}$$

we usually observe a pair of reduction and oxidation peaks with a peak position at V_{pc} and V_{pa}, respectively, which is independent of sweep rate and separated by

$$\Delta V = \left| V_{pa} - V_{pc} \right| = 2.303\frac{RT}{nF} \tag{19}$$

which is governed by the number of electrons, n, transferred in the reaction. The redox potential for the reaction can be determined by

$$V^{\circ} = \frac{V_{pa} + V_{pc}}{2} \tag{20}$$

For an ideal reversible reaction, we ought to expect that $I_{pa} = I_{pc}$. That said, the peak currents, I_{pa} and I_{pc}, if they are limited by diffusion of reactive species in the solution, will vary with sweep rate, ν, according to

$$I_p \propto \nu^{1/2} \tag{21}$$

Here we use the cyclic voltammograms of a poly-(MG) coated glassy carbon electrode cycled in the presence of a buffered NADH/NAD$^+$ solution as an example for illustration. Figure 8 presents a series of 15 cycles of cyclic voltammograms for such an experiment. The most prominent peak on the voltammogram is the NADH oxidation peak. The reduction peak, on the other hand, is a very vague shoulder on the onset of the two prominent MG reduction peaks [32,33,46]. The difference between the oxidation and reduction peak intensities is consistent with the ability of poly-(MG) to promote the oxidation of NADH to NAD$^+$, but not the reduction. The separation of the NADH/NAD$^+$ redox peaks is about 30 mV, which is characteristic of the two-electron charge transfer in this process. On the voltammograms, we also observed the presence of two couples of redox peaks, in the potential range of 0 to –0.5 V against Ag/AgCl reference. These two couples of redox peaks are signatures of the reversible MG intermediate formations [33,46], which are accompanied with the NADH oxidation. Interestingly, the presence of these two pairs of redox peaks suggests that the charge transfer of the NADH oxidation released a proton at the same time. The liberated proton may have reacted with the poly-(MG) chain and induced the redox reaction of the MG intermediate formations in the subsequent charge transfer reactions. The two conjugated redox peaks in each couple are separated by about 60 mV, which is also characteristic of the single-electron charge transfer process involved in each intermediate formation. The overall redox reaction involves MG is shown as follows [46]:

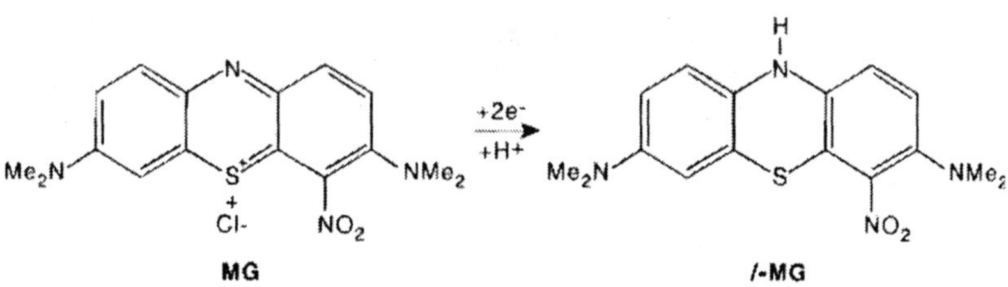

(22)

The CV in Figure 8 also reveals the formal potentials for the two charge transfer reactions of poly-(MG) on the glassy carbon electrode in the presence of the NADH/NAD$^+$ couple. In the solution environment of the experiment, the NADH/NAD$^+$ couple exhibits a formal potential at about 190 mV above the Ag/AgCl reference. The subsequent two charge transfer reactions of poly-(MG) occur at about –110 mV and –270 mV versus the same reference, respectively. In other words, the formal redox potentials for the two charge transfer reactions of poly-(MG) are 300 mV and 460 mV more positive than the NADH/NAD$^+$ couple, as shown in Figure 9. In this case, CV can be used to verify or derive thermodynamic information of the chemistry.

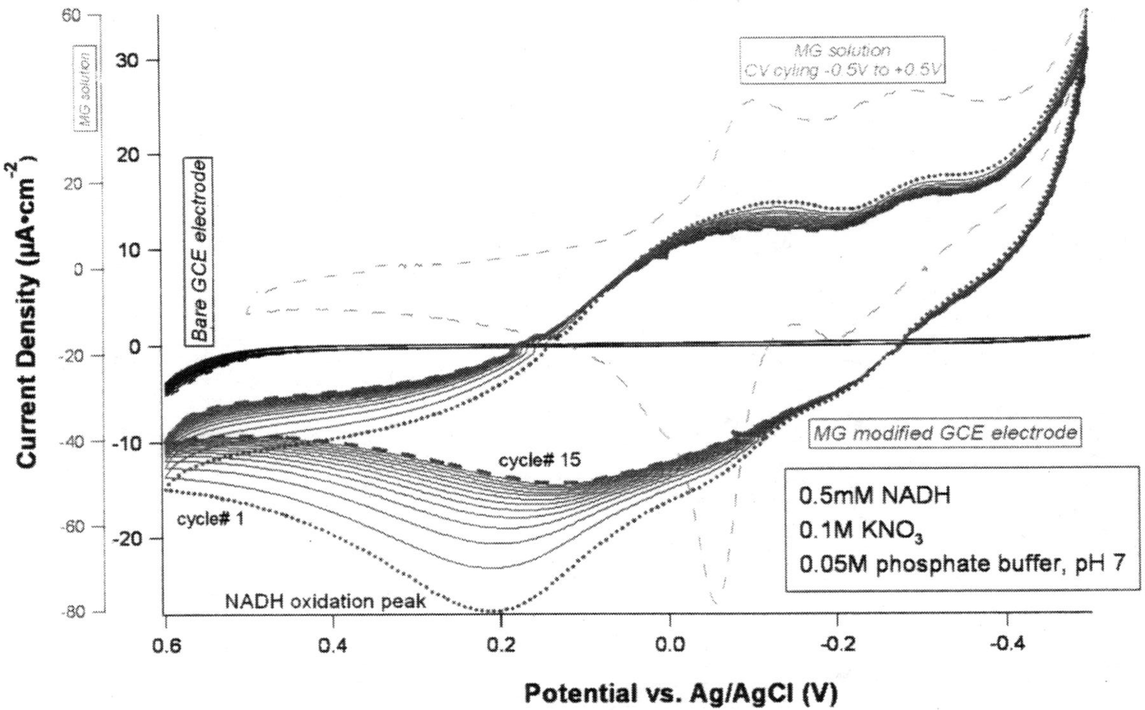

Figure 8. Cyclic voltammograms for NADH/NAD$^+$ redox reaction by poly-(methylene green) films on glassy carbon substrate. Conditions: NADH solution (0.5 mM), phosphate buffer (0.05 M), $NaNO_3$ electrolyte (0.1 M), sweep rate (10 mV/s), 15 cycles.

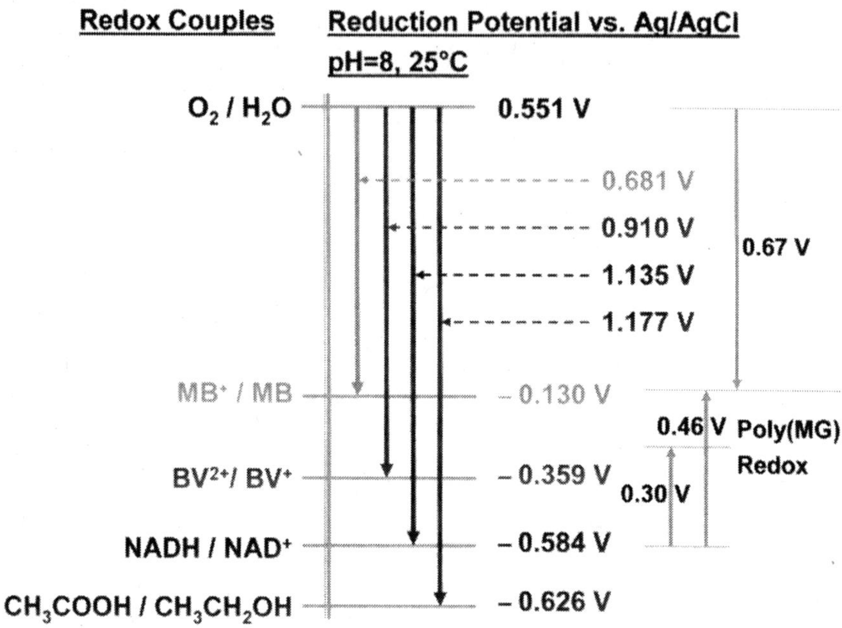

Figure 9. Formal redox potentials for poly-(MG) coated on glassy carbon electrode and the associated V°_{cell}.
(Reproduced with permission from the author. Copyright held by the author.)

Another interesting example is the ethanol enzymatic bio-fuel cell made of NAD^+-dependent ADH anode with poly-(MG) mediator and an oxygen air electrode as cathode. The cell open circuit voltage was measured to be ~0.67 V (see Figure 9), which is substantially lower than the conventional ethanol fuel cell, which should have an OCV close to 1.17 V. The disparity in the OCV's suggests that the final equilibrium on the anode is no longer represented by the half cell reaction based solely on the ethanol oxidation. The charge transfer involved in the ethanol oxidation, and NAD^+ reduction, as expressed in equations (9)-(11), apparently does not dictate the half-cell potential at the anode, either. The correct expression of the mediator involved NADH oxidation as depicted in equation (10) should be

$$\text{NADH} + \text{poly-(MG)} \rightarrow \text{NAD}^+ + l\text{-poly-(MG)} \qquad (23)$$

Here, analogous to MG in equation (22), poly-(MG) is positively charged when it dissociated in aqueous solution, as denoted in the equation. As such, the half-cell reaction on the anode is therefore different from that expressed in equation (11). The correct reaction is as follows:

$$\text{Ethanol} + \text{poly-(MG)} \rightarrow \text{Acetaldehyde} + H^+ + l\text{-poly-(MG)} \quad (24)$$

Equation (24) will account for the observed OCV, in which some of the free energy attainable from ethanol oxidation will be lost for the redox reactions involved in the co-factor and mediator for stripping protons and electrons from the hydrocarbon before any useful energy can be harnessed from the charge transfer. The determination of the formal potential of the poly-(MG) redox reactions allows us to derive the thermodynamics (cell open circuit voltage in this case) of the anode in terms of mechanistic steps involved in the half-cell reaction.

As we have discussed previously, the poly-(MG) film has an advantage over the use of benzyl viologen (BV) because of its tolerance to oxygen. The immobilized poly-(MG) film is also more effective in conducting charge transfer than the free flow BV, thus it should reduce the R_{cell} term. These benefits can be comprehended in terms of thermodynamic and kinetic considerations. In Figure 8 it can be seen that the oxidation peak occurs at roughly –150 mV which is a bit better than the Meldola's blue but still below that of BV. The question of whether (or not) the loss in electrochemical potential (e.g., V_{anode} is a little less negative) is balanced by the gains in kinetics due to the decrease in R_{cell} (e.g., the kinetics of electron transport is improved) is a matter of design consideration, as well as choice of optimization.

Electrochemical Impedance Spectroscopy (EIS)

EIS is a very powerful technique to study contributions of impedance in electrode kinetics [22,47]. In principle, the EIS is conducted on an electrode with a reference electrode, under a DC bias (or not), and with a superimposition of an AC perturbation signal over a range of frequencies, typically from several hundred kHz to a few mHz. The frequency scan allows us to separate contributions in impedance from different physical or chemical processes (since each of them exhibits a characteristic time constant in the frequency domain). The impedance spectra can be analyzed with an equivalent circuit model to yield specific parameters for the corresponding processes involved in the electrode system. The variation of the impedance spectra with operating conditions can provide valuable information regarding the electrode kinetics, such as possible rate limiting step and its kinetic characteristics as presented by the parameters derived from the model.

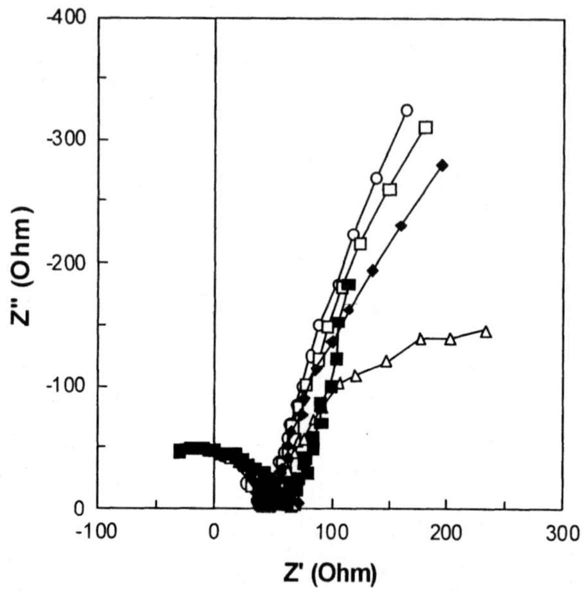

Figure 10. Nyquist plots for the Ni/PGC (O: –0.6 V, □: –0.5 V, and △: –0.4 V, vs. Ag/AgCl) and SS/PGC (■: –0.5 V and ◆: –0.3 V, vs. Ag/AgCl) electrodes with different DC biases. (Reproduced with permission from reference 45. Copyright held by author.)

EIS can be used to detect variations in dominant electrode behavior during the polarization process. The impedance data is then analyzed and visualized using software in graphical presentations such as the Nyquist (or Bode) plots. The variation of the frequency responses with polarization conditions are generally used to interpret possible limiting mechanisms associated with the overall charge transfer process. For example, Figure 10 presents the frequency response of the Ni/PGC and SS/PGC electrodes (see Figure 7) under three potentiostatic conditions in the hydrogen oxidation region [45]. As the Ni/PGC enzyme-immobilized electrode is polarized to the maximum hydrogen oxidation current (from –0.6 V to –0.4 V vs. Ag/AgCl at 75°C), the low frequency branch of the complex impedance curve becomes more defined as a semicircle arch with decreasing impedance, implying a radical departure in reaction mechanism. This was found to be consistent with the onset of nickel support oxidation, which was reinforced by DC-polarization and dynamic potentiometric data [45]. In the critical hydrogen oxidation region (–0.6 to –0.5 V vs. Ag/AgCl), both Ni/PGC and SS/PGC electrodes showed similar impedance behavior, which is independent of the DC bias (Figure 10). This behavior suggests that:

(1) The electrode supports (Ni or SS) did not affect the impedance behavior.
(2) The impedance behavior must be dominated by mass transport, not charge transfer. This conclusion was further supported by evidence obtained from DC polarization measurements and by mass transport calculations of limiting current densities.
(3) The result also implies that charge transfer is quite efficient by the adsorbed hydrogenase.

EIS has not yet received wide spread use in the investigations of bio-fuel cell applications. That said the broad utility and capability of EIS promises to be a useful tool for kinetic study in biological materials, films, and systems; e.g., [45-51].

Spectrophotometry

Spectrophotometry is generally used as a technique to determine enzyme activity and kinetics in free solution based on Michelis-Menten type models [52]. The spectrophotometric measurements of reagents exposed to immobilized enzyme electrodes have also been used to estimate the kinetic parameters (V_{max} and K_M) of bound enzymes on an electrode surface by the same analysis. In this case the performance of the electrode is affected by both enzyme kinetics and mass transfer throughout the electrode architecture; especially when porous medium transport is involved in the measurement. Ideally, one should estimate the kinetic parameters in a manner that separates out the effects of mass transport [52]. The issue lies on the concentration gradient created by the mass transport, largely due to diffusion and/or migration. A possible solution is to estimate the reactant concentration at the enzyme surface based on a diffusion model and the bulk concentration that can be measured by spectrophotometric detection *in situ* or, if necessary, *ex situ* by HPLC analysis. An example of this approach is illustrated in Figure 11 [52].

The mass transfer characteristics of the electrodes are estimated by comparison of experimental data against model predictions of system performance in the presence of continuous flow. Thereafter, a combined mass transfer parameter is calculated and the enzyme kinetic parameters evaluated. In the case of immobilized alcohol dehydrogenase, it can be shown that these terms can be related to the concentration of the NADH leaving the biofuel cell and the flow rate by [52]:

$$F \cdot NADH_b = \frac{V_{max}^{NAD^+} \cdot \left(NAD_{res}^+ - NADH_b - \dfrac{F \cdot NADH_b}{M_{NAD^+}} \right)}{K_M^{NAD^+} + \left(NAD_{res}^+ - NADH_b - \dfrac{F \cdot NADH_b}{M_{NAD^+}} \right)} \tag{25}$$

$$M(cm^3 / \min) = D_{eff} \cdot \frac{ASA \cdot 60}{L} \tag{26}$$

where F is the flow rate, the subscripts b refer to bulk solution and res to the reservoir, and M is the lumped mass transfer parameter. The LHS can be measured experimentally and the RHS can be plotted in the form of a Lineweaver-Burke plot, assuming a value for the combined mass transfer parameter M is used. If the assumption is made that the value for $K_M^{NAD^+}$ will not change from the K_M measured for the enzyme in free solution, then a value of M can be selected to yield a value for $K_M^{NAD^+}$ that is equivalent to the value determined in free solution (Figure 11).

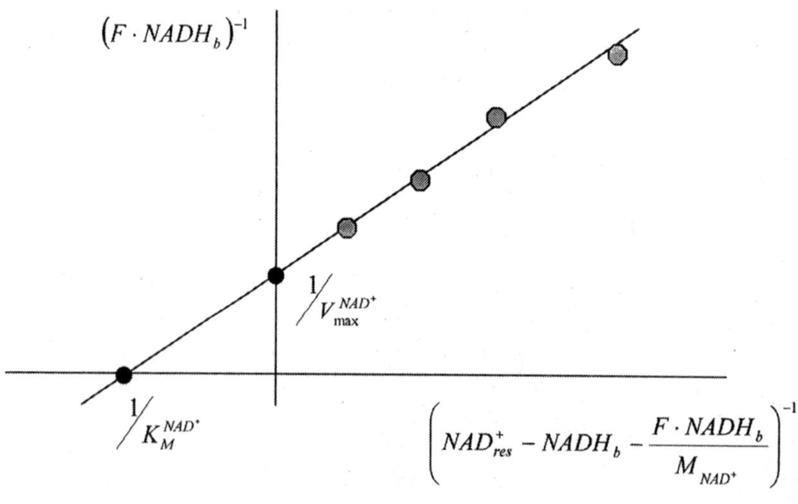

Figure 11. Lineweaver-Burke plot with a lumped mass transfer model [52] for immobilized enzymatic catalytic electrode, which can be used to estimate V_{max} and K_M for the specific enzymatic system.

In practice, we have found the reproducibility of the method to be quite good. If each determination of V_{max} and D'_{eff} has been executed using data from at least three experiments, with each experiment conducted at a constant flow rate and three different feed concentrations of NAD^+, consistent model fits for V_{max} and M were found for alcohol dehydrogenase immobilized on PGC electrodes (by direct adsorption) or within conductive polypyrrole films. The PGC was found to be 26 times more active than the polypyrrole/stainless steel electrode after correction for mass transfer effects (V_{max} of 0.062 versus 00025 μmol/min). While it is clear that the greater activity of the PGC electrode could have been due to a number of factors that the model is not detailed enough to predict, it can be stated that the measured activity has been separated from the effects of mass transfer. Hence, the lower measured activity (V_{max}) for the polypyrrole electrode is likely due to lower enzyme loading as opposed to mass transfer effects. While this model has its limitation, it does establish a simple method to address the need to eliminate the effect of mass transfer when using absorbance spectrophotometry to determine the kinetic parameters of bound enzymes.

Spectrophotometric measurements can also be combined with electrochemical characterizations to derive additional performance parameters. Usually, the catalytic behavior is measured through the absorbance of a cofactor (e.g., the disappearance or appearance of NAD^+ at 340 nm) or mediator that has been oxidized (e.g., methyl viologen). In the latter case, the actual enzyme activity can be measured, when coupled with electrochemical measurements of current, and used to estimate the charge transfer efficiency of bound enzyme [45]. To do this, the maximum theoretical current density is first calculated from the spectrophotometric determination of enzyme loading (and assuming perfect charge transfer efficiency), and then compared to the actual current density measured from the DC polarization experiments. The maximum theoretical current density is calculated from Faraday's law:

$$I = n \cdot F \cdot \frac{d\mathrm{H}/dt}{\mathrm{ASA}} \tag{27}$$

where (in the case of a hydrogenase enzyme electrode) the hydrogen oxidation rate per unit electrode surface area (dH/dt)/ASA is estimated by first measuring the activity per unit electrode surface area (U/cm^2) and then multiplying this term by the definition of hydrogenase enzyme activity (e.g., 1 unit of enzyme activity is equal to 1 μmole of hydrogen oxidized per minute). The charge transfer efficiency of the bound enzyme is then calculated by taking the ratio of the actual measured current density to the theoretical value, expressed as a percentage.

Imaging

Imaging techniques are used to characterize surface morphology of biocatalytic electrodes, which are often exposed to solution environments. There are many options available, including optical microscopy [53], scanning electron microscopy (SEM) [54], atomic force microscopy (AFM) [53-56], scanning tunneling microscopy (STM) [57], transmission electron microscopy (TEM) [58], scanning electrochemical microscopy (SECM) [59], acoustic microscopy [60], confocal microscopy [61,62], or imaging ellipsometry [63-65]. Most of these imaging techniques are typically performed *ex situ*, which may require that the samples are analyzed after certain preparation steps, e.g., dehydration; or under conditions that may introduce artifacts into the samples and lead to observations that might not reveal the true morphological features in the solution. Techniques that allow *in situ* observations are therefore more desirable in this case.

One of such *in situ* techniques is AFM because it allows imaging of the electrode surface in the liquid phase without the requirement of drying. When it is used in a tapping mode with a precisely balanced cantilever beam to trace out the surface of the sample in the solution, it can reveal detailed sample morphology. Figure 12 shows an example of a polypyrrole (PPy) film scanned for a 2 × 2 square micron area under a wet (right) and dry (left) condition, respectively. The images show that the morphology at about the same film location for both wet and dry conditions is about the same. The less discernable resolution in the height map of the wet condition indicates that the height variations are more in wet than in dry, a result of swelling.

Although *in situ*, imaging with AFM is tediously difficult to control and to yield high quality images. The scan rate for a large area is also slow, therefore time consuming. When a more informative and non-invasive *in situ* imaging technique is required for monitoring a dynamic system, for instance, to measure the growth of the immobilization matrix or polymerized mediator film *in situ* as it grows, a better approach has to be sought. Imaging ellipsometry is a technique that provides such capability for observing thin film growth or immobilization process, almost in real-time [54,63-65]. Consider the example of polypyrrole deposition from free solution onto platinum support. Figure 13 suggests how the initial film formation with different current density might have set the tune for the morphological development. As the current density increases, the film grows with increased roughness in the formation. On the other hand the film thickness increases with current density. Since the amount of charge passed in the deposition is the same, we would expect the mass deposited should be the same. This observation implies that the film gravimetric density might not be the same; thus, the higher current density will produce a thicker film with lower gravimetric density. Nonetheless, the thickness

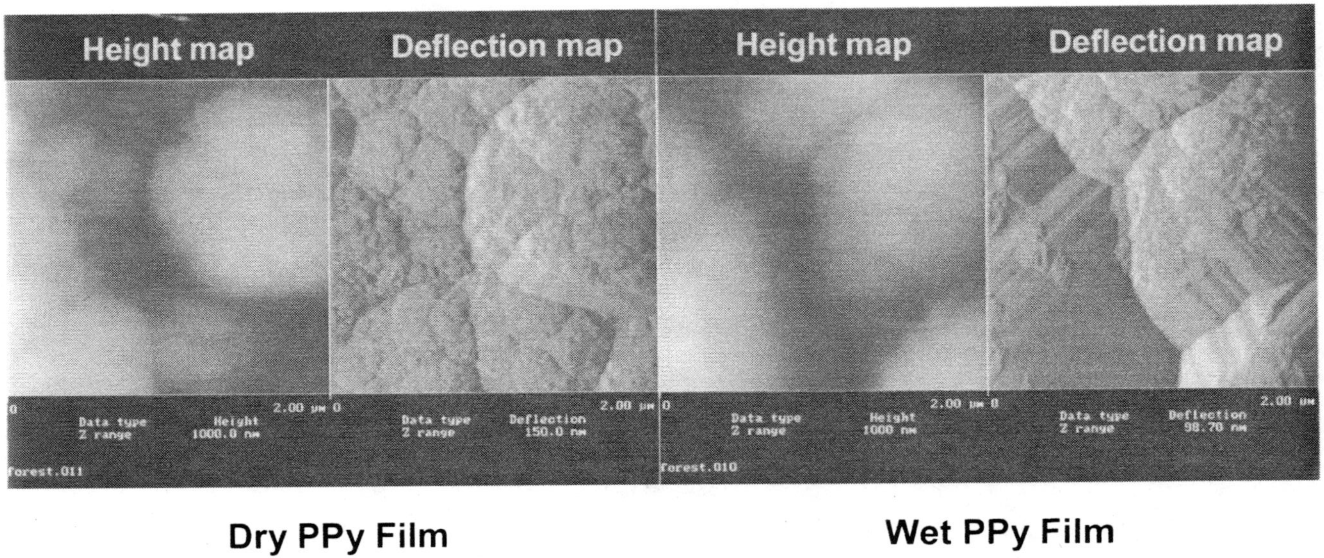

Figure 12. AFM images of a polypyrrole film taken in situ in solution (right) and in dry condition (left).

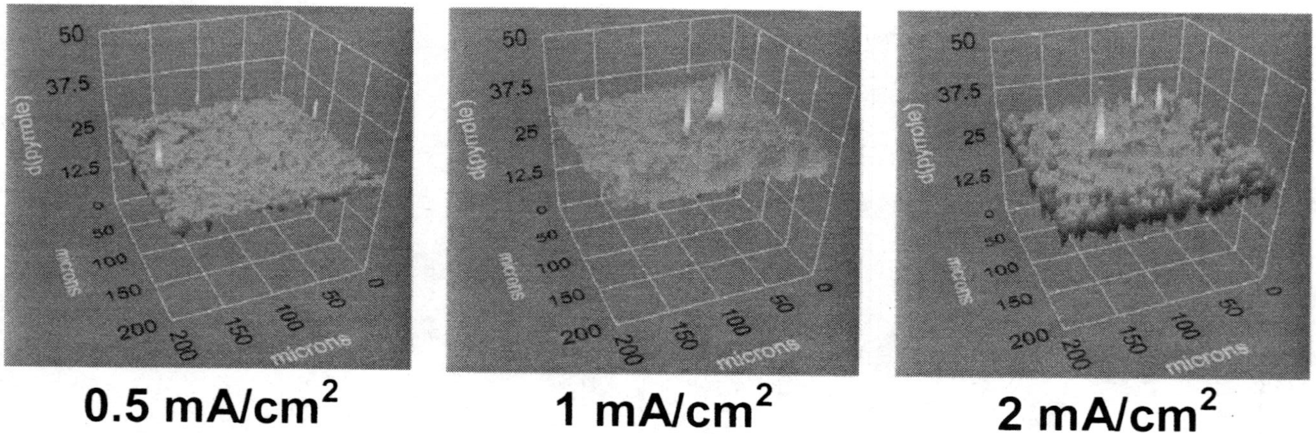

Figure 13. Morphology versus current density in the initial film formation of polypyrrole on Ni substrates.

estimates were however much lower than what we anticipated according to the amount of charge and the reported correlation of thickness with charge in the literature [66].

Characterizations of Physical Properties

Important physical properties of a fabricated electrode include porosity, pore size and distribution, geometric, total, and active surface area, thickness, and density. If the electrode is a (sub)-monolayer of enzyme immobilized onto a flat surface (e.g., on a glassy carbon or platinum electrode), some of the terms are either irrelevant (e.g., porosity, thickness, density) or quite easy to define and measure (e.g., for a smooth surface area with minimal roughness). In the case of an electrode with a thick, porous catalytic film on the surface (e.g., Figure 3), it will require more detailed characterization of its porosity, pore size and distribution, ASA, thickness, and gravimetric density, in order to correlate the catalytic properties (e.g., reaction kinetics and mass transport) with these physical parameters. Some of these properties are interdependent and also key parameters in defining performance. For instance, increasing the porosity of a film without changing the thickness or geometric surface area will decrease the gravimetric density but increase the ASA. Films with the same porosity but different pore sizes and distributions may exhibit very different transport and kinetic behavior in performance [6,67].

The latter concept underscores the impact of pore structure and distribution on the mass transfer properties of a porous electrode. The ability to control pore structure (architecture) permits the ability to control the R_{cell} term in equation (3) and consequently improve the current/power density that can be achieved for a given chemistry (V°_{cell}). The development of techniques to control pore structure and to accurately determine the physical properties of the porous catalytic film, especially in the micro- and nano-scale, is therefore highly desirable.

Porosity (Θ) is the ratio of void volume to total volume of the sample [68]:

$$\Theta = 1 - \frac{V_m\left(= V_t - V_e\right)}{V_t} \tag{28}$$

where V_m is the solid volume of the sample, whereas V_t is the geometrical total volume, and V_e is the empty void volume. V_m can be determined by the apparent densities method or the pycnometric method. In the apparent densities method V_m is given as $V_m = m_1/\rho_{rm}$, where m_1 is the mass of the dried sample and ρ_{rm} the density of the raw sample material which has been previously determined. Porosity can also be measured by the pycnometric method, which measures the mass m_2 of a pycnometer filled with a suitable fluid of known density and m_3 of the same pycnometer when it contains the sample.

$$\Theta = 1 - \frac{m_1 + m_2 - m_3}{V_t \cdot \rho_w} \tag{29}$$

where ρ_w is the density of the known fluid. It is important to note that the reliability of this method depends upon the ability of the fluid to wet the porous structure effectively. A more accurate variation of pycnometry uses mercury as the penetrating liquid, in the so called mercury-porosimetry. The mercury is forced to enter the pores under pressure in a calibrated pycnometer in a manner in which the intruded volume is registered at each pressure.

$$\Theta = \frac{V_{\text{intr}} \rho_{Hg}}{\left[\left(m_{c,Hg} - m_c\right) - \left(m_{c,s,Hg} - m_{c,s}\right)\right]} \tag{30}$$

where V_{intr} is the total volume intruded and m_c, $m_{c,Hg}$, $m_{c,s}$, and $m_{c,s,Hg}$ are the masses of the cell emptied, filled with mercury, filled with the sample and filled with the sample and mercury, respectively [68].

To get an estimate of pore size and total surface area, BET is the most often used technique. The BET technique measures both flow-through pores and blind pores in the porous medium in a more or less static mode (in equilibrium), which estimates the bulk pores size and total surface area available for adsorption of the gas, but does not yield an ASA, which represents the true surface area that is accessible by the reactants.

The gas permeation test is part of a larger group of tests that are applied to membranes (e.g., bubble point technique, solute transport, mercury intrusion, permporometry, and thermoporometry [69]). In gas penetration (or permeation) tests, a pure gas (e.g., nitrogen) is selected as a standard gas and the permeation flux through the membrane is measured at different transmembrane pressures. The mean pore size is calculated from the lineal plot of mean pressure and the following equation

$$d_g = \frac{16}{3}\left(\frac{S_o}{I_o}\right)\left(\frac{8 \cdot RT}{\pi \cdot M}\right)^{0.5} \mu \tag{31}$$

where S_o and I_o are the intercept and slope of the line between permeance and the mean pressure, respectively; R is the gas constant; T is the absolute temperature; M is the molecular weight of the gas; and μ is the gas velocity [55]. Unfortunately, such techniques useful for permeable membranes may not be directly applicable to biocatalytic electrodes in those cases where the immobilization matrices are made from the deposition of a polymer from solution (e.g., Nafion or polypyrrole) onto a flat surface. In these cases it can be difficult to either remove or isolate the membrane in a device or permit the application of transmembrane pressure across the polymer film matrix.

When defining the pore structure, the thickness of the polymer film is also an important parameter to consider. Several methods have been proposed to determine the thickness. In the case of conductive polypyrrole films [66], the film thickness per geometric area can be determined *in situ* by techniques such as ellipsometry [54,63-65], or by measuring the charge transferred and using the charge-thickness correlation reported in [66]. More recently, combining several techniques for *in situ* characterization has been favored. This approach can bring additional benefits for direct correlation of physical parameters in a system to gain better understanding and control of the process. For example, one can combine imaging ellipsometry and quartz crystal microbalance to characterize the deposition of polypyrrole. In Figure 14 the characteristic vibration frequency change of the quartz crystal electrode is plotted for the electrochemical deposition of polypyrrole from a 0.1 M pyrrole solution under a potentiostatic condition of 0.8 V vs. Ag/AgCl. The finished film reduced the crystal frequency by 1,542 Hz. According to the Sauerbrey equation, a polypyrrole film of 27.4 $\mu g/cm^2$ was deposited on the electrode. If the film thickness was estimated, according to $d = 2.5 \times C$ [66], to be 195 nm, the deposited polypyrrole film density would be 1.41 g/cm^3, which is close to but lower than the reported value of 1.58 g/cm^3 [66]. A possible source of error for the discrepancy may come from an inaccurate estimation of thickness using the literature value for much thicker films. If the film density would follow 1.58 g/cm^3 in the literature, the film thickness would be about 176 nm, instead of 195 nm. Using ellipsometry in combination with quartz crystal microbalance, one will have a much better methodology to investigate if the mass or the thickness attributed to the discrepancy.

Determination of Chemical Properties

Chemical composition can be determined by a number of techniques which include various forms of infrared or Raman spectroscopy [65,70,71], X-ray photoelectron spectroscopy (XPS) [72], electrochemical quartz crystal microbalance (EQCM) [73,74], pyrolysis-gas chromatography (GC)-mass spectroscopy [75], surface plasmon resonance [76], water contact angle or microscopic wetting [63,65]. The characterization of chemical properties along with the complementing physical properties can provide us needed information

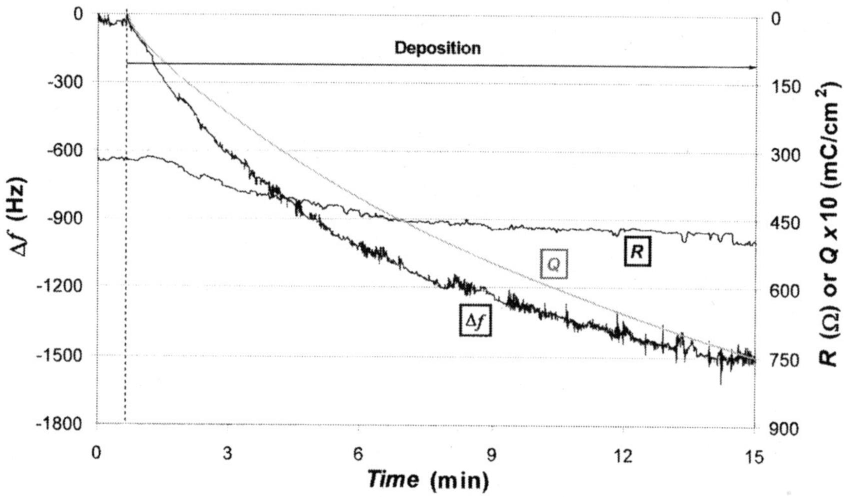

Figure 14. Quartz crystal microbalance result recorded in situ during a potentiostatic polypyrrole film deposition in aqueous solution. Correlation with charge is shown.

to define the local environment that the enzymatic catalysts embed with. This information is critical for the improvement of the enzyme activity and stability.

An interesting approach is to utilize the fluorescence tagging technique to label the enzymes or the adjacent chemical environment to reveal the distribution of the enzymes and their interactions with the local chemical environments. Fluorescence is the property of some atoms and molecules to absorb light at a particular wavelength and to subsequently emit light of longer wavelength after a brief interval, termed the fluorescence lifetime [77]. Through appropriate choice of a specific fluorophore, one can potentially select for probes that will emit at wavelengths that are unique to any other fluorescent material in the system. If this occurs, one can detect the presence of that specific probe and, if used in conjunction with confocal laser scanning microscope, actually focus measurements spatially.

As such, quantitative fluorescence spectroscopy can in principle be used to measure the amount of enzyme immobilized, to probe the regional pH or polarity of a polymer matrix, and to reveal dynamic aspects of the enzyme in the polymer matrix. For example, if the amount of fluorophore bound per enzyme can be controlled, then the measured intensity of fluorescence should be proportional to the amount of enzyme immobilized. If the fluorophore emission is sensitive to pH or polarity (as is the case with the dye Nile red), one can probe the micro-chemical environments immediately surrounding the immobilized probe [78,79]. If the probe is bound to the enzyme, fluorescence is a potentially powerful technique to probe how enzymes interact with their polymer matrix.

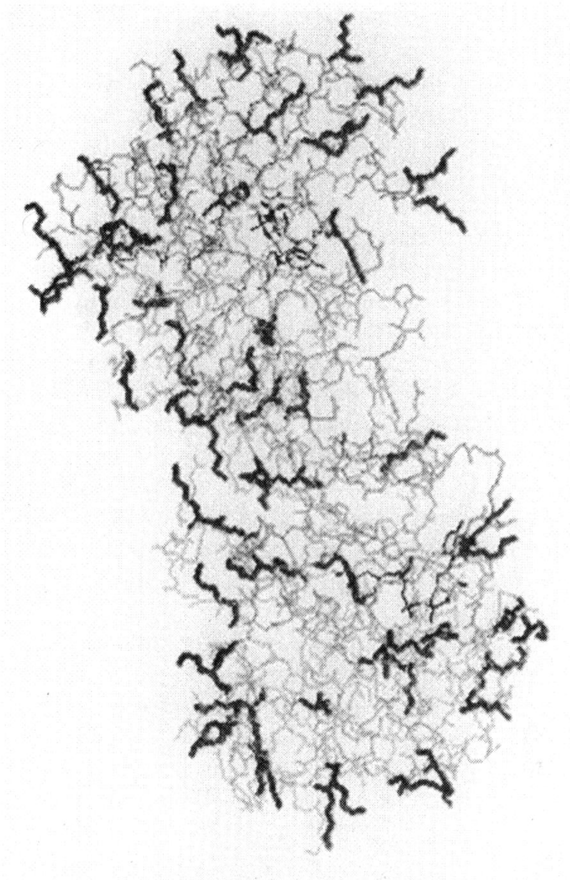

Figure 15. Lysine groups on YADH which are available for binding fluorescent probes (Reproduced with permission from the author. Copyright held by author.)

A working example can be found with yeast alcohol dehydrogenase (YADH) bound with different dyes to reveal their interactions with the hosting matrix, such as Nafion or Eastman AQ, which act as polymer backbones to host enzymes in their porous structure for immobilization [80]. Initially the selected probe (either amine reactive for binding to NH_2 groups or thyol reactive for binding to SH groups) is dissolved in an appropriate solvent (e.g., DMSO) and then mixed with the enzyme in buffer. The mixture is then incubated sufficiently long for the enzyme to bind to the target amino acid (typically lysine groups, Figure 15). In general one looks to have a 10 fold excess of dye to enzyme but this can vary depending upon the binding protocol which must be optimized for each enzyme and probe combination. The goal, however, is to achieve a dye to enzyme binding ratio that gives both a strong signal without deactivating the enzyme. In our work we have found, for example, that a binding ratio of 1:1 for Alexa 488 succinimidyl ester works well. In general, though, one will find that higher dye to enzyme binding ratios will lead to higher labeling of the enzyme but lower enzyme activity.

After incubation the unreacted dye must be separated from the labeled enzyme. Typically this can be achieved on a PD-10 column using 50 mM phosphate buffer as the effluent. The collected fractions should be verified for specific activity and the most active fractions combined and concentrated using microfilters (Millipore) and centrifuging at 14,000 rpm for about 20 minutes. Thereafter the labeled enzyme should be dialyzed for 24 hours against 50 mM TRIS buffer, pH 7. This should leave one with a pure solution of tagged enzyme absent of free probe, which is important for the subsequent microscopic imaging. One should confirm the successful labeling of enzyme by checking for an increase in the polarization of the dye. Finally, the degree of labeling can be determined by taking absorption spectrum of the product to estimate the concentration of both protein and dye using the extinction coefficient. Typical protocols can be found through the manufacturer (Molecular Probes, Invitrogen, Inc.). It is worth noting that, in reality, this calculated value is an approximation of a Poisson distribution.

In our work we have entrapped tagged YADH in both Nafion and Eastman AQ polymer films and imaged the composite with both wide field epifluorescence and confocal laser scanning microscopy [81]. Wide field epifluorescence microscopy has not been proven particularly useful since the resulting image is a gross measurement of fluorescent intensity (i.e., the 3-D distribution mapped down to a 2-D surface). Unless one were able to slice the composite films down into ultra thin cross sections it is difficult to extract useful information. By contrast we have found that confocal microscopy to promise powerful results. While a description of confocal microscopy and how it works is beyond this presentation, it is worth noting that confocal imaging permits the viewing of isolated slices or sections of the sample and these images can be viewed and analyzed using appropriate software. Each slice can then be restacked to reconstruct the volume using appropriate software. Figure 16 [81] shows two such images for Alexa 488 tagged YADH entrapped in Eastman AQ. Image A (left) shows a typical 2-D slice taken near the surface of the film. One can see that the enzyme is sequestered to circular pores. In Image B (right), it shows the 3-D reconstructed map in which Image A was just one of many slices. In this image it becomes clear that the pores extend deep into the film suggesting that the entrapped enzyme has been entrapped in deep circular pores and are not distributed homogeneously throughout the polymer matrix.

We are currently investigating the role of the probe in directing the enzyme distribution and interaction with the polymer. As with all characterization

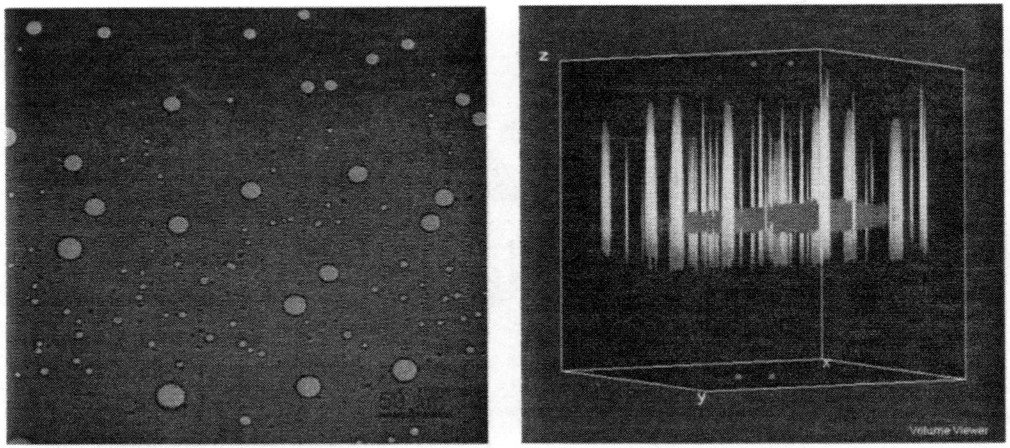

Figure 16. A slice and volume reconstruction of the stack of confocal images obtained with Eastman AQ 55 mixed with Alexa 488 labeled YADH. Volume reconstruction was performed using ImageJ volume viewer plug-in.

techniques, it is important to clarify how the measurement technique impacts the result. In our work we have shown that enzymes can be tagged with more than enough probe to be visualized but without loss of enzyme activity. We are now looking into the effect of probes that possess positive, negative, and neutral charge to understand how they interact with the chemical environment of the hosting matrix. Clarifying these effects will permit the expanded use of fluorescence to measure the distribution of enzyme in polymer matrices, as well as probing the micro-chemical environments immediately surrounding the enzyme.

Standardization of Electrochemical Measurements

As a last comment the authors would like to note an important aspect of, and the perceived difficulty in, standardizing electrochemical measurements. Most such characterizations of electrodes or cells are difficult to replicate across labs, in part because the lack of a consistent physical geometry used for test cells makes it exceedingly difficult to compare performance. Although this is a common issue among many disciplines, in the field of biofuel cells it is more sensitive and vulnerable to this concern due to the low level of signals. For instance, as the electrolyte solutions used in the investigations are often buffered, the solution resistance to ion transport can significantly alter signals in electrochemical measurements. Given this impact, the placement of electrodes (i.e., the test cell configuration and the associated field) becomes critical in shaping the electrical signals in the measurements. Even small variations in the placement can alter the signals noticeably, making it difficult to discern any relative performance deviations of an electrode or cell. Therefore, it is sensible to conduct testing in common test cell geometry across experiments and labs in order to compare experimental results and facilitate meaningful discussions. Currently, the authors and their MURI colleagues are engaged in an exercise to determine to what extent using test cells of fixed geometry can be employed for cross laboratory comparison and validation of experimental data using identical electrode fabrication and test protocols. A prototype test cell being used in this exercise is presented in Figure 17A. Figure 17B presents an extended version that permits experiments where solution and headspace aeration with inert gas is required.

Conclusion

Power generation in bio-fuel cells involves complex optimization of electrode design and fabrication. Enhancing enzyme loading and prolonging their stability are critical issues, which require careful engineering of

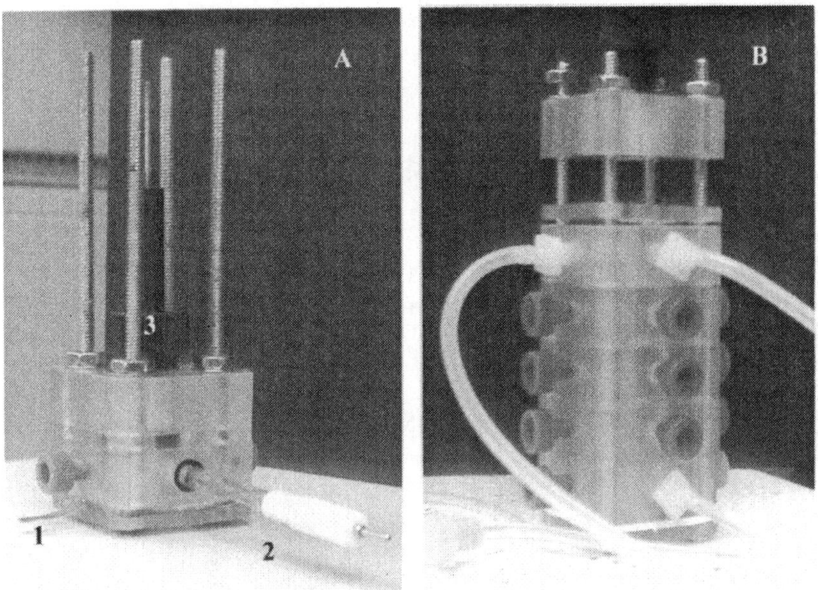

Figure 17. Modular stack cell. In image (A), (1) the platinum counter electrode, (2) reference electrode, and (3) working electrode.

immobilization processes, the creation of favorable porous structure to host the enzymes and to retain their activity and stability, and the provision of multidirectional and multidimensional pore channels that promote mass transfer of fuel. To quantify and characterize improvements in these aspects, it is important to employ various *in situ* analytical techniques that can characterize properties on scales ranging from microscopic (atomic level) to macroscopic (size of cell) in order to provide temporal and spatial correlations. In this manner, one can begin to understand the intricacy of the enzyme's interaction with its local environment and ability to effectively conduct catalytic behavior to generate power with high efficiency. It is also important to finally test performance in a well defined test cell geometry.

We have discussed several important aspects in the optimization of power generation and cell performance. The aspects of thermodynamics and kinetics in the selection of cell chemistry and design have been presented in terms of the basic power generation concept and requirements. These considerations dictate the characterization and comparison methods for electrode and cell performance. We outlined a number of characterization techniques, particularly for *in situ* observations, to help characterize the material, process, component, and system performance of the bio-fuel cells.

Using a few interesting illustrations, we showed that, for instance, the use of electrochemical techniques to yield macroscopic performance characteristics *in situ*, and imaging ellipsometry and fluorescence to reveal microscopic properties of the electrode and its porous matrix. In addition, we were able to correlate performance characteristics with detailed physical properties in the microscopic dimensions and with improved temporal and spatial resolutions. Although it is important to realize the advantages of each technique, one should also understand the limitations.

In the area of physical characterizations, we often need to have dimensional resolutions, particularly in the imaging application, on the order of nanometers. Among several useful techniques available to date, including high resolution SEM and TEM, AFM seems to be the choice for *in situ* measurements. However, operating AFM *in situ* in solutions is very challenging and tedious; therefore it is difficult to yield useful temporal information in transient studies. There exists a disparity in the limitations of spatial resolution in the imaging techniques. Typical optical characterizations, including confocal microscopy and surface imaging in ellipsometry; have limitations in the spatial resolution. Due to constraint in the optics, they only can achieve the best in the submicron range. Although high resolution SEM and TEM can provide better resolutions, they often provide *ex situ* information, therefore presenting limited value. Imaging ellipsometry provides very good resolution for *in situ* thin film and interface study, therefore its potential needs to be fully explored. We believe that there is a need to develop more powerful imaging techniques in the meso-scale range for *in situ* studies.

Most importantly, it is conceivable to integrate several non-invasive, *in situ* characterization techniques, with high spatial and temporal resolutions, to allow us to monitor and characterize the biocatalyst and microelectrode behavior. It is lucrative enough to believe that such well-correlated characterization is the most direct method to fully understand the biocatalysis and to effectively optimize its use. It is also important to integrate from both macro and micro scales a collective understanding of material, electrode, cell, and system performance.

Acknowledgments

We would like to thank funding support for our enzymatic biofuel cell work from the Office of Naval Research (Contract No. N0014-01-1-0928) under the Hawaii Environmental and Energy Technology (HEET) Initiative projects (Richard Rocheleau, PI), from the Director of Central Intelligence under the Postdoctoral Fellow Research Program (National Geospatial Intelligence Agency, Contract No. HN1582-04-1-2013), and from the Air Force Office of Scientific Research under a Multidisciplinary University Research Initiative (MURI) program project awarded to the University of New Mexico (Plamen

Atanassov, PI). Many of our students and research staff co-workers have contributed to the work presented herein. In particular, Dr. Wayne Johnston contributed much to our early work on hydrogenase. Dr. Vojtech Svoboda has contributed to the work related to the polypyrrole and methylene green polymer deposition, alcohol ADH cell study, and in situ QCM and imaging ellipsometric work. Dr. Anastassija Konash has contributed to the study of enzyme polymer interaction using chemical fluorescence. The Pacific Biology Research Center at the University of Hawaii has helped us in the SEM imaging. Dr. Pavel Zinin helped in the imaging of polypyrrole using AFM. We would also like to thank our collaborators, including those in the MURI program, Professors Plamen Atanassov, Shelley Minteer, Scott Calabrese Barton, Scott Banta, and Sanjeev Mukerjee, and Dr. Christopher Apblett of Sandia, for their inspiring discussions and intellectual exchanges.

References

1. Katz, E., A.N. Shipway, I. Willner, (2003). *Biochemical fuel cells*, in: Handbook of Fuel Cells- Fundamentals, Technology and Applications. W. Vielstich, H.A. Gasteiger, and A. Lamm, eds., John Wiley & Sons, Ltd, New York.
2. Barton, S.C., J. Gallaway, P. Atanasov, (2004). *Biofuel cells for implantable and microscale devices*. Chemical Reviews, **104**: p. 4867.
3. Chen, T., S.C. Barton, G. Binyamin, Z. Gao, Y.C. Zhang, H.H. Kim, A. Heller, (2001). *A miniature biofuel cell*. J. Am. Chem. Soc., **123**: p. 11917.
4. Kukul'a, R., P. Hasal, T. Schultz, T. Schröder, K. Sundmacher, (2006). *Can a fuel-cell stack and an enzyme electro-membrane reactor be combined into a multi-functional unit?* Catalysis Today, **118**: p. 104.
5. Reiner, A., B. Steiger, G.G. Scherer, A. Wokaun, (2006). *Influence of the morphology on the platinum electrode surface activity*. J. Power Sources, **156**: p. 28.
6. Arbizzani, C., S. Beninati, E. Manferrari, F. Soavi, M. Mastragostino, (2006). *Electrodeposited PtRu on cryogel carbon-Nafion supports for DMFC anodes*. J. Power Sources, **161**: p. 826.
7. Hughes, M., G.Z. Chen, M.S.P. Shaffer, D.J. Fray, A.H. Windle, (2004). *Controlling the nanostructure of electrochemically grown nanoporous composites of carbon nanotubes and conducting polymers*. Composites Science and Technology, **64**: p. 2325.
8. Invnitski, D., B. Branch, P. Atanassov, C. Apblett, (2006). *Glucose oxidase anode for biofuel cell based on direct electron transfer*. Electrochemistry Communications, **8**: p. 1204.
9. Ho, C.-C., A.L. Zydney, (2000). *Measurement of membrane pore interconnectivity*. J. Membrane Science, **170**: p. 101.

10. Asberg, P., O. Inganas, (2003). *Hydrogels of a conducting conjugated polymer as 3-D enzyme electrode.* Biosensors & Bioelectronics, **19**: p. 199.
11. Kim, P., H. Kim, J.B. Joo, W. Kim, I.K. Song, J. Yi, (2005). *Preparation and application of nanoporous carbon templated by silica particle for use as a catalyst support for direct methanol fuel cell.* J. Power Sources, **145**: p. 139.
12. Delhaes, P., M. Couzi, M. Trinquecoste, J. Dentzer, H. Hamidou, C. Vix-Guterl, (2006). *A comparison between Raman spectroscopy and surface characterizations of multiwall carbon nanotubes.* Carbon, **44**: p. 3005.
13. Imageolcová, O., H. Imagenajdaufová, P. Schneider, (2003). *Liquid-expulsion perm-porometry for characterization of porous solids.* Microporous and Mesoporous Materials, **65**: p. 209.
14. Kim, T.-W., R. Ryoo, M. Kruk, K.P. Gierszal, M. Jaroniec, S. Kamiya, and O. Terasaki, (2004). *Tailoring the pore structure of SBA-16 silica molecular sieve through the use of copolymer blends and control of synthesis temperature and time.* J. Phys. Chem. B, **108**: p. 11480.
15. Thommes, M., B. Smarsly, M. Groenewolt, P.I. Ravikovitch and A.V. Neimark, (2006). *Adsorption hysteresis of nitrogen and argon in pore networks and characterization of novel micro- and mesoporous silicas.* Langmuir, **22**: p. 756.
16. Jena, A. and K. Gupta, (2004). *Characterization of pore structure of fuel cell components containing hydrophobic and hydrophilic pores.* Presented in the 41st Power Sources Conference, June 14-17, 2004, Philadelphia, PA.
17. Falk, B., S. Garramone, and S. Shivkumar, (2004). *Diffusion coefficient of paracetamol in a chitosan hydrogel.* Materials Letters, **58**: p. 3261.
18. Madihally, S.V., H.W.T. Matthew, (1999). *Porous chitosan scaffolds for tissue engineering.* Biomaterials, **20**: p. 1133.
19. S.D. Minteer, B.Y. Liaw, and M.J. Cooney (2007). *Enzyme-based biofuel cells.* Current Opinion in Biotechnology, 18(3): in press.
20. Liu, Y., M. Wang, F. Zhao, Z. Xu, and S. Dong, (2005). *The direct electron transfer of glucose oxidase and glucose biosensor based on carbon nanotubes/chitosan matrix.* Biosensors & Bioelectronics, **21**: p. 984.
21. Zhang, M., A. Smith, and W. Gorski, (2004). *Carbon nanotube-chitosan system for electrochemical sensing based on dehydrogenase enzymes.* Analytical Chemistry, **76**: p. 5045.
22. Bard, A.J., L.R. Faulkner, (2001). *Electrochemical Methods.* John Wiley & Sons, New Jersey.
23. Palmore, G.T., H. Bertschy, S.H. Bergens and G.M. Whitesides, (1998). *A methanol/dioxygen biofuel cell that uses NAD^+-dependent dehydrogenase as catalysts: application of an electro-enzymatic method to regenerate nicotinamide adenine dinucleotide at low overpotentials.* J. Electroanalytical Chemistry, **443**: p. 155.
24. Gorton, L. and E. Dominquez, (2002). *Electrochemistry of $NAD(P)^+$ /NAD(P)H.* Encyclopedia of Electrochemistry, A.J. Bard and M. Stratmann,

eds., Bioelectrochemistry, Vol. 9, G.S. Wilson, ed., Wiley-VCH Verlag GmbH & Co., KGaA, D-69496 Weinheim, FRG, p. 67.

25. Prieto-Simon, B. and E. Fabregas, (2004). *Comparative study of electron mediators used in the electrochemical oxidation of NADH.* Biosensors and Bioelectronics, **19**: p. 1131.
26. Santos, A.S., R.S. Freire and L.T. Kubota, (2003). *Highly stable amperometric biosensor for ethanol based on Meldola's blue adsorbed on silica gel modified with niobium oxide.* J. Electroanalytical Chemistry, **547**: p. 135.
27. Yabuki, S., H. Shinohara, Y. Ikariyama and M. Aizawa, (1990). *Electrical activity controlling system for a mediator-coexisting alcohol dehydrogenase-NAD conductive membrane.* J. Electroanalytical Chemistry, **277**: p. 179.
28. Persson, B., H.L. Lan, L. Gorton, Y. Okamoto, P.D. Hale, L.I. Boguslavasky, and T. Skotheim, (1993). *Amperometric biosensors based on electrocatalytic regeneration of* NAD^+ *at redox polymer-modified electrodes.* Biosensors and Bioelectronics, **8**: p. 81.
29. Nagy, G., I. Kapui and L. Gorton, (1995). *Effect of surface-active agents on amperometric NADH measurements with chemically modified electrodes.* Analytica Chimica Acta, **305**: p. 65.
30. Grundig, B., G. Wittstock, U. Rudel and B. Strehlitz, (1995). *Mediator-modified electrodes for electrocatalytic oxidation of NADH.* J. Electroanalytical Chemistry, **395**: p. 143.
31. Mao, L. and K. Yamamoto, (2000). *Glucose and choline on-line biosensors based on electropolymerized Meldola's blue.* Talanta, **51**: p. 187.
32. Svoboda, V., M.J. Cooney, C. Rippolz, and B.Y. Liaw, (2007). *In-Situ characterization of electrochemical polymerization of methylene green on platinum electrodes.* J. Electrochemical Society, **154**: p. D113.
33. Zhou, D.-M., H.-H. Fang, H.-Y. Chen, H.-X. Ju, and Y. Wang, (1996). *The electrochemical polymerization of methylene green and its electrocatalysis for the oxidation of NADH.* Analytica Chimica Acta, **329**: p. 41.
34. Akers, N.L., C.M. Moore, and S.D. Minteer, (2005). *Development of alcohol/*O_2 *biofuel cells using salt-extracted tetrabutylammonium bromide/Nafion membranes to immobilize dehydrogenase enzymes.* Electrochimica Acta, **50**: p. 2521.
35 Liu, Y., M. Wang, F. Zhao, Z. Guo, H. Chen, S. Dong, (2005). *Direct electron transfer and electrocatalysis of microperoxidase immobilized on nanohybrid film.* J. Electroanalytical Chemistry, **581**: p. 1.
36. Gavalas, V.G., S.A. Law, C. Ball, R. Andrews, L. Bachas, (2004). *Carbon nanotube aqueous sol-gel composites: enzyme-friendly platforms for the development of stable biosensors.* Analytical Biochemistry, **329**: p. 247.
37. Klotzbach, T., M. Watt, Y. Ansari, S.D. Minteer, (2006). *Effects of hydrophobic modification of chitosan and Nafion on transport properties, ion-exchange capacties, and enzyme immobilization.* J. Membrane Science, **282**: p. 276.

38. Qian, L., X. Yang, (2006). *Composite film of carbon nanotubes and chitosan for preparation of amperometric hydrogen peroxide biosensor.* Talanta, **68**: p. 721.
39. Liu, Y., X. Qu, H. Guo, H. Chen, B. Liu, S. Dong, (2006). *Facile preparation of amperometric laccase biosenser with multifunction based on the matrix of carbon nanotubes-chitosan composite.* Biosensors & Bioelectronics, **21**: p. 2195.
40. Wei, Y., J.-J. Xu, Q. Feng, H. Dong, M. Lin, (2000). *Encapsulation of enzymes in mesoporous host materials via the nonsurfactant-templated sol-gel process.* Materials Letters, **44**: p. 6.
41. Singh, K.V., R.R. Pandey, X. Wang, R. Lake, C.S. Ozkan, K. Wang, M. Ozkan, (2006). *Covalent functionalization of single walled carbon nanotubes with peptide nucleic acids: Nanocomponents for molecular level electronics.* Carbon, **44**: p. 1730.
42. Ge, J., H. Liu, (2007). *A three-dimensional mathematical model for liquid-fed direct methanol fuel cell.* J. Power Sources, **163**: p. 907.
43. Perrin, C.L., P.M.J. Tardy, K.S. Sorbie, J.C. Crawshaw, (2006). *Experimental and modeling study of Newtonian and non-Newtonian fluid flow in pore network micromodels.* J. Colloid and Interface Science, **295**: p. 542.
44. Barton, S.C., M. Pickard, R. Vazquez-Duhalt, A. Heller, (2002). *Electroreduction of O_2 to water at 0.6V (SHE) at pH 7 on the 'wired' Pleurotus ostreatus laccase cathode.* Biosensors and Bioelectronics, **17**: p. 1071.
45. Johnston, W., M.J. Cooney, B.Y. Liaw, R. Sapra, and M.W.W. Adams, (2005). *Design and characterization of redox enzyme electrodes: new perspectives on established techniques with application to an extremophilic hydrogenase.* Enzyme and Microbial Technology, **36**: p. 540.
46. Akkermans, R.P., S.L. Roberts, F. Marken, B.A. Coles, S.J. Wilkins, J.A. Cooper, K.E. Woodhouse, R.G. Compton, (1999). *Methylene green voltammetry in aqueous solution: studies using thermal, microwave, laser, or ultrasonic activation at platinum electrodes.* J. Physical Chemistry B, **103**: p. 9987.
47. Scully, J.R., D.C. Silverman, M.W. Kendig, eds. (1993). Electrochemical Impedance: Analysis and Interpretation. ASTM International, Philadelphia, PA.
48. Wang, C., C. Yang, Y. Song, W. Gao, and X. Xia, (2005). *Adsorption and direct electron transfer from hemoglobin into a three-dimensionally ordered macroporous gold film.* Adv. Functional Materials, **15**: p. 1267.
49. Yan, J., S. Dong, J. Li, and W. Chen, (1997). *Formation of self-assembled monolayers on gold electrodes with inclusion complexes of cyclodextrins and viologens.* J. Electrochemical Society, **144**: p. 3858.
50. Zhao, G., J.-J. Xu, and H.-Y. Chen, (2006). *Fabrication, characterization of Fe_3O_4 multilayer film and its application in promoting direct electron transfer of hemoglobin.* Electrochemistry Communications, **8**: p. 148.

51. Zhao, G.-C., L. Zhang, X.-W. Wei, and Z.-S. Yang, (2003). *Myoglobin on multi-walled carbon nanotubes modified electrode: direct electrochemistry and electrocatalysis*. Electrochemistry Communications, **5**: p. 825.
52. Johnston, W., N. Maynard, B.Y. Liaw, and M.J. Cooney, (2006). *In situ measurement of activity and mass transfer effects in enzyme immobilized electrodes.* Enzyme and Microbial Technology, **39**: p.131.
53. Miles, M.J., W.T. Smith, and J.S. Shapiro, (2000). *Morphological investigation by atomic force microscopy and light microscopy of electropolymerised polypyrrole films.* Polymer, **41**: p. 3349.
54. Zhou, Y., O. Anderson, P. Lindberg, and B. Liedberg, (2004). *Reversible hydrophobic barriers introduced by microcontact printing: application to protein microarrays.* Microchimica Acta, **146**: p. 193.
55. Khayet, M. and T. Matsuura, (2003). *Determination of surface and bulk pore sizes of flat-sheet and hollow fiber membranes by atomic force microscopy, gas permeation and solute transport methods.* Desalination, **158**: p. 57.
56. Muller, D.J. and D.J. Anderson, (2002). *Biomolecular imaging using atomic force microscopy.* Trends in Biotechnology, **20**(8 (Suppl.)): p. S45.
57. Czajka, R., C.G.J. Koopal, M.C. Feiters, J.W. Gerritsen, R.J.M. Nolte, and H. Van Kempen, (1992). *Scanning tunneling microscopy study of polypyrrole films and of glucose oxidase as used in a third-generation biosensor.* Electrochemistry and Bioenergetics, **29**: p. 47.
58. Iriyama, Y., T. Abe, M. Inaba, and Z. Ogumi, (2000). *Transmission electron microscopy (TEM) analysis of two-phase reaction in electrochemical lithium insertion within alpha-MoO_3.* Solid State Ionics, **135**: p. 95.
59. Barker, A.L., P.R. Unwin, J.W. Gardner, and H. Rieley, (2004). *A multi-electrode probe for parallel imaging in scanning electrochemical microscopy.* Electrochemistry Communications, **6**: p. 91.
60. Ngwa, W., R. Wannemacher, and W. Grill, (2004). *Phase-sensitive acoustic microscopy of polymer thin films.* Ultrasonics, **42**: p. 983.
61. Amatore, C., F. Bonhomme, J.-L. Bruneel, L. Servant, L. Thouin, (2000). *Mapping concentration profiles within the diffusion layer of an electrode. Part I. Confocal resonance Raman microscopy.* Electrochemistry Communications, **2**: p. 235.
62. Cannan, S., I.D. Macklam, and P.R. Unwin, (2002). *Three-dimensional imaging of proton gradients at microelectrode surfaces using confocal laser scanning microscopy.* Electrochemistry Communications, **4**: p. 886.
63. Hamelmann, F., U. Heinzmann, U. Siemeling, F. Bretthaurer and J. Vor der Bruggen, (2004). *Light-stimulated switching of azobenzene-containing self-assembled monolayers.* Applied Surface Science, **222**: p. 1.
64. Winsel, K., D. Honig, K. Lunkenheimer, K. Geggel and C. Witt, (2003). *Quantitative Brewster angle microscopy of the surface film of human bronch-alveolar lavage fluid.* European Journal of Biophysics, **32**: p. 544.

65. Zhou, Y., R. Valiokas, and B. Liedberg, (2004). *Structural characterization of microcontact printed arrays of hexa(ethylene glycol)-terminated alkanethiols on gold.* Langmuir, **20**(15): p. 6206.
66. Diaz, A.F., J.I. Castillo, J.A. Logan, and W.Y. Lee, (1981). *Electrochemistry of conducting polypyrrole films.* J. Electroanalytical Chemistry, **129**: p. 115.
67. Rao, V., P.A. Simonov, E.R. Savinova, G.V. Plaksin, S.V. Cherepanova, G.N. Kryukova, U. Stimming, (2005). *The influence of carbon support porosity on the activity of PtRu/Sibunit anode catalysts for methanol oxidation.* J. Power Sources, **145**: p. 178.
68. Palacio, L., P. Pradanos, J.I. Calvo, and A. Hernandez, (1999). *Porosity measurements by a gas penetration method and other techniques applied to membrane characterization.* Thin Solid Films, **348**: p. 22.
69. Nakao, S.I., (1994). *Determination of pore size and pore size distribution 3. Filtration membranes.* J. Membrane Science, **96**: p. 131.
70. Aurbach, D. and A. Zaban, (1994). *Impedance spectroscopy of lithium electrodes. 3. The importance of electrode surface preparation.* J. Electroanalytical Chemistry, **41**: p. 365.
71. Agosti, E. and G. Zerbi, (1996). *Conducting polymer-protein interactions in composite films: a spectroscopic study.* Synthetic Metals, **79**: p. 107.
72. Aurbach, D., M.L. Daroux, P. Faguy, and E. Yeager, (1987). *Identification of surface films formed on lithium in propylene carbonate solutions.* J. Electrochemical Society, **134**: p. 1611.
73. Naoi, K., M. Mori, Y. Naruoka, W.M. Lamanna, and R. Atanasoski, (1999). *The surface film formed on a lithium metal electrode in a new imide electrolyte, lithium bis(perfluoroethylsulfonyl)imide [$LiN(C_2F_5SO_2)_2$].* J. Electrochemical Society, **146**: p. 462.
74. Snook, G.A., G.Z. Chen, D.J. Fray, M. Hughes, and M. Shaffer, (2004). *Studies of deposition of and charge storage in polypyrrole-chloride and polypyrrole-carbon nanotube composites with an electrochemical quartz crystal microbalance.* J. Electroanalytical Chemistry, **568**: p. 135.
75. Mogi, R., M. Inaba, Y. Iriyama, T. Abe, and Z. Ogumi, (2003). *Study of the decomposition of propylene carbonate on lithium metal surface by pyrolysis-gas chromatography-mass spectroscopy.* Langmuir, **19**: p. 814.
76. Raitman, O.A., V.I. Chegel, A.B. Kharitonov, M. Zayats, E. Katz, and I. Willner, (2004). *Analysis of NAD(P)+ and NAD(P)H cofactors by means of imprinted polymers associated with Au surfaces: A surface plasmon resonance study.* Analytica Chimica Acta, **504**: p. 101.
77. Jameson, D.M., J.C. Croney, P.D. Moens, (2003). *Fluorescence: basic concepts, practical aspects, and some anecdotes.* Methods in Enzymology, **360**: p. 1.
78. Cooksey, K.E., J.B. Guckert, S. Williams, P.R. Callis, (1987). *Fluorometric determination of the neutral lipid content of microalgal cells using Nile Red.* J. Microbiological Methods, **6**: p. 333.

79. Greenspan, P., S.D. Fowler, (1985). *Spectrofluorometric studies of the lipid probe, Nile Red.* J. Lipid Research. **26**: p. 781.
80. Konash, A., E. Magner, (2006). *Characterization of an organic phase peroxide biosensor based on horseradish peroxidase immobilized in Eastman AQ.* Biosensors & Bioelectronics, **22**: p. 116.
81. Konash, A., M.J. Cooney, B.Y. Liaw, D.M. Jameson, (2006). *Characterization of Polymer-Enzyme Interactions using Fluorescence*, J. Materials Chemistry, **16**: p. 4107.

Chapter 20

Soybean Oil Biofuel Cell: Utilizing Lipoxygenase Immobilized by Modified Nafion® to Catalyze the Oxidation of Fatty Acids for Biofuel Cells

Jeanne L. Kerr and Shelley D. Minteer

Department of Chemistry, Saint Louis University, 3501 Laclede Avenue, St. Louis, MO 63103

The goal of this research was to build a biofuel cell that produces power from lipoxygenase enzymes catalyzing the oxidation of the main components of soybean oil. In this research, Nafion® was modified by ammonium salts to immobilize the lipoxygenase enzyme and allow facile transport of the fuel through the membrane to the enzyme. Modified Nafion® retains the original properties of Nafion and improves the surface conditions of the electrode for an enzyme. Bioanodes are fabricated by coating carbon supports with lipoxygenase immobilized within the modified Nafion®. Lipoxygenase catalyzes the oxidation of the fatty acid components of soybean oil. Ultimately, the goal will be to optimize the soybean oil bioanode to optimize power output and produce a stable bioanode for the fuel cell.

Biofuel cells are being developed as an alternative to traditional fuel cells. A biofuel cell is similar to a traditional fuel cell except at least one of the anodic or cathodic catalysts is a biomolecular catalyst, such as an enzyme or a living cell *(1)*.

The goal of this project is to build a biologically active fuel cell that produces power from enzymes catalyzing the oxidation of soybean oil. Traditional batteries and fuel cells are made of metals that are not biodegradeable and many produce an environmental hazard due to the possibility of these metals leaching into the ground water. This research describes a fuel cell made of soybean oil fuel, carbon conductor, and enzyme catalysts is far more environmentally friendly and renewable than other fuel cells and batteries.

In this research, commercial Nafion® is modified by quaternary ammonium bromide salts to immobilize the chosen enzyme and allow facile transport of the fuel through the membrane to the enzyme. Quaternary ammonium bromide salt-treated Nafion® membranes provide an almost ideal environment for enzyme immobilization *(2)*. Modified Nafion® retains the original electrical properties of Nafion®, increases the mass transport of ions and neutral species, lowers the acidity of Nafion® polymer to neutral pH, and increases the size of the pore for easier immobilization of large and complex molecules such as enzymes *(2-4)*.

This research developed lipoxygenase bioanodes which are utilized in a soybean oil biofuel cell. Lipoxygenase catalyses the oxidation of certain fatty acids found in soybean oil. Soybeans are grown in bulk quantities in the United States and soybean oil is easily extracted from the soybeans as a byproduct of soy-protein/soy-meal extraction. Although the individual components of soybean oil vary by source, most soybean oil contains linoleic acid, linolenic acid, oleic acid, stearic acid, and palmitic acid. Myristic acid also sometimes appears in soybean oil. Since enzymes are a renewable resource, and lipoxygenase can also be extracted from the soybean plant itself among other plant and animal source tissues, this electrocatalytic system is more environmentally friendly than traditional metallic systems. Lifetime studies were conducted in this research to determine how long lipoxygenase can be immobilized at the surface of an electrode without any significant degradation in power performance. So far, the results of the lifetime studies have been promising and they are still continuing. Ultimately, the goal will be to optimize the soybean oil-based lipoxygenase bioanode to optimize power output and produce a stable anode for the fuel cell.

Background

Enzyme Immobilization Method

In this research, lipoxygenase is immobilized within hydrophobically modified Nafion membranes. Our research group originally developed this method of enzyme immobilization for use in immobilizing dehydrogenase enzymes for fuel cells involving such fuels as ethanol, glucose and fructose (5). Within the past couple of years, other classes of enzymes that were immobilized have also shown success within this immobilization system. An ion exchange polymer, Nafion®, was chosen because of its unique micellar structure, its' useful physical properties, and ability for facile mass transport for ions and neutral species. While researching Nafion®, it was found that the polymer was acidic and was limited on mass transport due to it's small pore sizes. It was determined through experimentation that the polymer could be modified by exchanging the protons on the sulfonic acid side chains with quaternary ammonium salts to decrease acidity to a near neutral pH and increase pore sizes to enhance mass transport, while maintaining useful physical properties and lifetime of the polymer (5). Specifically, quaternary ammonium bromide salts were chosen because they are commercially available in a wide variety of alkyl chain lengths. Tetrabutylammonium bromide was utilized to modify Nafion® in order to immobilize alcohol dehydrogenase and bilirubin oxidase for the anode and the cathode of the ethanol biofuel cell (5). Triethylhexylammonium bromide modified Nafion® was also utilized to immobilized the two enzymes and both stability and increased mass transport were noted in both enzymatic systems (5).

Soybean Oil as a Fuel

Soybean oil is composed from five primary fatty acids: linoleic acid, oleic acid, linolenic acid, palmitic acid, and stearic acid (6). Linoleic acid usually composes the majority of the soybean oil with a minimum of half of the composition. Oleic acid usually composes around twenty percent of the soybean oil. The remaining fatty acids (linolenic, stearic, palmitic, and trace acids such as myristic) compose the remainder of the soybean oil (6). The composition of soybean oil actually varies within the individual plants grown in various farmer's fields. Linolenic and linoleic acids are both polyunsaturated fatty acids, oleic acid is a monounsaturated fatty acid, and the remaining stearic and palmitic acids are saturated fatty acids. DuPont has been working on optimizing the oleic acid content of soybean oil so as to produce a potentially healthier product. The results of DuPont's work comparing the standard composition of soybean oil to the oleic acid enhanced composition of soybean oil are shown in Figure 1 (6).

Lipoxygenase

Lipoxygenases constitute a family of non-heme iron enzymes that catalyze the dioxygenation of polyunsaturated fatty acids. Their recommended names are listed as: lipoxygenase [E.C. 1.13.11.12] (LOX), arachidonate 5-lipoxy-

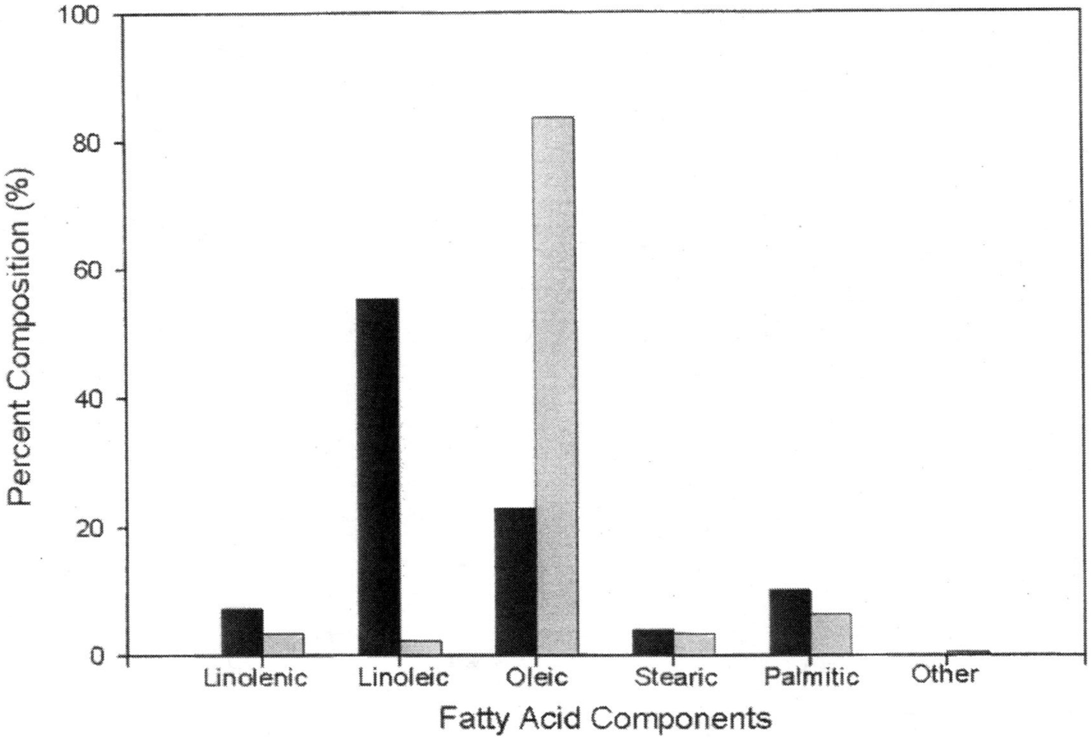

Figure 1. Soybean Oil Fatty Acid Percent Composition. Standard soybean oil is shown in black and DuPont's optimized soybean oil is shown in grey.

geanse [E.C. 1.13.11.34] (5-LOX), arachidonate 8-lipoxygenase [E.C. 1.13.11.40], linoleate 11-lipoxygeanse [E.C. 1.13.11.45], arachidonate 12-lipoxygenase [E.C 1.13.11.31] (12-LOX), and arachidonate 15-lipoxygenase [E.C 1.13.11.33] (15-LOX). The recommended and common names are generally determined by which carbon within the carbon-chain on archidionic acid is oxidized. Some lipoxygenases will almost exclusively produce one product, while other lipoxygenases will produce a mixture of products. The standard lipoxygenase tends to be more selective towards the thirteenth carbon in the fatty acid chain (7). In polyunsaturated fatty acids, the double bond located near the carbon effected by the oxidative process is shifted one bond closer to the carbonyl and changes from a cis-formation to a trans-formation. Lipoxygenase is sometimes used in the process to synthesize certain anti-inflammatories. In these studies, lipoxygenase is utilized to aid in the break down of soybean oil into smaller components. Our studies primarily focused on the use of lipoxygenase-1 from glycine max. Lipoxygenase-1 from glycine max was selected due to its' selectivity in the carbons affected and its commercial availability. This particular lipoxygenase has shown that it is stable when immobilized in a modified Nafion® membrane and literature has shown that it consistently produces the same products from linoleic and linolenic acids.

Experimental Procedure

Materials

The following materials were used for the fuel cell: Nafion™ 117 membrane (DuPont Company), platinum cathode (E-Tek ELAT Electrode, 20% Pt on Vulcan XC-72, 0.5mg/cm^2 Pt), carbon cloth (AvCarb™ Grade P50 Carbon Fiber Paper) with modifications as listed below.

The following reagents were used to prepare the fuels: potassium phosphate, sodium hydroxide, sodium chloride, soybean oil, linoleic acid, oleic acid, linolenic acid, stearic acid, palmitic acid, and myristic acid.

The following materials were used to modify the anode: Nafion® perfluorinated ion-exchange resin (5 wt.% solution in lower aliphatic alcohols/H_2O mix), lipoxygenase (EC 1.13.11.12, 70,800 Units/mg), tetrabutylammonium bromide (TBAB), triethylhexylamonium bromide (TEHA), trimethylhexylammonium bromide (TMHA), trimethyloctylammonium bromide (TMOA), triethyldecylammonium bromide (TMDA), trimethyldodecyl-ammonium bromide (TMDDA), and trimethyltetradecylammonium bromide (TMTDA). All materials listed were obtained from Sigma-Aldrich unless otherwise noted. Data was obtained with a CH Instruments Electrochemical

Analyzer Model 611B, and a CH Instruments Electrochemical Analyzer/ Workstation Model 684. At least three replicates of each type of biofuel cell are tested.

The following materials were used to perform the enzymatic assays: 200 mM borate buffer (B-0252), 1 M sodium hydroxide, 95% (v/v) ethanol, linoleic acid (L-1376), Nafion® perfluorinated ion-exchange resin (5 wt.% solution in lower aliphatic alcohols/H_2O mix), lipoxygenase (EC 1.13.11.12, 70,800 Units/mg), tetrabutylammonium bromide (TBAB), triethylhexylamonium bromide (TEHA), trimethylhexylammonium bromide (TMHA), trimethyloctyl-ammonium bromide (TMOA), triethyldecylammonium bromide (TMDA), trimethyldodecylammonium bromide (TMDDA), and trimethyltetradecyl-ammonium bromide (TMTDA). Data was obtained with a Cary 50 Bio UV-Vis Spectrophotometer. All experiments were performed in triplicate.

Preparation of Anode

Each ammonium salt is co-cast with 5% by weight Nafion® suspension into a weighing boat. All mixture-casting solutions were prepared so the concentration of ammonium salts is in a three-fold excess of the concentration of sulfonic acid sites in the Nafion® suspension. The membranes were allowed to dry overnight in a low humidity environment. The excess bromide salts were extracted by placing 18 MΩ water into the weigh boat and allowing the membranes to soak overnight. The water was removed and the membranes were rinsed thoroughly (three times) with 18 MΩ water. The salt-extracted membranes were resuspended in a lower aliphatic alcohol (ethanol). This will result in an ammonium modified Nafion® membranes. The enzyme/Nafion® casting solutions were prepared by adding 2.1 mg of enzyme to 5 mL of buffer pH 7.15 phosphate buffer. Then, equal amounts of enzyme solution and modified Nafion® suspensions (50/50) were mixed. The resulting solution was pipetted onto the surface of a carbon support (carbon cloth) and allowed to dry thoroughly. The electrodes were stored at room temperature before use.

Preparation of Fuels

All fuels were buffered with pH 7.15 phosphate buffer. The fuels were stored at room temperature and used within a week of preparation. Saturated solutions of soybean oil in pH 7.15 phosphate buffer were prepared. The fuel solution was sonicated for approximately five minutes to aid in dispersion of the soybean oil throughout the fuel solution. One millimolar fatty acid solutions were also formed in pH 7.15 phosphate buffer.

Testing the Biofuel Cell

A U-shaped glass test cell with Nafion® 117 membrane separating the anode and cathode compartments was constructed, as shown in Figure 2. The catholyte solution contains phosphate buffer (pH ~7.15). The analyte solution contains the soybean oil or fatty acid fuel.

Performing Enzymatic Assays

A standard UV/Vis spectrophotometric assay was used for lipoxygenase using a Cary 50 UV/Vis spectrophotometer.

Preparation of Reagents

Cuvettes were prepared by pipetting 100 μL of 50/50 enzyme/modified Nafion® suspension per 1 cm^2 onto the base of the cuvettes. Cuvettes were then allowed to air dry overnight. All enzyme/modified Nafion® solution samples (TBAB, TEHA, TMHA, TMOA, TMDA, TMDDA, and TMTDA) were prepared in triplicate.

Three reagents were prepared for the assay reaction. Reagent A was a 200mM borate buffer adjusted to pH 9.0 at 25°C with 1 M NaOH. Reagent B was 95% (v/v) ethyl alcohol (EtOH) and readily available in the laboratory. Reagent C was a 0.017% (v/v) linoleic acid solution prepared by combining 0.05 mL of reagent B and 0.05 mL of linoleic acid (Sigma-Aldrich) into a suitable container, vortexing completely, then adding reagent A while stirring until the solution was homogeneous until a total of 50.0 mL was obtained. Five milliliters of the linoleic acid/EtOH solution was combined with 20.0 mL of reagent A and 5.0 mL of 18 MΩ water. The solution was then mixed by stirring.

Procedure for UV Assay

The UV assay reaction was performed at room temperature (~25°C), pH 9.0, with a light path of 1 cm. The lipoxygenase enzymatic reaction was begun with 0.50 mL of reagent A and 1.00 mL of reagent C were added to the previously prepared and thoroughly dried cuvettes. The initial absorbance reading was taken immediately after pipetting the reagents into the cuvette at a wavelength of 234 nm. A second absorbance reading was taken after five minutes at a wavelength of 234 nm. The final absorbance reading was taken at a wavelength of 234 nm after 24 hours.

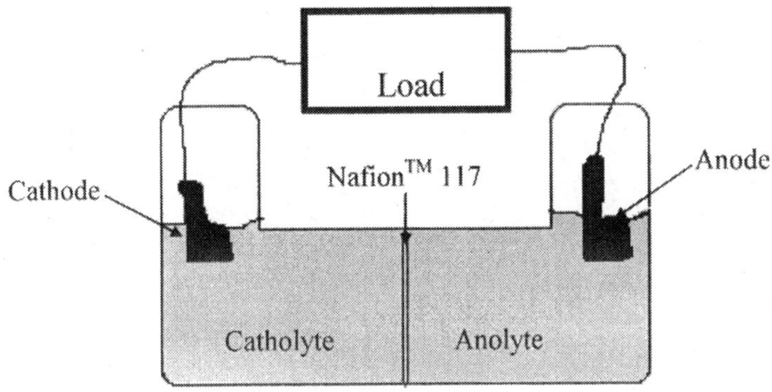

Figure 2. Biofuel Cell Test System

Results and Discussion

Several ammonium salts were used to modify the Nafion® membrane including: tetrabutylammonium bromide (TBAB), triethylhexylamonium bromide (TEHA), trimethylhexylammonium bromide (TMHA), trimethyl-octylammonium bromide (TMOA), triethyldecylammonium bromide (TMDA), trimethyldodecylammonium bromide (TMDDA) and trimethyltetradecyl-ammonium bromide (TMTDA). These modified Nafion membranes were mixed with lipoxygenase and coated on a carbon electrode. As shown in Figure 3, the modified Nafion® pores are significantly larger to promote easier immobilization of larger biomolecules such as enzymes and encourage mass transport of fuels. The various ammonium salts were utilized to produce the optimal size and shape for the biomolecule employed to oxidize the fuel solution. The modifications produced fewer, but larger pores that ultimately yielded better mass transport for the fuels utilized in these studies.

These electrodes were used as anodes for a soybean oil-based biofuel cells. Lipoxygenase is not the only enzyme to catalyze the oxidation of soybean oil components, such as: linoleic acid, linolenic acid, and oleic acid, but lipoxygenase is the only commercially available enzyme to oxidize the species. Figure 4 shows a general schematic of the catalytic event at the anode for a fatty acid oxidized by lipoxygenase within soybean oil.

Lipoxygenase Activity Assays

Figure 5 shows the results of the UV/Vis enzyme activity tests for the enzyme in the modified Nafion membrane. Figure 5 shows that TMDDA is

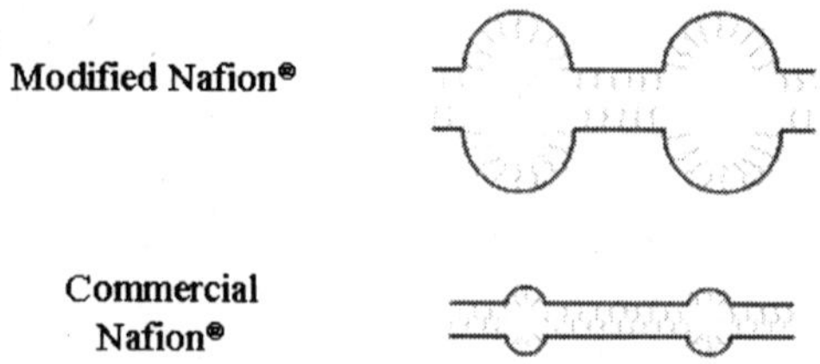

Figure 3. Comparison of Commercial Nafion® Structure Serses Modified Nafion® Structure (figures not to scale)

clearly the choice polymer for optimizing lipoxygenase activity while immobilized within a polymer. TMDA is another viable polymer choice for lipoxygenase immobilization and activity. TBAB, TEHA, TMHA, TMOA, and TMTDA perform approximately the same in this particular activity assay. Note: TMTDA has the largest error bar, but this is not surprising as it is difficult to get the polymer to resuspend in solution resulting in uneven polymer distribution on the electrode. While pipeting the TMTDA modified Nafion®/enzyme solution onto the base of the cuvette, it was noted that the polymer formed an uneven distribution before and after drying. All other polymers tended to form a relatively smooth surface on the bases of the cuvettes. Relative standard deviations were calculated for all of the polymers so as to ascertain the consistency of the results. The relative standard deviations for all of the modified Nafion® polymers with the exception of TMTDA was less than 5%, while the relative standard deviation of TMTDA was 17%. TMTDA was noted during this portion of the research to have various problems throughout the experiment. Problematic issues with TMTDA will be further discussed in later sections covering fuel cell performance and lifetime studies on the TMTDA polymer used within a soybean oil fuel cell system.

Nafion® Modification by Ammonium Salts for Biofuel Cell Optimization

Table 1 shows the results of soybean oil fueled biofuel cells utilizing lipoxygeanse (LOX) with each of the ammonium salt modified membranes. Figure 6 is a representative power curve for a soybean oil/oxygen biofuel cells employing the TMDDA modified Nafion® membrane to immobilize lipoxygenase at the carbon paper electrode surface. The peak of the power curve is used to determine the maximum power density, the maximum current density is determined as the current density when a load is applied to the cell to produce a potential of 0.1V, and the open circuit potential is measured at no load. It is important to note that although there is little change in the open circuit potentials

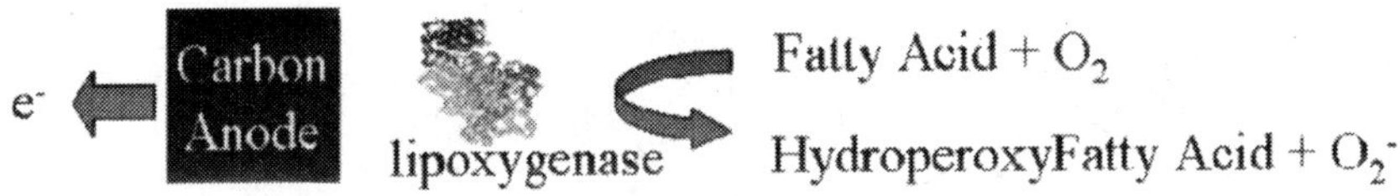

Figure 4. Lipoxygenase Catalytic Event Schematic

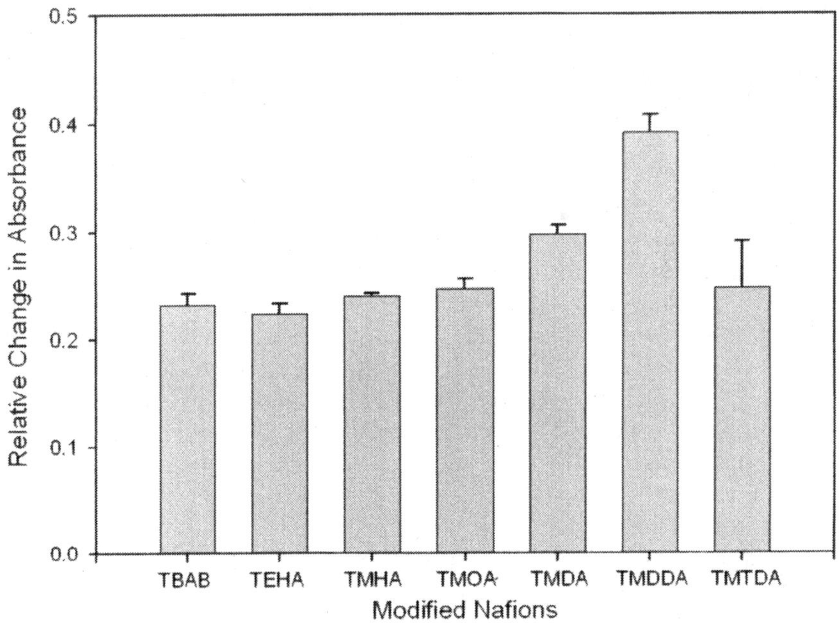

Figure 5. Results of Enzymatic Assay for Lipoxygenase in a Series of Modified Nafion® Membranes

for biofuel cells fabricated from each membrane, there are changes in the maximum current and maximum power. These membranes are stable for greater than one year and lifetime tests are continuing. Previous work in ethanol/oxygen biofuel cells performed by our research group has shown best performance for TBAB modified Nafion® membranes, but it is clear from both the current density and power density data that the TMDDA is the optimal membrane for immobilization of lipoxygenase at the surface of a carbon graphite support. TMTDA has slightly larger pores, and yet it yields lower power densities.

TMDDA modified Nafion® membranes tend to form smooth power curves as shown in Figure 6. Power curves for other modified polymers tend to be similar, but lower in peak current and power densities. TMTDA power curves tend to not be as smooth in appearance as the other modified polymers discussed in this research.

Biocatalysis of Soybean Oil Fatty Acid Component Studies

Table 2 shows the results of linolenic acid, linoleic acid, and oleic acid fuel cells employing lipoxygenase immobilized within TBAB and TMDDA membranes. The data shows increased open circuit potentials for TMDDA membranes that are statistically higher than open circuit potentials for TBAB membranes. The maximum current densities and maximum power densities for TMDDA membranes are also statistically higher than their corresponding TBAB membranes. These results are expected because the larger pore structure of the TMDDA membranes should provide increased mass transport of the fuel (each acid) through the membrane to the enzyme. It is important to note that even though the enzyme has different turnover rates for each acid, there is no statistical difference between the fuel cell performance of any of the acid components of soybean oil. This is indicative of a transport-limited system. These studies indicated that oleic acid was a substrate of lipoxygenase and that a reaction was catalyzed despite a variety of literature debating whether oleic acid is oxidized by lipoxygenase.

Further tests were done to determine if lipoxygenase was capable of oxidizing all components of soybean oil. Soybean oil consists of: linoleic acid, oleic acid, linolenic acid, stearic acid, palmitic acid, and sometimes myristic acid. Soybean oil composition varies depending on the source plant. Therefore it was important to determine whether lipoxygenase was able to oxidize the fuel regardless of the source of the soybean plant. Literature shows that lipoxygenase oxidizes linoleic acid and linolenic acid. This research shows that lipoxygenase is also capable of oxidizing unsaturated (oleic acid) and saturated (stearic, palmitic, and myristic acids). The products of the oxidations of the fatty acids that are mono-unsaturated and saturated are unknown, but the electrochemistry indicates that an oxidative process is taking place. Figure 7 shows the results of

Table 1. Results of Soybean Oil Biofuel Cell with Lipoxygenase Immobilized in Nafion® Modified with Various Ammonium Salts

	Open Circuit Potential (Volts)	Maximum Current (mA/cm^2)	Maximum Power (mW/cm^2)	Lifetime (Years)
TBAB	0.96	7.67	3.78	1.5+
TEHA	0.91	9.18	3.85	1.5+
TMHA	0.90	9.53	3.89	1.5+
TMOA	0.91	8.23	3.54	1.5+
TMDA	0.91	9.23	3.95	1.5+
TMDDA	0.97	10.6	4.39	1.5+
TMTDA	0.96	8.83	4.14	1.0+

Table 2. Results of tests on TBAB verses TMDDA on the primary acid components of soybean oil

	Linoleic Acid TBAB	Linolenic Acid TBAB	Oleic Acid TBAB	Linoleic Acid TMDDA	Linoleic Acid TMDDA	Oleic Acid TMDDA
Open Circuit Potential (Volts)	0.70	0.75	0.73	0.91	0.95	0.95
Maximum Current (mA/cm^2)	3.72	3.73	4.07	7.22	8.02	6.59
Maximum Power (mW/cm^2)	1.32	1.66	1.47	3.21	2.81	2.88

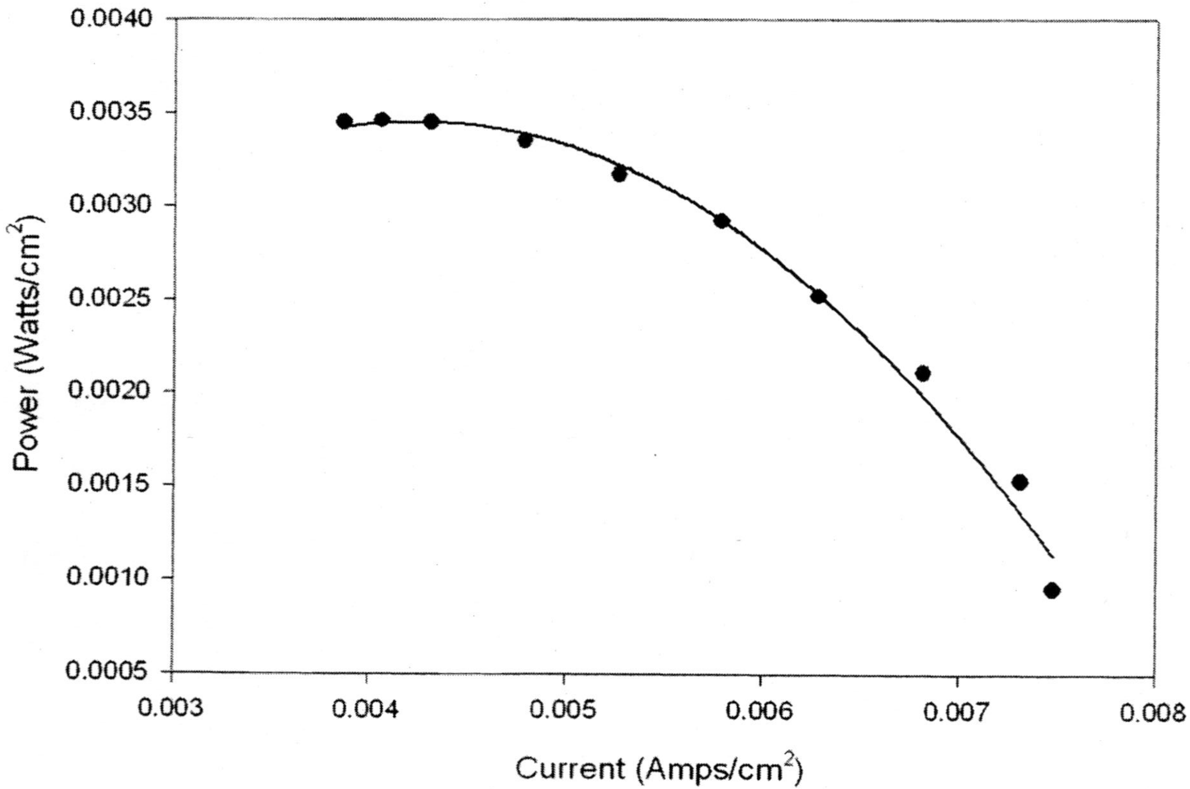

Figure 6. Representative Power Curve for TMDDA modified Nafion® Anodes in Soybean Oil.

these tests to determine whether lipoxygenase oxidizes all fatty acid components of soybean oil. As you can see the highest current and power densities are produced by the unsaturated fatty acids, while the saturated fatty acids produced smaller current and power densities.

At first glance one would think that stearic acid would outperform palmitic and myristic acids because it is an 18-carbon chain fatty acid like linolenic, linoleic, and oleic acids. However, upon closer inspection of those fatty acid chains, as depicted in Figure 8, it was noted that the unsaturated fatty acids curved in on themselves to have a more bent shape that is advantageous for the enzyme binding site. This would explain why the 18-carbon chain of the stearic acid would produce lower current and power densities than that of the 16-carbon chain palmitic acid and the 14-carbon chain of the myristic acid.

Lipoxygenase Modified Nafion Anode Lifetime Studies

The largest problem that has plagued enzymatic biofuel cells has been limited lifetime. For this reason, the lifetime of the soybean oil biofuel cells has been studied. Lifetime is defined as a reproducible 20% degradation in power output. Although there have been daily fluctuations in the power density due to changes in room temperature (18-27°C), all of the lifetime studies are still ongoing because the cells have not shown degradation in performance. The TBAB, TEHA, TMHA, TMOA, TMDA, and TMDDA-based soybean oil biofuel cell has been tested for over one and a half years with no degradation in power. The TMTDA-based soybean oil biofuel cell has been tested for seventeen months without any degradation in power. This is an increase from other biofuel cells (sugar based) in literature that are only stable for up to 10 days *(8-10)*.

Figure 9 demonstrates that no significant power degradation has taken place over one year. One can easily see on this particular day TBAB actually performed better than it did the previous year, while the other modified polymer results are very similar to the results of the previous year. Results actually fluctuate daily with humidity and temperature fluctuations within the laboratory. TMTDA appears to have the most random daily fluctuations of power and current densities during the lifetime studies performed over the past seventeen months. It has been observed that TMTDA power densities can range from approximately 0.5 mW/cm^2 to 4 mW/cm^2 reguardless of fuel cell conditions and room conditions. All other modified polymer electrodes tested tend to give relatively consistant results. Further studies will be done to determine exact influences of each of the fuel cell and room condition factors in the future.

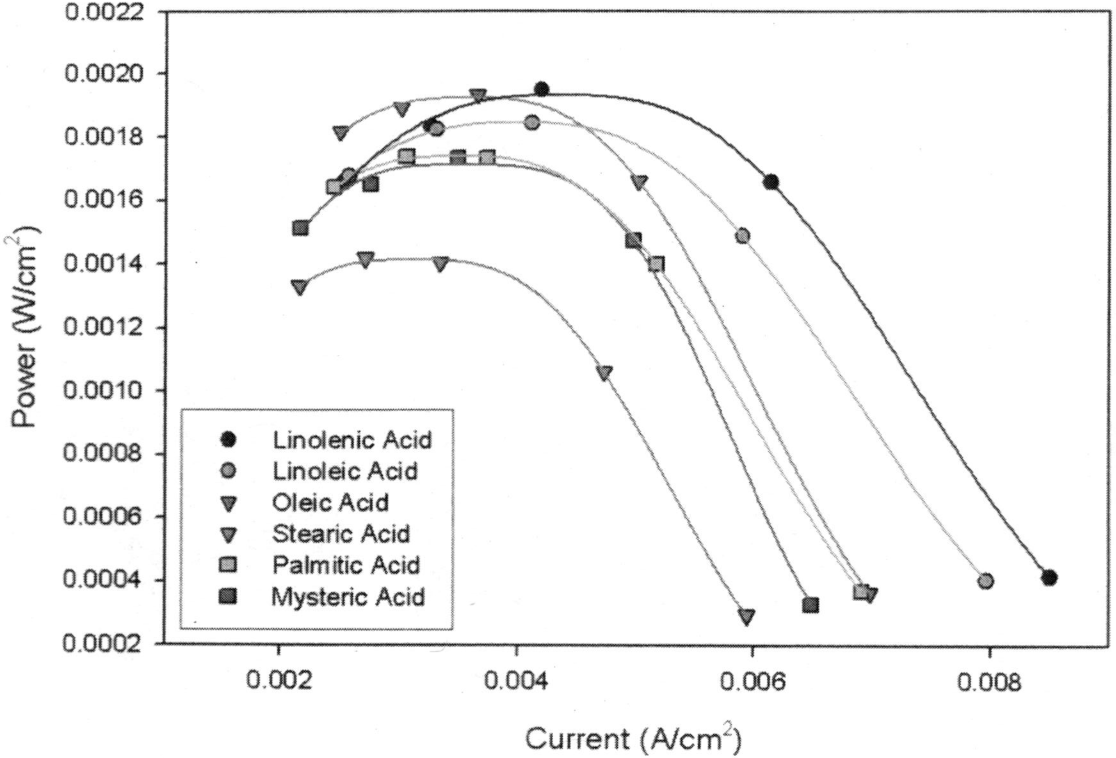

Figure 7. Representative Power Curves for TMDDA modified Nafion® Anodes in the Fatty Acid Components of Soybean Oil.

Conclusions

Soybean oil can be oxidized by lipoxygenase to produce energy. A lipoxygenase-based biofuel cell has an advantage over other fuel cells in that it does not require a coenzyme and yields similar or better results than many of the alcohol and sugar fuel cells *(8-11)*.

Trimetheyldodecylammonium bromide (TMDDA) modified Nafion® is clearly the choice polymer to optimize enzymatic activity of lipoxygenase. TMDA is also a possible candidate for lipoxygenase immobilation at the surface of an electrode utilized within a fuel cell. TMTDA is a poor choice for consistant activity of the lipoxygenase enzyme in a polymer utilized for immobilization.

Larger pore sizes have, so far, yielded higher power and current densities in a soybean oil biofuel cell. Lipoxygenase has shown in literature to oxidize linoleic and linolenic acids, and even more recently oleic acid. This research has indicated that lipoxygenase is capable of oxidizing all of the 6 fatty acids found in lipoxygenase: linolenic acid, linoleic acid, oleic acid, stearic acid, palmitic acid, and myristic acid.

Acknowledgements

This research could not have been done without financial support from the the United Soybean Board, the Saint Louis University Department of Chemistry, and the Saint Louis University Graduate School.

References

1. Palmore, G.T.R.; Whitesides, G.M. *ACS Symposium Series* **1994**, *566*, 271.
2. Moore, C.M.; Akers, N.L.; Hill, A.D.; Johnson, Z.C.; Minteer, S.D. *Biomacromolecules* **2004**, *5*, 1241.
3. Schrenk, M.; Villigram, R.; Torrence, N.; Brancato, S.; Minteer, S.D. *J. Mem. Sci.* **2002**, *205*, 3.
4. Thomas, T.J.; Ponnusamy, K.E.; Chang, N.M.; Galmore, K.; Minteer, S. D. *J. Mem. Sci.* **2003**, *213*, 55.
5. Akers, N.L.; Minteer, S.D. *Fuel Cell Sci. Eng. And Tech.* **2004**, *2*, 497.
6. Payne, J.H. Dupont Petition 97-008-01p, 2001.
7. Samuelsson, B.; Dahlen, S.E.; Lindgren, J.A.; Rouzer, C.A.; Serhan, C.N. *Science* **1987**, *237*, 1171.

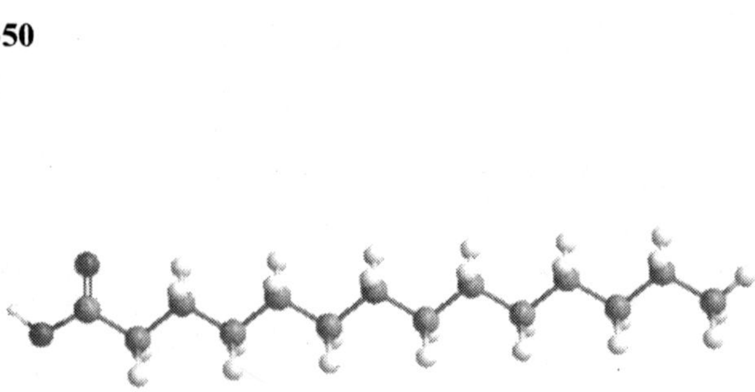

a) Myristic Acid

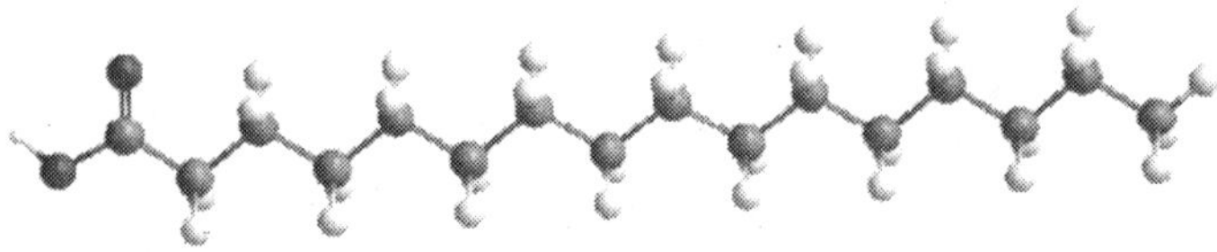

b) Palmitic Acid

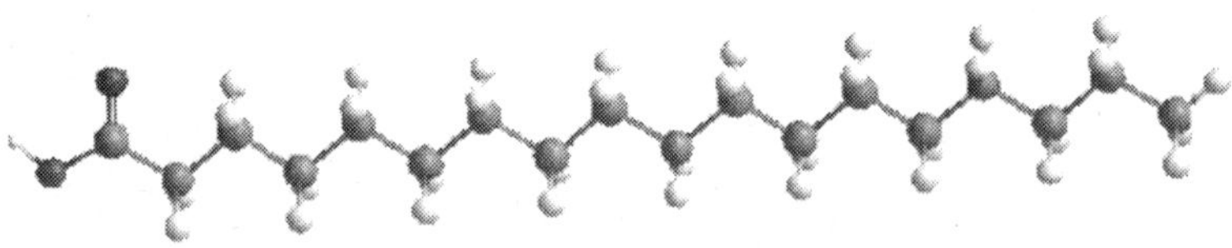

c) Stearic Acid

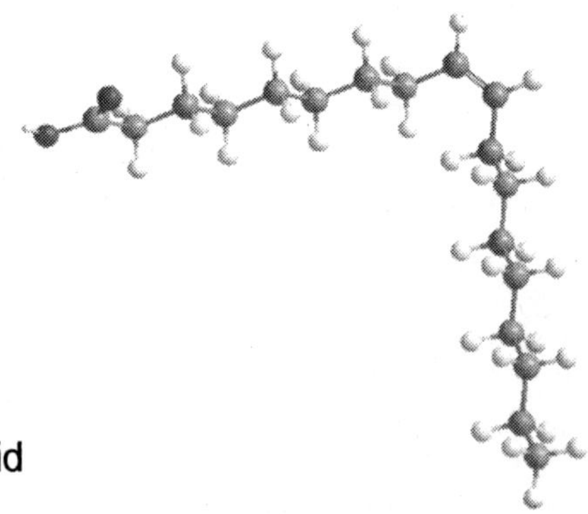

d) Oleic Acid

Figure 8. Structure Fatty Acid Components of Soybean Oil

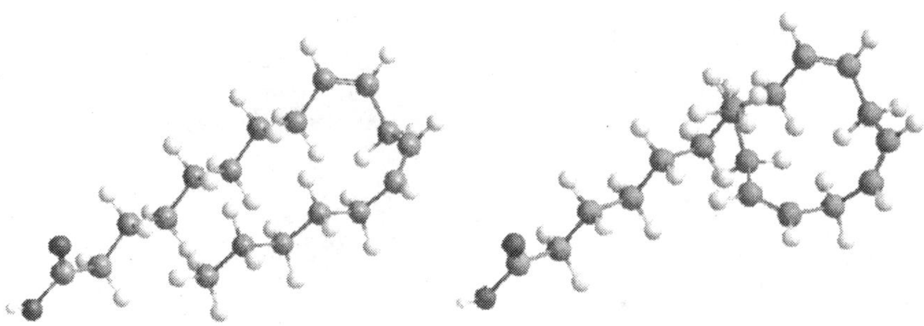

e) Linoleic Acid   f) Linolenic Acid

Figure 8. Continued.

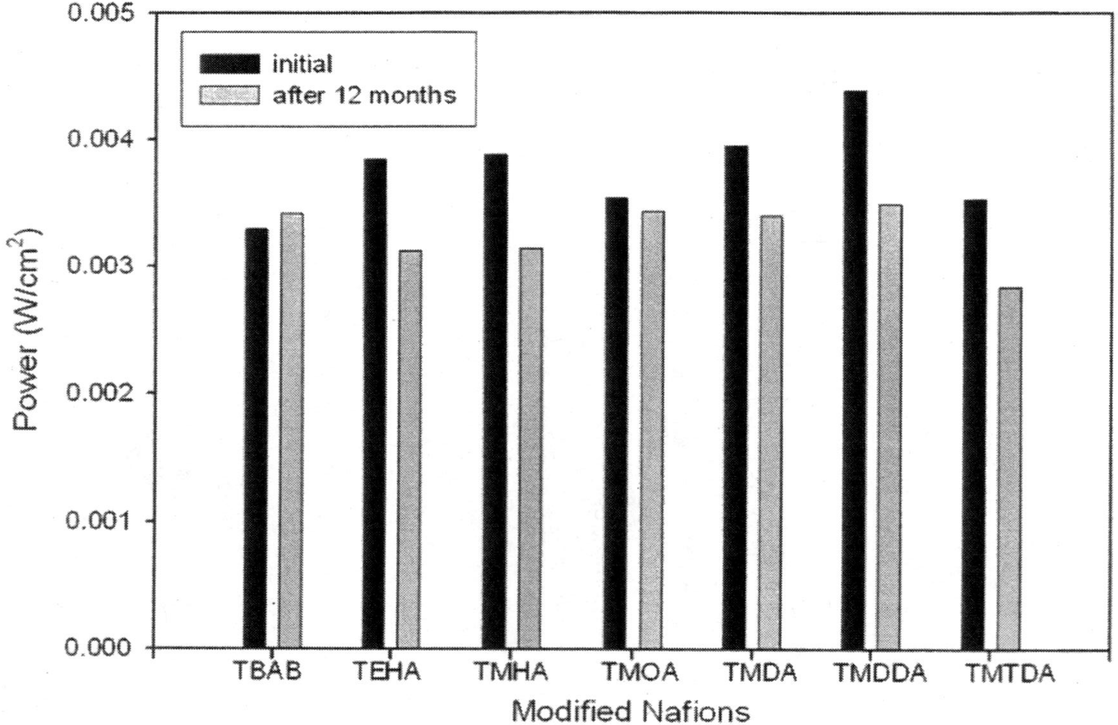

Figure 9. Comparison of power densities over a year period.

8. Mano, N.: Mao, F.; Heller, A. *J. Am. Chem. Soc.* **2003**, *125*, 6588.
9. Katz, E.; Willner, I. *J. Am. Chem. Soc.* **2003**, *125*, 6803.
10. Mano, N.; Mao, F.; Shin, W.; Chen, T.; Heller, A. *Chem. Communications* **2003**, 518.
11. Akers, N.L.; Moore, C.M.; Minteer, S.D. *Electrochimica Acta* **2005**, 50, 2521.

Chapter 21

Surface Patterning and Functionalization for Biomolecular Motor Nanotechnology

WenLiang He, Thorsten Fischer, and Henry Hess*

Department of Materials Science and Engineering, University of Florida, Gainesville, FL 32611–6400

The successful integration of biomolecular motors into synthetic environments enables the design of active nanostructures and nanodevices, such as the "molecular shuttle". Over the past few years, continuous efforts have been devoted to guide the movement of such shuttles consisting of surface-adhered motors transporting functionalized filaments. Both, controlled placement of motors and guiding obstacles, have been achieved, employing diverse surface modification techniques including microcontact printing, laser ablation, plasma deposition, photolithography, E-beam and nanoimprint lithography. This paper reviews recent developments in this area, with emphasis on techniques and the corresponding materials used to achieve this goal.

Introduction

Organizing transport on the nanoscale is a challenge for nanotechnology. Transport by diffusion is effective on small scales but requires concentration gradients as driving forces. A "molecular shuttle" offers an alternative means to transport nanoscale cargos under user control *(1)*.

The idea of the molecular motor shuttle is inspired by nature, where nanoscale transport systems are already realized *(2)*. In cells, motor proteins powered by hydrolysis of adenosine triphosphate (ATP) transport cellular cargos throughout the cytoplasm by moving along stationary parts of the cytoskeleton, in particular actin filaments and microtubules *(3,4)*. Molecular shuttle systems often utilize the inverted geometry, where motors are adsorbed to the surface and transport the corresponding filaments, and have found applications as nanoscale sensors and probes *(5)*, in single-molecule studies *(6,7)*, molecular assembly *(8)*, lab-on-a-chip systems *(9,10)*, and nanofluidics *(11,12)*. The primary topic of this paper is to review various means of achieving directed shuttle movement along predetermined tracks.

Motor Proteins and Their Related Filaments

Kinesin and myosin are two families of molecular motors which have found application in such hybrid systems. Kinesin proteins are eukaryotic microtubule-associated motor proteins that are typically composed of three parts: (i) two large N-terminal globular "heads" (head region) that allow it to attach to and move along microtubules (MTs), (ii) a central coiled-coil region, and (iii) a C-terminal tail region termed light-chains, which bind kinesins to the cargo *(13)*. Kinesin molecules are about 80 nm in length, and move unidirectional along MTs towards the fast-growing plus end in a hand-over-hand fashion with 8 nm per step *(14)*. Individual kinesin can generate force up to 6 pN *(15)*, and can be adsorbed on surfaces at 10^3 motors/μm^2, theoretically producing forces on the order of nN per μm^2. Microtubules are hollow dimeric tubular protein filaments, self-assembled from α and β tubulins. They are 25 nm in diameter, and range in length from 1 to 100 µm, depending on polymerization conditions *(16)*.

Skeletal myosin, termed Myosin II, also contain two globular motor heads, a neck, and a tail domain. Myosin II binds to actin filaments, which are composed of two strands of actin monomer wound around each other in a helical arrangement with a repeat distance of 72 nm are 8 nm in diameter and typically 1 to 10 µm in length. *(16)*

Motility Assays

Motility assays have been used extensively to determine the mechanisms of molecular motor movements and interactions *(17-19)*. Two distinct configurations are commonly used: the bead motility assay and the gliding motility assay, also known as the "inverted assay", as shown in Figure 1. The bead assay mimics the way motor proteins and filaments interact in nature: the filaments act as static components while motor proteins step along them. In the inverted gliding assay a carpet of motor proteins coats the surface while the head groups propel the filaments. The inverted gliding assay is more popular, due to the possibility of producing extended networks and complex patterns of tracks.

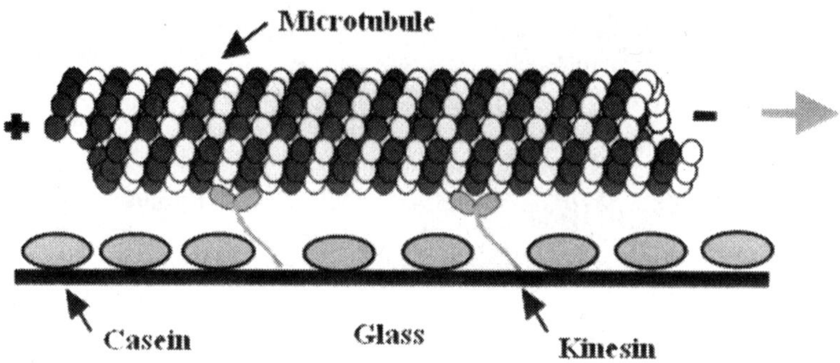

Figure 1. Illustration of the gliding motility assay for the kinesin-MT system: casein is pre-coated on glass to support kinesin function, and MT is transported towards its minus end as kinesin moves towards the plus end.

Confinement and Motility Regulation

In early studies of both kinesin-MT *(20)* and myosin-actin systems *(18)*, filaments moved over smooth glass surfaces with no preferred orientation. However, meaningful work cannot be extracted from such systems unless filament motion can be controlled. For the rest of this paper, the topic of filament control will be termed "confinement".

Confinement can be achieved via chemical or topographic methods, or by a combination of the two. **Chemical confinement** is achieved by selectively coating a surface, resulting in regions with high functional motor density and regions with no motor functionality, as seen in Figure 2A. Filaments are confined to move only in regions where motors are available to bind to and transport them. Since a single kinesin motor can move a MT *(21)*, high motility contrast among surfaces cannot be obtained unless the active motor density on the motility blocking regions is decreased to a degree that the average distance between two available motors is smaller than the average length of the filaments being transported. **Physical confinement** is obtained by physically imposing barriers to confine filament motility, as shown in Figure 2B. If the walls are made of non-fouling materials, filaments will travel only along the bottom motor protein channel, which gives rise to the **combined confinement** as shown in Figure 2C.

Within the channels, filaments are transported in both directions. In order to achieve unidirectional transportation, patterned structures can be added to regulate filament directionality, as shown by Figure 3.

For both chemical and combined confinements, protein adsorption is the primary concern and established strategies to prevent it are employed *(26)*. An ethylene glycol based polymer $(-CH_2CH_2O-)_n$ is commonly used to create a non-fouling surface. This material is also called poly(ethylene glycol) (PEG) when n

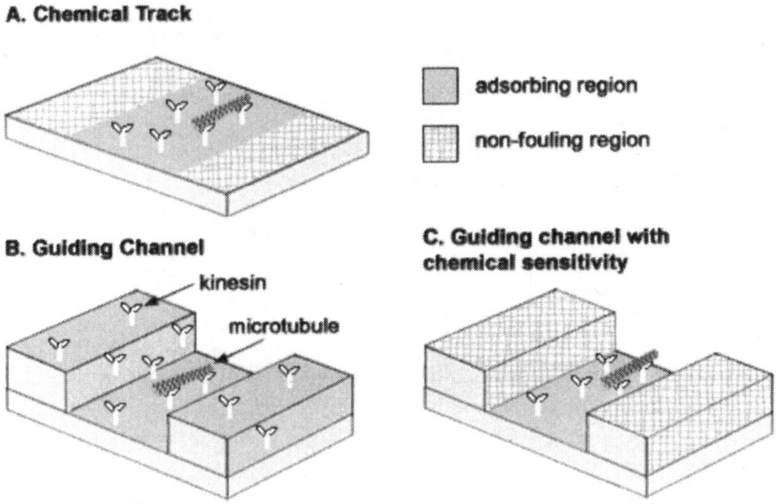

Figure 2. Three basic ways to confine filament motility: A. chemical confinement – motility contrast is obtained by contrast in active surface motor density among surfaces; B. physical confinement – filaments can move on both upper and lower surfaces; C. combined confinement – filaments only move within channels. Reproduced with permission from Nano Lett. 2003, 3, 1651-1655. Copyright 2003 Am. Chem. Soc.

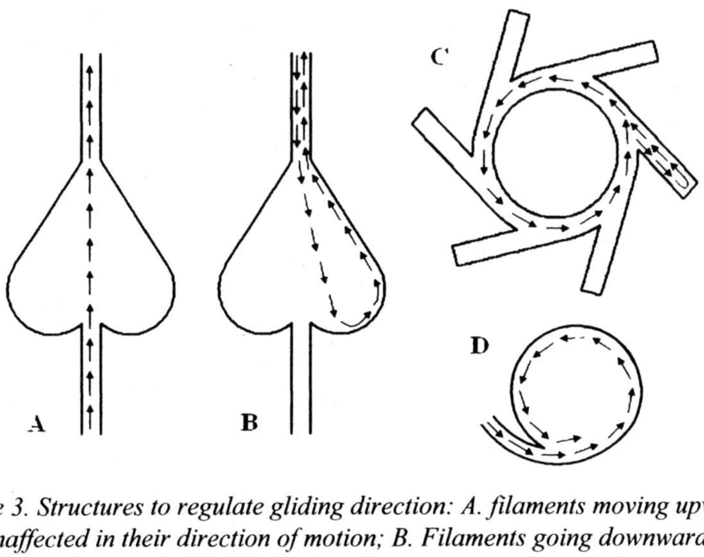

Figure 3. Structures to regulate gliding direction: A. filaments moving upwards are unaffected in their direction of motion; B. Filaments going downwards are redirected upwards (22,23); C. a ratchet pattern used to obtain counter-clockwise motion (24,25); D. a structure that can concentrate gliding filaments (25).

is around 15-3500, while the term poly(ethylene oxide) (PEO) is designated when n is greater than 3500. Two main classes of PEG-based materials have been utilized in this field: plasma polymerized tetraglyme (pp4G) coating *(27,28)*, and various kinds of poly(ethylene glycol) self-assembled monolayers (PEG-SAM) *(23,29,30)*.

Common Confinement Strategies

Kinesin-MT System overview

In general, chemical confinement is of very limited use for the kinesin-MT system *(24)*, due to the high rigidity of the MTs. Successful guidance can only occur at shallow approach angles (defined as the angle between a MT and the sidewall when they were first in contact), since the guiding probability decreases exponentially with approach angle *(28)*, indicating that only narrow tracks can assure desirable filament transportation *(27)*. Since chemical confinement relies on the Brownian motion of the swiveling tip of a gliding microtubule to bind to an off-axis motor, the minimum radius of a curving track is on the order of the persistence length of a trajectory of an unconfined microtubule (0.1 mm *(31)*). For the actin-myosin system, however, chemical confinement is much more successful due to the more than hundred-fold higher flexibility of actin filaments compared to microtubules *(32)*.

Physical barriers can be used to transform the pushing forces of kinesins into bending forces, which makes open channels an attractive means to direct MTs on smaller scales. But confinement by open channels *(33,34)* also has its own problems *(28,35)*: motor proteins are uniformly adsorbed on all the surfaces, and thus filaments are often observed climbing up the walls, since active motors attached on sidewalls can transport them to the upper surfaces. In fact, for MTs moving parallel to the edges of the walls, the probability of successful confinement is only about 60%, thus the probabilities for climbing up walls or continuing to travel along the channels are almost equal. The guiding probability of physical confinement decreases linearly from about 60% at low angles to almost 0% at 90° angles *(28,35)*. Unlike for chemical confinement, "the narrower the better" does not apply for the width of the channel, but an intermediate value is optimal for directed transport *(35)*. This is explained as a compromise between the favorable reduction of approach angles and the unfavorable increase in the number of collision per distance covered. However, with the introduction of semi-enclosed *(36)* and totally enclosed channels *(37)*, 100% confinement can be easily achieved, since microtubules are prevented from climbing the overhanging sidewall. But for such restricted structures, two problems may occur: (i) the supply of reagents, e.g. ATP, is restricted and (ii) the introduction of solutions without creating trapped air bubbles is challenging.

Combined confinement is a very successful method to control filament movement *(22,28,29,38)*, since by creating non-fouling sidewalls microtubules cannot bind to motors and escape. The probability of successful confinement is typically over 90%.

Myosin-Actin System overview

For the myosin-actin system, it has been shown that only cationic or hydrophobic surfaces can adsorb and support myosin motor function *(12,39-43)*, while negatively charged or hydrophilic surfaces are usually used to bring motility contrast to the system. Among the photoresists, MRL-6000 1.XP *(11,12,44)* is identified as a superior material supporting motility, while treatment of glass surfaces with hexamethyldisilazane (HMDS) *(45)*, trimethylchlorosilane (TMCS) *(41)* and positively charged poly(allylamine hydrochloride) (PAH) *(42)* also supported myosin-induced actin motility. E-beam treated tBuMA & MMA (4:1) *(39)* and PMMA 950 *(11,12,44)* are photoresists that disable myosin motor function and this effect can be further enhanced by O_2 plasma etching *(12)*. Bovine serum albumin (BSA) *(43,45)* and negatively charged PEBSS *(42)* as well as poly(styrene sulfonic acid) (PSS) are reagents that help to block motor protein adsorption onto surfaces.

Compared to the kinesin-MT system, actin filaments move much faster (typically more than 4 μm/s; MTs typically move less than 1 μm/s) and can pass through smaller curvatures (through circles 2 μm in diameter), which enables fast transportation and miniaturization of devices. However, much narrower channels (usually down to 100 - 500 nm) are also required to effectively confine actin filament motility and to prevent it from taking U-turns. The requirement of a hydrophobic surface to support myosin motor function may further impose some difficulty with respect to solution introduction – air bubbles can be trapped in the channels due to the unfavorable interactions between the hydrophobic surface and the aqueous buffer solutions.

In the first motility assays *(18)*, nitrocellulose was spread on a glass slide to produce a hydrophobic surface that supported myosin function. However, nitrocellulose lacks the ability to be lithographically patterned, and it does not facilitate production of films with controlled uniform thickness. Later studies focused on finding effective methods to confine and regulate actin filament motility.

Methods to Achieve Chemical Confinement

Thermal Stretching of PTFE

Through mechanical deposition of PTFE on glass a layer of polymer with ridges and grooves along the sheer-axis can be produced for both kinesin-MT *(1)* and myosin-actin systems *(46)*. The dimensions of these ridges are randomly distributed; they are 10 - 20 nm in height, 10 - 100 nm in width and 10 - 100 nm apart. Although MTs were observed to become oriented toward the shear-axis after several directional changes the oriented motion was maintained only at moderate motor densities. For lower motor density, MTs tended to detach, while for higher motor density, MTs began moving randomly as they switched among adjacent lines. Similarly, at high motor density areas, orientated actin filament motility would be lost, and actin filaments tended to switch among lines and to take U-turns.

E-beam Treatment of tBuMA:MMA (4:1) Copolymer

Under E-beam exposure, the copolymer of tert-butyl-methacrylate (tBuMA) and methyl methacrylate (MMA) (4:1) tends to become more hydrophilic as the contact angle is dropped from 87° to 82° *(40)*. Thus, contrast in protein adsorption can be produced by selectively exposing the photoresist, which creates hydrophilic regions surrounded by hydrophobic regions and discourages adsorption of functional myosin. Unfortunately, the mobility confinement obtained by this method was not very effective; both the exposed hydrophilic regions and the unexposed hydrophobic regions absorbed motor proteins. Filaments moving on myosin-rich regions tended to be either immediately reflected or gradually deflected when they encountered the edges, while filaments on myosin-poor regions were not guided by the edges *(40)*.

Glow Charge Plasma Deposition of pp4G

Tetraglyme, CH_3-O-($CH_2CH_2O)_4$-CH_3, also known as tetraethylene glycol dimethyl ether, can be used to deposit a "PEG-like" coating by glow charge plasma deposition. This coating strongly exhibits the hydrophilic character of PEG, and thus it discourages kinesin motor adsorption effectively.

As shown in Figure 4, selective coating of polymerized tetraglyme on glass can be obtained by a succession of UV lithography, plasma charge deposition and final lift-off process *(27)*. Through this method, Lipscomb et al. *(27)* have demonstrated that narrow and smooth tetraglyme tracks can significantly improve MT motility performance under chemical confinement. When MTs moved along rough and wide (125 μm) tracks, they couldn't trace the edges but tended to detach when they reached the edges. On the other hand, MTs could be reflected by the edges and several of them traveled as long as 250 μm along smooth and narrow (2 μm) lines. Clemmens et al. *(28)* also showed that about 88% of MTs would detach as they encountered the edges, and chemical confinement was effective only at very shallow approach angles (< 10°, almost 100%), while it became almost useless at larger angles (> 30°, almost 0%).

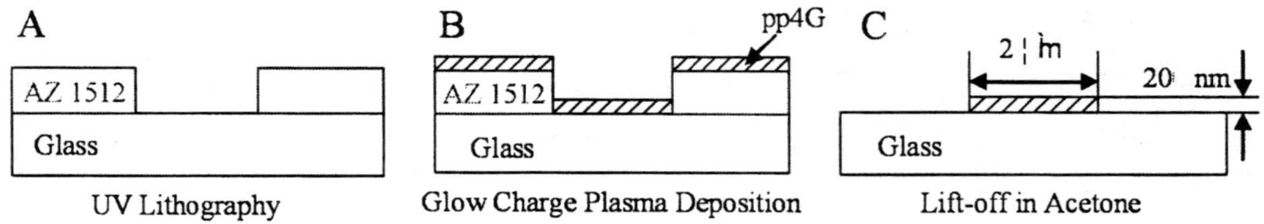

Figure 4. The process of glow charge deposition of pp4G: A. UV lithography is used to selectively coat AZ 1512 on glass; B. plasma deposition of pp4G through vapor phase ensures that tetraglyme can reach every small corner of the exposed surface; C. the final lift-off process exposes glass surfaces that are surrounded by non-fouling PEG based regions (~ 20 nm in thickness).

Microcontact Printing & Biotin-Streptavidin Interactions

Highly effective chemical confinement can be achieved by taking advantage of the very strong and specific biotin-streptavidin interactions *(43)*, as shown in Figure 6. No filaments were observed moving outside the MHA-treated regions (where his-tagged biotin adsorbs), and those traveling within the regions bounced back when they encountered the edges. In contrast to the behavior of MTs under chemical confinement, successful guidance occurred at all approach angles.

PDMS Stamp Shielding & Pluronic Treatment

Pluronic F108 is a triblock copolymer (PEO_{129}-PPO_{56}-PEO_{129}) that is comprised of hydrophobic poly(propylene oxide) (PPO) domains and hydrophilic poly(ethylene oxide) (PEO) domains. In the presence of hydrophobic substrate, the PPO domain of the molecule will bind to the surface, leaving the hydrophilic PEO domains extending out towards the aqueous environment; Pluronic F108 thus can be used to confer hydrophobic surface non-fouling properties.

By pressing a PDMS stamp against glass surfaces, the exposed hydrophilic glass areas can be transformed into hydrophobic regions by silane treatment (tridecafluoro - 1,1,2,2, - tetrahydrooctyl - 1 - trichlorosilane) *(24)*, as shown by Figure 5. The hydrophobic regions can be further transformed to be non-fouling surfaces through Pluronic F108 treatment.

Methods to Achieve Physical Confinement

Replica Molding to Produce Open Channels

Replica molding is a simple and convenient way to produce topographic patterns *(47)*. Through standard UV lithography, SU8 photoresist can be patterned on silicon wafers. PDMS precursor is then applied to the patterned surface. After curing with heat, a PDMS mold is formed which can be used to produce a PU replica by imprinting it into the polyurethane precursor that can be cured under UV light as seen in Figure 7. These PU replicas are reproduced from the originally SU8 patterned silicon wafer with high fidelity. The steepness of the sidewalls obtained by this method is usually about 80°. PDMS molds can be used multiple times before being discarded.

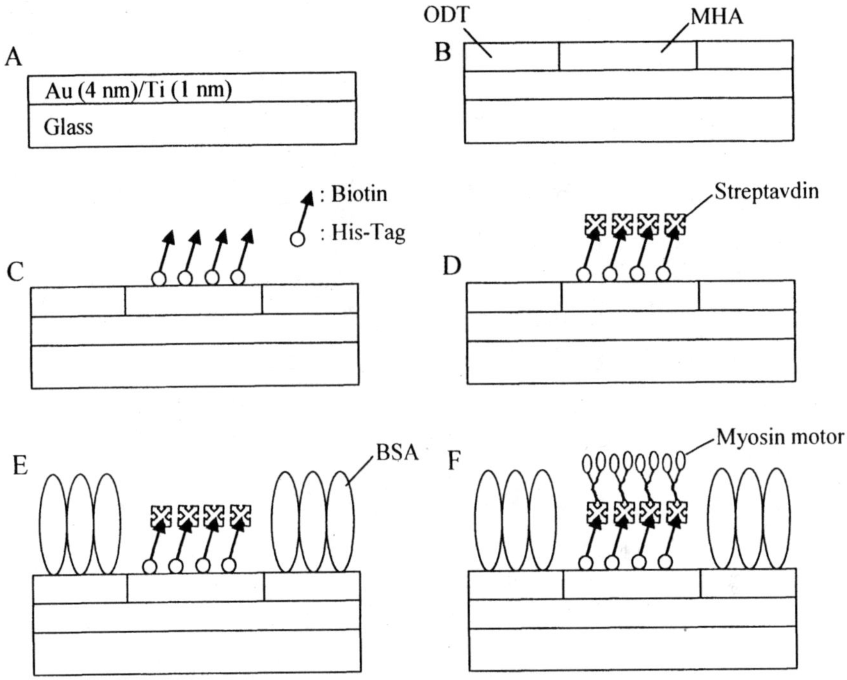

Figure 6. Chemical confinement through biotin-streptavidin interactions: A. Au/Ti (4.2 nm Au on 1.4 nm Ti) bilayer deposited on glass substrate; B. 16-mercaptohexadecanoic acid (MHA) is coated on Au/Ti through microcontact printing, and the rest of the surface is coated with 1-octadecanethiol (ODT) by dipping the patterned surface in its solution; C. The his-tag groups of the biotinylated his-tags will tightly bind to the negatively charged MHA regions; D. Streptavidin is introduced to selectively bind to the biotinylated his-tags; E. Bovine serum albumin (BSA) is used to block the hydrophobic ODT surfaces from further myosin adsorption; F. The biotin groups on the biotinylated myosin motors bind to the streptavidin coated regions.

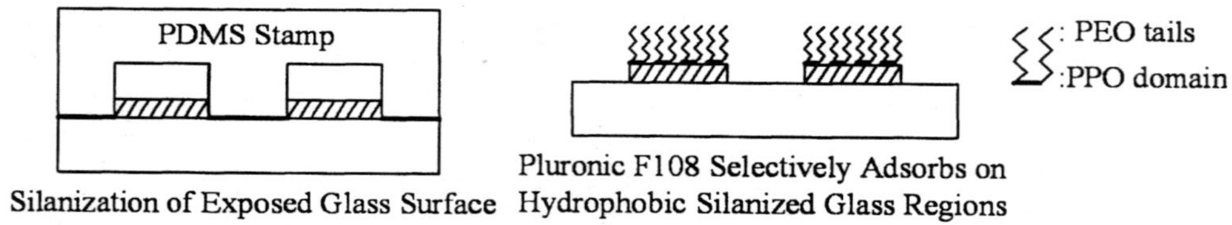

Figure 5. PDMS stamp shielding: silanization of unprotected glass regions (left). After removing the stamp, the silanized regions are rendered non-fouling by Fluronic treatment (right).

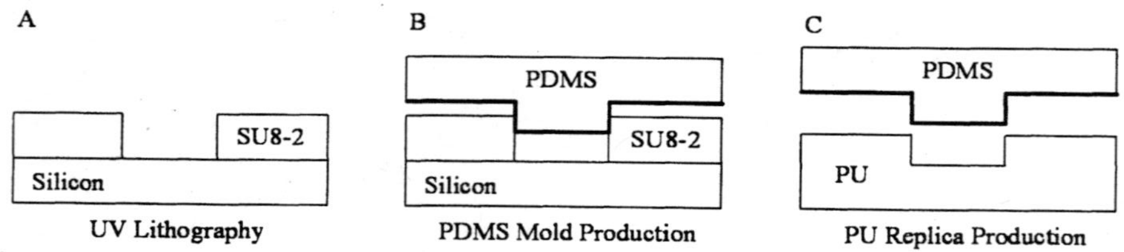

Figure 7. The process of producing a PU replica: A. A UV lithographically patterned master; B. Pattern transfer from the master to a PDMS mold; C. A replica reproduced from the PDMS mold.

UV Lithography to Produce a Semi-Enclosed Structure

Semi-enclosed channels, as shown in Figure 8, were applied to confine MT motility by forming undercut regions *(36)*. The formation of the undercut region is due to underexposure that fails to strongly cross-link the exposed resists. The dose levels used are typical of those for semiconductor surfaces which readily reflect exposure light back to the resists. Glass, as used in this case, allows most of the UV light to pass through it and thus decreases the exposure doses compared to the standard situation.

Although incompatible with rhodamine-labeled MT observation due to its strong fluorescence *(22,38)*, AZ 5214 photoresist was used here in combination with Oregon-green-labeled MTs. AZ 5214 surfaces supported motility and thus no chemical confinement was achieved, but the confinement due to the undercut was still excellent – no MTs were observed moving out of the channels, and those moving in the channels showed the same average gliding velocity as the ones moving over the upper open plateaus. This indicated that no shortage of ATP supply was experienced in the spatially constricted undercut regions. Due to their high rigidity, MTs traveling on the upper plateaus can move unaffected by the gaps if the MTs are longer than 5 µm and the gaps are 0.5 µm in width and at least 1 µm deep, leading towards a functional multilevel transportation system *(36)*.

Totally Enclosed Structures

Huang et al. *(37)* produced totally enclosed channels, where selective adsorption of motor proteins is no longer needed for confinement purpose, and 100% confinement can be achieved. Two ways have been explored to cap the open channels. One simple and low cost method is to use Riston, a photoimagable dry-film. A 40 µm thick Riston film is brought into contact with a UV lithographically patterned SU-8 photoresin coated substrate with 5 µm wide channels. One drawback of such technique is the thickness of the Riston film. The minimum film thickness of 40 µm may render this method unusable for light microscopy.

The other way to realize enclosed channels is schematically shown in Figure 9. A trilayer of benzocyclobutene (BCB) (1 µm thick) coated on Au modified silicon wafer is brought into contact with UV lithographically patterned SU-8 channels. Under moderate pressure, the sample is cured at 200°C for 2 hours. Because gold bonds poorly to silicon, the capped structure can be easily peeled off from the silicon wafer due to the strong adhesion between BCB and SU-8 surfaces. A post-etching process is performed to remove the attached Au layer. It should be noted that the thin BCB layer will collapse if the width of the SU-8 channels is greater than 10 µm. One problem related to BCB is that it is strongly

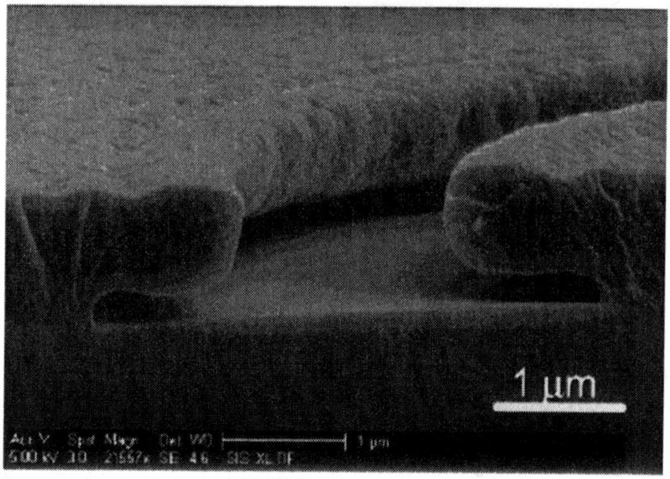

Figure 8. The structure of the semi-enclosed channels: A layer of negative photoresist AZ 5214 is spin-coated on glass substrate. After a 2.3 s exposure under UV light, the sample is baked at 110°C for 50 s. A final aggressive flood exposure is then delivered for 45 s to the sample. When the sample is developed the undercut regions (200 nm high, 1 µm deep into the sidewall) will appear and the size of the undercut areas is the function of development time. Reproduced with permission from Nano Lett. 2003, 3, 1651-1655. Copyright 2003 Am. Chem. Soc.

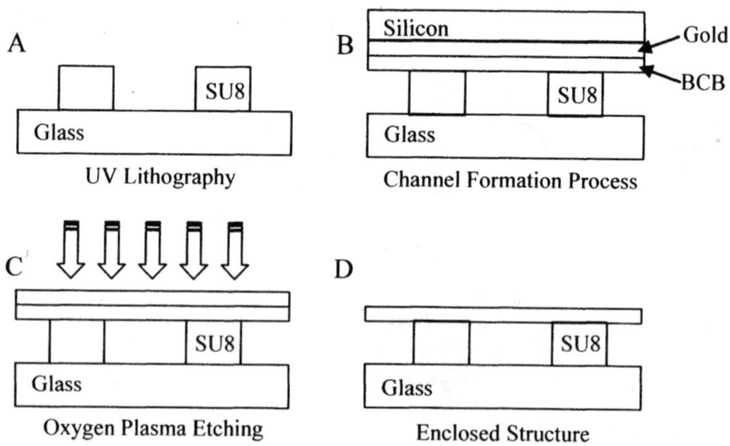

Figure 9. The process of fabricating totally enclosed structures: A. UV lithographically patterned structure; B. The structure is brought in contact with a layer of BCB coated on Au/Si; C. After peeling off the silicon wafer, oxygen plasma etching is used to remove the Au layer; D. The final enclosed structure.

auto-fluorescent, making it difficult to image the samples. These channels are 5 μm high and less than 15 μm wide, and microfluidic connections are needed for sample introduction and outflow.

Methods to Achieve Combined Confinement

UV Lithography and E-beam Lithography

Only two photoresists have successfully been used to achieve combined confinement of kinesin-MT systems. Both, SAL601 *(22)* and SU8-2 *(38)*, have been used in conjunction with the non-ionic detergent Triton X100 (Figure 10), as well as a buffer with increased ionic strength. Triton X100, which adsorbs selectively to the hydrophobic photoresist, is hypothesized to either discourage protein adsorption *(22)* or denature adsorbing motors *(38)*. High ionic strength of the solution seems to decrease the electrostatic interactions between protein and substrate, and thus helps to discourage non-specific protein binding.

By photolithography, SAL 601 and SU8 can be easily patterned on glass, and MTs can be guided at all approach angles. Moorjani et al. *(38)* quantitatively described the behavior of MTs when they encountered sidewalls: for 87

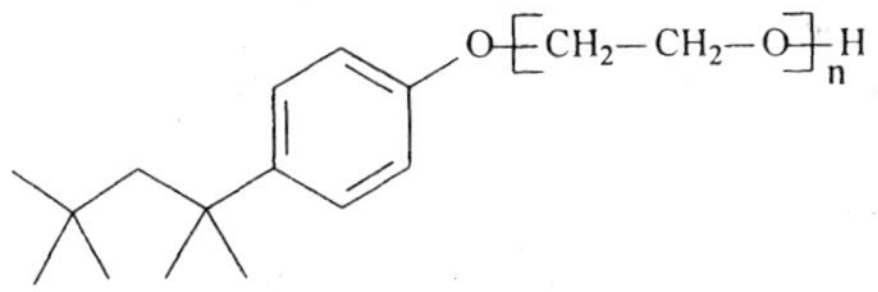

Figure 10. The structure of Triton X100 ($C_{34}H_{62}O_{11}$ for n = 10): The hydrophobic moiety of the Triton X100 structure enables its adsorption to surfaces through hydrophobic interactions, while the "PEG-like" hydrophilic part of the molecule exhibit non-fouling characteristics.

observed collations, 81 (93%) were properly guided, 3 (4%) stalled, and 3 (4%) diffused away, but no MTs climbed up the walls. Moreover, among the 81 successful guidance incidents, 86% of MTs traced the edge of walls and moved parallel to them, while 14 % were reflected back to the channels, which indicated that MTs tended to trace the edges of walls. An alternative to Triton X100 is the surfactant Pluronic F108 *(28)*.

Van den Heuvel et al. developed a combined confinement approach using e-beam lithography (23). As shown in Figure 11, PMMA is selectively coated on parts of the SiO_2/Si substrate. Plasma etching is performed to etch both the PMMA and the SiO_2 surfaces, and thus the structured PMMA profile is transferred about 800 nm deep into the SiO_2. The substrate is subsequently

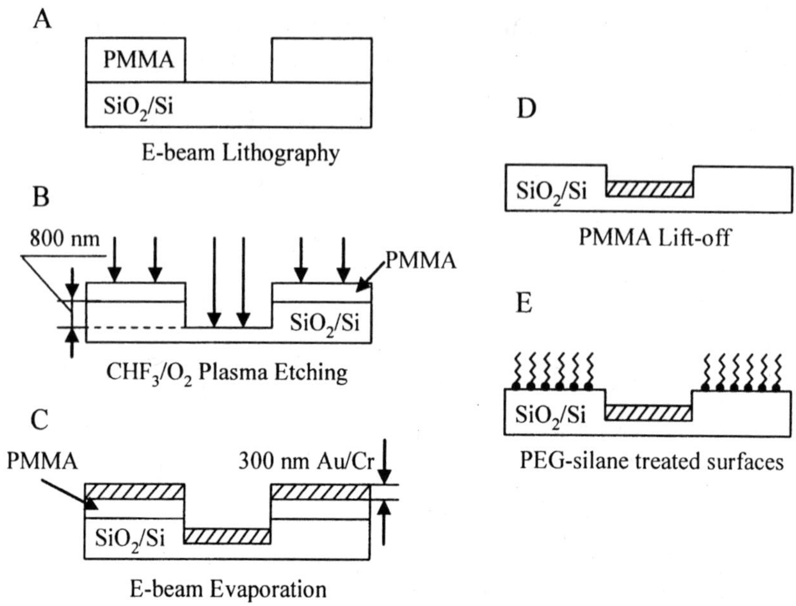

Figure 11. E-beam lithography and plasma etching: A. E-beam lithographily produced structure; B. Oxygen plasma etching to transform the pattern into the SiO_2 layer; C. E-beam evaporation of Au and Cr; D. Lift-off the PMMA layer; E. Non-fouling modification of SiO_2 surface by forming a PEG-silane monolayer.

coated with chrome (5 nm) and gold (30 nm). After PMMA lift-off, the underlying glass surfaces are exposed. PEG-silane is then applied to the substrate, binding only to the glass regions and rendering them non-fouling. Since MTs were observed to move only on the recessed gold regions but never on SiO_2 plateaus, combined confinement was achieved. Guiding was more successful for high MT densities, and guiding efficiency began to decrease from near 100% for approach angles over 50°-60° to about 50% for orthogonal collisions.

For the myosin-actin systems, E-beam lithography is commonly used, since narrower channels are required to prevent actin filaments from taking U-turns due to their high flexibility. The negative photoresist MRL-6000.1 XP was identified as an excellent motility supporting material, while polymethyl-methacrylate (PMMA-950) was selected as a motility suppressing material *(11)*. PMMA-950 can be selectively patterned on a MRL-6000 layer coated onto a SiO_2/Si substrate by E-beam lithography. The resulting patterns consist of MRL-6000 channels (100-200 nm wide and 140 μm long) that were surrounded by PMMA-950 sidewalls (300 nm high). The density of moving actin filaments within the channels was about 20 times greater than that from the outside.

Although no U-turns were observed within the narrow channels, the filaments were able to climb up the sidewalls and escape from the channels.

Nanoimprint Lithography

Compared with conventional photolithography or E-beam lithography, nanoimprint lithography (NIL) *(48,49)* is considered to be a less time-consuming process which enables mass production. For the kinesin-MT system, Cheng et al. *(29)* were able to generate structured patterns on the sub-micro scale with NIL, as shown in Figure 12. By pressing a mold into a layer of perfluoro-based polymer CYTOP precoated on glass, 350 nm-wide channels can be easily generated with steep sidewalls 350 nm in height. Finally the hydrophobic CYTOP surface is treated with Pluronic F108 resulting in a non-fouling surface. This combined confinement system resulted in almost 100% successful confinement, and the average MT traveling distance was about 2.5 mm, approximately 70 times longer than that obtained by physical confinement on PU surfaces *(35)*. In a standard NIL process, O_2 plasma etching is carried out to remove the organic residues. However, as O_2 plasma etching would render the CYTOP surface hydrophilic, oxygen was replaced by sulfur hexafluoride (SF_6) for the etching process. SF_6 was able to enhance surface hydrophobicity of some polymers, e.g. polyurethane, silicone and PTFE *(50)*.

The same combination of materials, MRL-6000.1 XP and PMMA-950, can also be used to produce patterned structure through NIL for the myosin-actin system *(44)*. A SiO_2/Si stamp was pressed into a 120 nm thick PMMA layer and

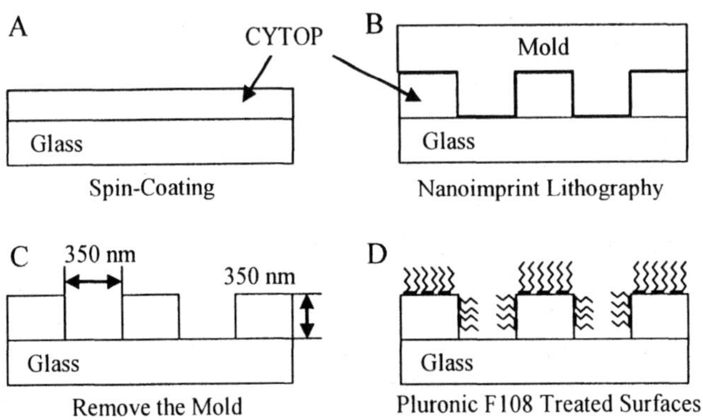

Figure 12. Patterned structures produced by nanoimprint lithography (29): A. spin-coat a layer of CYTOP on glass; B. nanoimprint lithography; C. the resulting structure after removing the mold; D. non-fouling modification of CYTOP surfaces after Pluronic treatment.

100 nm-deep indentations were formed. O_2 plasma etching was then performed to reduce the thickness of the PMMA layer until the underlying motility-supporting MRL surface was exposed. The patterned channels were 100-400 nm wide and 150 nm deep. This patterned structure showed good motility contrast and no U-turns were observed. During the etching process, both PMMA and MRL surfaces became more hydrophilic, and consequently, the corresponding average velocity of actin filaments was reduced to ~2 μm/s, which was only 35% of that measured on the untreated hydrophobic MRL surface.

Polymer-on-polymer Stamping

The isoelectric point of myosin is 5.4, which indicates that it will become negatively charged in imidazole buffer solutions (pH 7.4), and thus tends to adsorb onto cationic surfaces. Based on this idea, Jaber et al. *(42)* tested three materials for their ability to adsorb and support myosin motors. Although all three could adsorb myosin motors, only the positively charged poly(allylamine hydrochloride) (PAH) coated surface supported protein function.

By alternately applying glass coverslips to polyelectolyte solutions of positively charged PAH and negatively charged polystyrene sulfonate (PSS), polyelectrolyte multilayers (PEMUs) - having the positive PAH layer always as the terminating surface - could be produced. Then, polymer-on-polymer stamping (POPS), as developed from microcontact printing techniques *(51,52)* was used to selectively pattern protein-inactivating poly(styrene sulfonate)-block-poly-(ethylene-ran-butylene)-block-poly(styrenesulfonate) (PEBSS) regions on PAH surface. PEBSS was applied to a poly(dimethyl siloxane) (PDMS) stamp with a cotton swab, and then the stamp was pressed against the PEMUs surface for 7 s. The patterned surface consisted of PAH channels 3.9 μm wide with 132-nm-high PEBSS sidewalls. Longer filaments (> 4.5 ± 0.5 μm) traveled along the tracks without changing lanes or making U-turns, while only 40% of the smaller filaments (~ 2 μm) could maintain their direction during an observation period of 26 s, and 11.5% filaments diffused away after collision with sidewalls.

Combined Confinement with Semi-Enclosed Structure

In most studies, actin filaments could enter and escape the channels at random locations in an uncontrolled fashion due to the poor contrast in myosin adsorption and thus filament motility. Within the channels, filament transport in such conditions could hardly be sustained for more than a few tens of micrometers, which would severely hinder the application of myosin-actin system on large scale transportation. By inventing another method to produce channels with undercut regions, as shown in Figure 13, Bunk et al. *(12)*

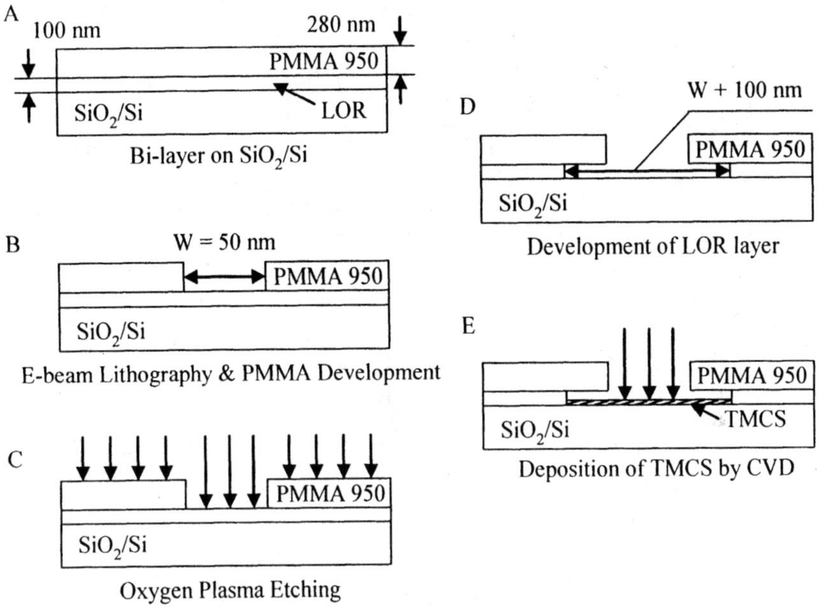

Figure 13. Combined confinement with a semi-enclosed structure: A. Photoresist LOR and PMMA is spin-coated on a SiO_2/Si wafer; B. By E-beam lithography, the top PMMA layer is exposed and developed first; C. O_2 plasma etching is employed to remove the organic residues and to render the top PMMA surface hydrophilic that will inactivate motor function; D. The exposed regions of the lower LOR layer are subsequently developed; E. By chemical vapor deposition (CVD), TMCS is selectively deposited only onto the SiO_2 surface of the bottom channels. TMCS strongly enhances the hydrophobicity of SiO_2 surfaces, which will further support robust filament motility.

developed a structure that allowed almost 100% confinement while still retaining high motor efficiency in terms of filament gliding velocity.

High contrast of filament mobility was observed between the top polymethylmethacrylate (PMMA) surface (almost no filaments were found) and the trimethylchlorosilane (TMCS) treated channel surfaces. It was seen that the motility environment along the channel edges under the PMMA overhangs was superior to the top or bottom of the channels as the filaments preferred to travel along the edges. This edge-tracing effect was explained in terms of the high affinity of LOR070A resist walls to myosin motor adsorption. Filament motion within the channels was not very smooth and the average velocity was 4.3 ± 2 μm/s. Filament transport was observed in all channels with varying widths, and no actin filament was observed to escape the channels. U-turns were not observed in channels narrower than 500 nm.

Outlook

The combination of biological nanomachines with micro- and nanofabrication methods enables the design of active nanostructures. However, this synthesis requires extensive experimentation, since critical interactions between motor proteins and synthetic materials are not well understood. This review presented the fabrication approaches taken in this field, and their relative success in balancing sometimes competing demands, such as low autofluorescence of photoresists for high contrast fluorescence observation and high hydrophobicity for efficient surfactant adhesion.

At this point, we have not arrived at one overarching solution, but rather a number of approaches tailored to the properties of motors, filaments and specific experimental situations. An emerging trend is the increase in the physical confinement of moving filaments, and we expect that the geometry of axons (tubes with submicron diameter and up to meter length) will eventually be employed. The competition between the two systems, microtubules transported by kinesin motors and actin filaments transported by myosins, has been extremely beneficial in clarifying the physical properties of biological and synthetic components which determine the efficiency of guiding. It will be interesting to see if two or three motors from the main families (kinesins, dyneins, myosins) can eventually be employed simultaneously in the same system. The ultimate challenge is the placement of individual motors in defined locations, a feat not achieved by any of the existing approaches.

Motor proteins, which undergo large conformational changes during their mechanochemical cycle and are consequently very sensitive to their immediate environment, can serve as "canaries in the coalmine" for a range of bio-nano interfacing approaches. Furthermore, in particular for the kinesin-microtubule system, motor coverages of a hundredth percent of a monolayer can cause observable changes in the system behavior, placing extreme requirements on the quality of non-fouling surfaces. In combination, this makes the interfacing of motor proteins with synthetic surfaces a challenging test for engineered biocompatible surfaces, and lends relevance to the results presented here beyond the biomotor community.

References

1. Dennis, J. R.; Howard, J.; Vogel, V. *Nanotechnology* **1999**, *10*, 232-236.
2. Hess, H.; Vogel, V. *Reviews in Molecular Biotechnology* **2001**, *82*, 67-85.
3. Howard, J. *Mechanics of Motor Proteins and the Cytoskeleton*; Sinauer: Sunderland, MA, 2001.
4. Hirokawa, N.; Takemura, R. *Nature Reviews Neuroscience* **2005**, *6*, 201-14.
5. Hess, H.; Clemmens, J.; Howard, J.; Vogel, V. *Nano Letters* **2002**, *2*, 113-116.

6. Hess, H.; Howard, J.; Vogel, V. *Nano Letters* **2002**, *2*, 1113-1115.
7. Diez, S.; Reuther, C.; Dinu, C.; Seidel, R.; Mertig, M.; Pompe, W.; Howard, J. *Nano Letters* **2003**, *3*, 1251-1254.
8. Hess, H.; Clemmens, J.; Brunner, C.; Doot, R.; Luna, S.; Ernst, K.-H.; Vogel, V. *Nano Letters* **2005**, *5*, 629-633.
9. Ramachandran, S.; Ernst, K.-H.; Bachand, George D.; Vogel, V.; Hess, H. *Small* **2006**, *2*, 330-334.
10. Bachand, George D.; Rivera, Susan B.; Carroll-Portillo, A.; Hess, H.; Bachand, M. *Small* **2006**, *2*, 381-385.
11. Bunk, R.; Klinth, J.; Montelius, L.; Nicholls, I. A.; Omling, P.; Tagerud, S.; Mansson, A. *Biochemical Biophysical Research Communications* **2003**, *301*, 783-8.
12. Bunk, R.; Sundberg, M.; Mansson, A.; Nicholls, I. A.; Omling, P.; Tagerud, S.; Montelius, L. *Nanotechnology* **2005**, *16*, 710-717.
13. Vale, R. D.; Milligan, R. A. *Science* **2000**, *288*, 88-95.
14. Schliwa, M.; Woehlke, G. *Nature* **2003**, *422*, 759-65.
15. Visscher, K.; Schnitzer, M. J.; Block, S. M. *Nature* **1999**, *400*, 184-9.
16. Alberts, B.; Johnson, A.; Lewis, J.; Raff, M.; Roberts, K.; Walter, P. *Molecular Biology of the Cell. Fourth Edition.*; 4 ed.; Garland: New York, 2002.
17. Spudich, J. A.; Kron, S. J.; Sheetz, M. P. *Nature* **1985**, *315*, 584-586.
18. Kron, S. J.; Spudich, J. A. *Proc. Natl. Acad. Sci. USA* **1986**, *83*, 6272-6.
19. Harada, Y.; Noguchi, A.; Kishino, A.; Yanagida, T. *Nature* **1987**, *326*, 805-808.
20. Vale, R. D.; Reese, T. S.; Sheetz, M. P. *Cell* **1985**, *42*, 39-50.
21. Howard, J.; Hudspeth, A. J.; Vale, R. D. *Nature* **1989**, *342*, 154-158.
22. Hiratsuka, Y.; Tada, T.; Oiwa, K.; Kanayama, T.; Uyeda, T. Q. *Biophys J* **2001**, *81*, 1555-61.
23. van den Heuvel, M. G.; Butcher, C. T.; Smeets, R. M.; Diez, S.; Dekker, C. *Nano Lett* **2005**, *5*, 1117-22.
24. Hess, H.; Clemmens, J.; Matzke, C. M.; Bachand, G. D.; Bunker, B. C.; Vogel, V. *Appl. Phys. A* **2002**, *75*, 309-313.
25. Clemmens, J.; Hess, H.; Doot, R.; Matzke, C. M.; Bachand, G. D.; Vogel, V. *Lab on a Chip* **2004**, *4*, 83-86.
26. Hoffman, A. S. *J Biomater Sci Polym Ed* **1999**, *10*, 1011-4.
27. Lipscomb, R. C.; Clemmens, J.; Hanein, Y.; Holl, M. R.; Vogel, V.; Ratner, B. D.; Denton, D. D.; Böhringer, K. F. In *Second International IEEE-EMBS Special Topic Conference on Microtechnologies in Medicine & Biology*; IEEE: Madison, Wisconsin, 2002; pp 21-26.
28. Clemmens, J.; Hess, H.; Lipscomb, R.; Hanein, Y.; Boehringer, K. F.; Matzke, C. M.; Bachand, G. D.; Bunker, B. C.; Vogel, V. *Langmuir* **2003**, *19*, 10967-10974.
29. Cheng, L. J.; Kao, M. T.; Meyhofer, E.; Guo, L. J. *Small* **2005**, *1*, 409-414.
30. Ionov, L.; Stamm, M.; Diez, S. *Nano Lett* **2005**, *5*, 1910-1914.

31. Nitta, T.; Hess, H. *Nano Letters* **2005**, *5*, 1337-1342.
32. Gittes, F.; Mickey, B.; Nettleton, J.; Howard, J. *J. Cell Biol.* **1993**, *120*, 923-934.
33. Stracke, P.; Bohm, K. J.; Burgold, J.; Schacht, H. J.; Unger, E. *Nanotechnology* **2000**, *11*, 52-6.
34. Hess, H.; Clemmens, J.; Qin, D.; Howard, J.; Vogel, V. *Nano Letters* **2001**, *1*, 235-239.
35. Clemmens, J.; Hess, H.; Howard, J.; Vogel, V. *Langmuir* **2003**, *19*, 1738-1744.
36. Hess, H.; Matzke, C. M.; Doot, R. K.; Clemmens, J.; Bachand, G. D.; Bunker, B. C.; Vogel, V. *Nano Letters* **2003**, *3*, 1651-1655.
37. Huang, Y. M.; Uppalapati, M.; Hancock, W. O.; Jackson, T. N. *IEEE Transactions on Advanced Packaging* **2005**, *28*, 564-570.
38. Moorjani, S. G.; Jia, L.; Jackson, T. N.; Hancock, W. O. *Nano Letters* **2003**, *3*, 633-637.
39. Suzuki, H.; Yamada, A.; Oiwa, K.; Nakayama, H.; Mashiko, S. *Biophysical Journal* **1997**, *72*, 1997-2001.
40. Nicolau, D. V.; Suzuki, H.; Mashiko, S.; Taguchi, T.; Yoshikawa, S. *Biophys J* **1999**, *77*, 1126-34.
41. Sundberg, M.; Rosengren, J. P.; Bunk, R.; Lindahl, J.; Nicholls, I. A.; Tagerud, S.; Omling, P.; Montelius, L.; Mansson, A. *Anal Biochem* **2003**, *323*, 127-38.
42. Jaber, J. A.; Chase, P. B.; Schlenoff, J. B. *Nano Letters* **2003**, *3*, 1505-1509.
43. Manandhar, P.; Huang, L.; Grubich, J. R.; Hutchinson, J. W.; Chase, P. B.; Hong, S. H. *Langmuir* **2005**, *21*, 3213-3216.
44. Bunk, R.; Klinth, J.; Rosengren, J.; Nicholls, I.; Tagerud, S.; Omling, P.; Mansson, A.; Montelius, L. *Microelectronic Engineering* **2003**, *67-8*, 899-904.
45. Mahanivong, C.; Wright, J. P.; Kekic, M.; Pham, D. K.; dos Remedios, C.; Nicolau, D. V. *Biomedical Microdevices* **2002**, *4*, 111-116.
46. Suzuki, H.; Oiwa, K.; Yamada, A.; Sakakibara, H.; Nakayama, H.; Mashiko, S. *Jap. J. Appl. Phys. Part 1* **1995**, *34*, 3937-3941.
47. Xia, Y. N.; Rogers, J. A.; Paul, K. E.; Whitesides, G. M. *Chemical Reviews* **1999**, *99*, 1823-1848.
48. Chou, S. Y.; Krauss, P. R.; Renstrom, P. J. *Science* **1996**, *272*, 85-87.
49. Chou, S. Y.; Krauss, P. R. *Microelectronic Eng.* **1997**, *35*, 237-240.
50. Rangel, E. C.; Bento, W. C. A.; Kayama, M. E.; Schreiner, W. H.; Cruz, N. C. *Surface and Interface Analysis* **2003**, *35*, 179-183.
51. Jiang, X. P.; Zheng, H. P.; Gourdin, S.; Hammond, P. T. *Langmuir* **2002**, *18*, 2607-2615.
52. Jiang, X. P.; Hammond, P. T. *Langmuir* **2000**, *16*, 8501-8509.

Chapter 22

Bionanotransporters

Evgeny Katz

Department of Chemistry and Biomolecular Science, Clarkson University, Potsdam, NY 13699-5810

Recent advances in the design of nanomachines performing transporting functions using driving forces generated by biomolecular units, such as kinesin and myosin motor proteins, are reviewed. The bionanotransporters are capable of translocating a nano-cargo in various nano-devices and they represent important parts of future nanomachines.

Bioelectronics is a multi-facet scientific and technological area that includes electronic (or optoelectronic) coupling of biomaterials (enzymes, DNA, recognition proteins, their integrated assemblies or, eventually, biological cells or their fragments) with electronic devices (*1*). Interfacing of biomaterials and electronic devices can be used to transduce chemical signals generated by biological components into electronically readable signals, or to activate the biomaterials by applying electronic (or optical) signals, thus resulting in the switchable/tunable performance of the biological components. The bioelectronic (optobioelectronic) systems can be used to develop sensing devices (*2*) (enzyme-based biosensors, DNA sensors, immunosensors, etc.) and to develop biofuel cells (*3*) (implantable biofuel cells for biomedical applications, self-powered biosensors, autonomously operated devices, etc.). New methods and new materials (functionalized nanoparticles (*4*), carbon nanotubes (*5*), etc.) developed due to the tremendous recent success in nanotechnology pave the way for the novel possibilities of coupling biomaterials and electronic transducers. Recent cooperative efforts in bioelectronics and nanotechnology resulted in the formulation of the novel scientific direction named "bionanotechnology". Ingenious methods were developed to design biomaterial-functionalized nano-objects, such as metallic or semiconductive nanoparticles and carbon nanotubes and to couple them with various electronic transducers. Dimensional similarity of biomolecules and nano-objects allows their functional coupling, thus providing effective chemical/electronic signal transduction in the hybrid systems. The bionano-hybrid systems keep great promise for the development of novel biosensors, biofuel cells and biocomputing elements.

A special field of research, within the general frame of bionanotechnology, is directed to the design of bionanomechanical devices, namely bionanomotors and bionanotransporters. This research is inspired by studying natural biomechanical assemblies, and the aim of this research is the functional coupling of natural biomechanical assemblies with artificial nano-objects to develop bionanomechanical devices that could be integrated in future with nano-biofuel elements and nano-biosensors to perform bionanorobotic functions.

Natural engines, motor proteins, that transport molecular cargos over long distances are far superior to all synthetic miniaturized transporting systems (*6,7*). Motor proteins demonstrate much better efficiency comparing to artificial micro- and nano-motors, they consume less energy, generate more power and have much smaller dimensions than any artificially engineered device. Because of the unique transport and power generating capabilities of motor proteins, there is a considerable interest in incorporating them into nano- and microscale hybrid devices (*8,9*). Isolation of the motor proteins from cells and then reconstitution of their activity in nano-engineered systems could provide a powerful approach to the transporting of a nano-cargo in various nano-devices.

Three major groups of the motor proteins are used in natural biological systems to provide a molecular-cargo transport over long distances. Two of them: kinesin and dynein move along microtubules toward plus-end and minus-end directions, respectively, using the microtubules paths similarly to rails for a train (*10,11*). The building blocks of microtubules are dimers of α- and β-tubulin interconnected in a head-to-tail mode to yield protofilaments. In natural systems, each microtubule consists of 13 protofilaments, but microtubules generated *in vitro* can include different numbers of the protofilament units, thus having different lengths (typically of 2-10 μm). Microtubules represent hollow tubes with a 24 nm diameter and a thickness of the walls of ca. 6 nm. The directionality of the microtubules is provided by the polarity of the dimeric tubulin subunits. The β-tubulin subunit of the αβ-dimer points to the plus-end side of the microtubule, while the opposite side of the microtubule is called the minus-end side and it is terminated by the α-tubulin subunit (Figure 1A).

The third motor protein - myosin moves along actin filaments that consist of two helically wound rows of actin monomers (*12*). The important feature of actin filaments is their dynamic mobility: the filaments are easily assembled and disassembled upon interaction with special enzymes. The motor proteins have two main parts: twin "heads" connected together at a "stalk" part that move along the molecular paths and a "tail" part that attaches and transports a molecular cargo (Figure 1B) (*7*). The motor proteins mechanistically represent stepping motors that move along the paths provided by microtubules or actin

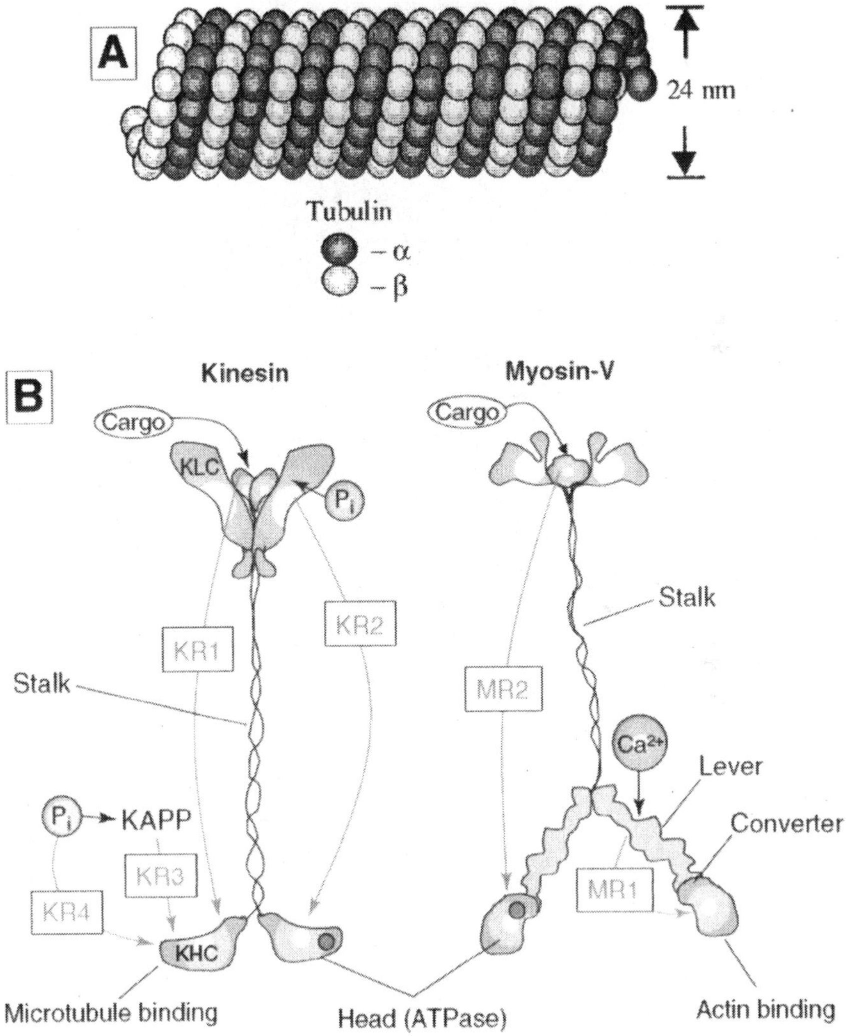

Figure 1. (A) Microtubule structure. (B) Schematic structure and operation of motor proteins kinesin-1 and myosin-V. (Part A is adapted with permission from reference 20, Figure 1. Copyright 2001 American Chemical Society. Part B is adapted with permission from reference 7, Figure 2. Copyright 2004 Elsevier Ltd.)

filaments in a stepwise mode. The energy needed for the motion is supplied to the system upon hydrolysis of adenosine triphosphate (ATP) that is a universal biological energy supplier. For example, kinesin performs an 8 nm step along the microtubule path, a distance corresponding to the size of the tubulin subunit, upon hydrolysis of one ATP molecule. The force of ca. 6-7 pN with the power efficiency of ca. 50-70% is generated by kinesin molecule upon each step, as derived from load-velocity measurements (*13,14*).

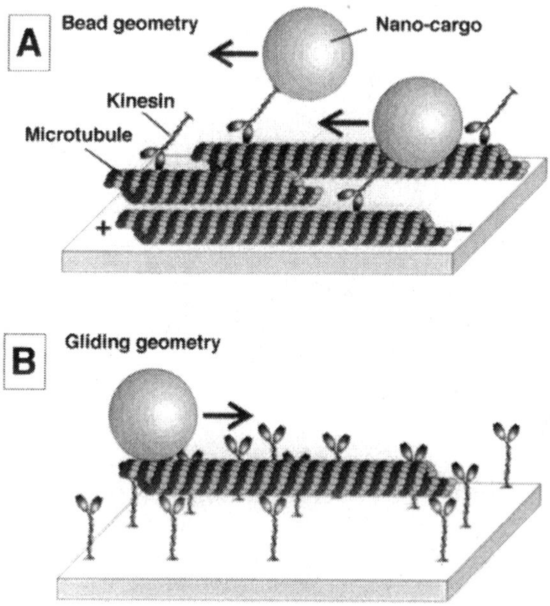

Figure 2. Motor protein-based nano-transporting systems: (A) Bead geometry. (B) Gliding geometry. (Adapted with permission from reference 9. Copyright 2004 Springer.)

Two different approaches to the nano-transporting systems were developed (*9*): (*i*) Motor protein (e.g. kinesin) units loaded with a nano-cargo can move along the respective paths (e.g. microtubules) that are bound to a surface support (transporting systems with a bead geometry) (Figure 2A). (*ii*) Fragments of paths (e.g. short fragments of microtubules) loaded with a nano-cargo can move on a solid support functionalized with many motor protein units (transporting systems with a gliding geometry) (Figure 2B). In this configuration, the front end of the microtubule is free from the surface and randomly searches over the surface for new motor units to bind. This results in the random move of the microtubules over the motor-functionalized surface. It should be noted, however, that the gliding geometry of the nano-transporting systems has obvious disadvantages: In order to provide continues movement of the paths fragments along the functionalized surface, the motor protein units should be assembled in a coherent structure on the solid support with the optimal distances between the units equal to the step-length of the motor. In the bead geometry of the nano-transporting systems, the motor units are free moving components and they can easily adjust their steps to the paths immobilized on the surface.

In order to design the nanotransporting systems with the bead geometry, various approaches were utilized to attach microtubules to a solid support and to organize the kinesin movement along them. For example, the negatively charged microtubules were electrostatically bound to a positively charged amino-silane functionalized surface (*15*), or microtubules were linked to surfaces through oligonucleotide bridges (*16*). The surface-confined microtubules were used as paths for kinesin driven transport of beads (*17*), Au nanowires (*18*), or even micron-size silicon microchips (*19*). As a step toward kinesin-powered microtransporting devices, microchips were functionalized by kinesin units and loaded onto a surface modified by aligned microtubules (*19*). In the presence of ATP, the microchips were translated, rotated and flipped over demonstrating the moving ability of the system. A specific placement of microtubules on a patterned surface was based on their linkage through hybridization of oligonucleotides attached to the tubulin units and complementary oligonucleotides bound to the solid support (*16*). A glass slide was lithographically patterned by neutravidin using a microstamp printing technique, and then biotinilated oligonucleotide (**1**) was linked to the patterned surface through the neutravidin units (Figure 3A). Also the microtubule was covalently labeled by biotin units that were further reacted with neutravidin. A complementary biotinylated oligonucleotide (**2**) was bound to the microtubule through the neutravidin/biotin bridges. The hybridization of the surface-confined oligonucleotide (**1**) and the microtubule-bound oligonucleotide (**2**) yields the surface-bound microtubilins that are linked to the oligonucleotide-patterned domains of the surface (Figure 3B). The oligonucleotide-bound microtubule was used to direct the translocation of kinesin loaded with a nano-cargo within the patterned domains. The movement of each individual kinesin occurred along a path provided by an individual microtubule (Figure 3C). It should be noted, however, that the microtubules bound to the surface were not aligned and the translocation of all kinesin motors in the modified domains was random. Similarly, Au nanowires (200 nm diameter and 6 μm length) functionalized by kinesin motor-units were deposited on a solid support modified by microtubules and transported over a long distance (50 μm) with a mean speed of 0.6 $\mu m\ s^{-1}$ (*18*). However, this movement was also random on a long distance, since the microtubules, which served as the directing tracks, were randomly immobilized on the surface.

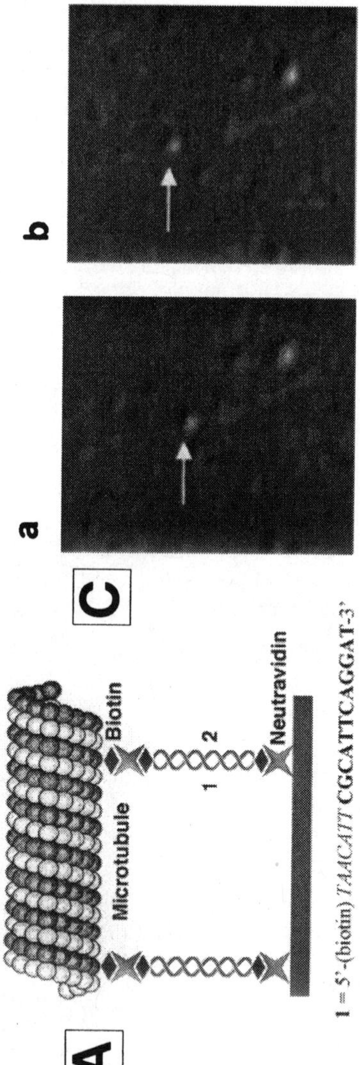
A
Microtubule
Biotin
1
2
Neutravidin
1 = 5'-(biotin) TAACATT CGCATTCAGGAT-3'
2= 5'-(biotin) TAACATT ATCCTGAATGCG-3'
C
a
b

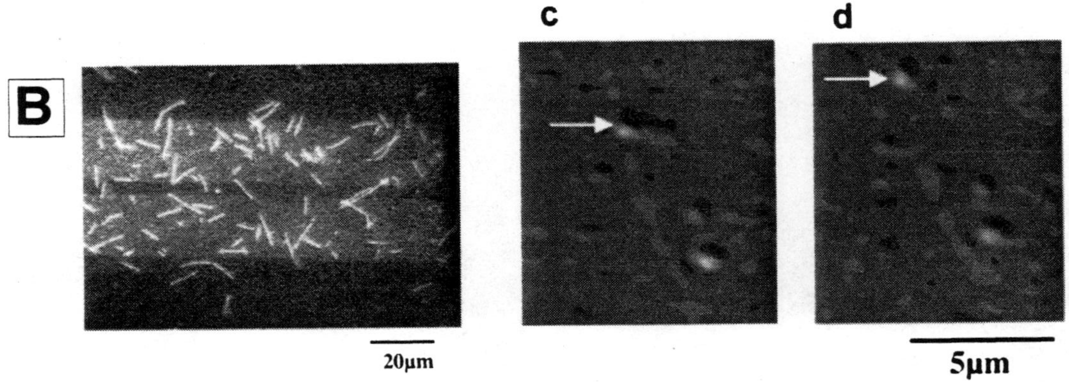

Figure 3. (A) Binding of micritubulin to a solid support using complementary oligonucleotides as linkers. (B) Assembly of microtubulines on the DNA-functionalized micropatterned array. (C) Kinesin-functionalized beads moving along microtubules immobilized on a patterned surface via DNA hybridization. The pictures are taken 1 sec apart. In (d), the bead has reached the end of the microtubule and is diffusing away. (Adapted with permission from reference 16, Figures 6b and 7. Copyright 2004 American Chemical Society.).

Since the translocation of the motor proteins along the microtubules is vectorial (kinesin moves to the plus-end, while dynein steps to the minus-end), the direction of the nano-cargo transportation is controlled by the orientation of the microtubules. Vectorial alignment of microtubules on a surface could be achieved by their anchoring at one end and further stretching by a liquid flow (*20*). An antibody (VαNT1) specific to the α-tubulin unit exposed only at the minus-end of microtubules was generated by a phage display technique. The recombinant VαNT1 antibody was produced with a C-terminal His-tag that allowed the antibody anchoring on the surface. Glass slides were coated by a monolayer of Pluronic F108-NTA-metal-chelating surfactant (NTA - nitrilotriacetic acid) with chelated Ni^{2+}, and then the VαNT1 antibodies were anchored to the Ni^{2+} ions on the surface through the His-tag units (Figure 4). Microtubules were bound to the antibody-modified surface at their minus-ends, and then they were vectorially aligned by application of a liquid flow in the lateral to the surface direction. Approximately 90% of the microtubules were aligned along the stream direction with the anchored minus-ends pointing upstream. The aligned microtubules could be used for the directional translocation of kinesin along them.

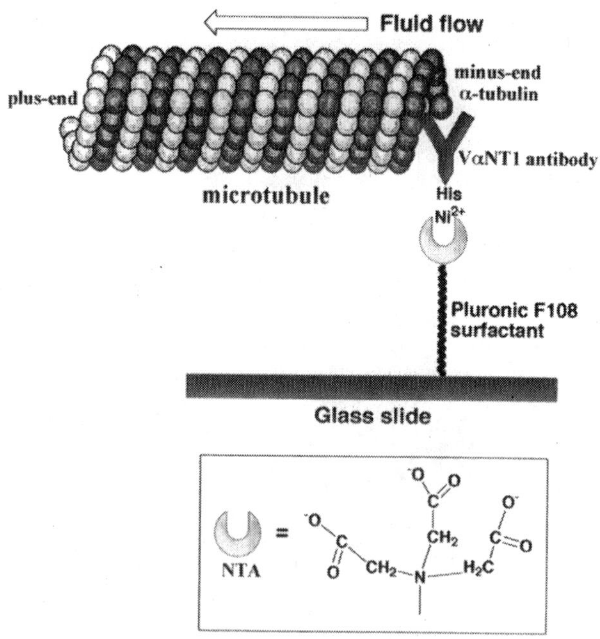

Figure 4. Anchoring of the microtubule at the minus-end on the surface through α-tubulin-specific VaNT1 antibody and the flow-induced alignment of the microtubule.

Despite the obvious disadvantages of the gliding geometry (compared to the bead geometry) for the nanotransporting systems, where microtubules are propelled across a surface functionalized with motor protein units (e.g. kinesin) (*17,21*), most of the research was performed in this direction. For example, microtubules loaded with CdSe nanoparticles as nano-cargos were moved over a kinesin-functionalized surface in the presence of ATP as a fuel component (*22*). Similarly to the microtubules moving over surfaces functionalized with kinesin motor units, another nanotransporting system based on the movement of actin filaments over myosin-functionalized surface was studied (*23*). The actin/Au-nanowire/actin filament generated by the stepwise polymerization of Au nanoparticle-functionalized g-actin, followed by the polymerization of naked g-actin at the nanoparticle-filament ends, and the subsequent enlargement of the Au nanoparticles to a continues Au bar was used as a nanotransporting system. Towards this goal the actin/Au-nanowire/actin filament was rigidified with phalloidin and deposited on a myosin interface linked to a glass support. Upon addition of the ATP "fuel" to the system, the actin-linked Au nanorods randomly moved on the surface, and the motility of the nano-objects was followed by reflectance microscopy. The Au nanorods were found to move at a speed of 250 nm s^{-1}.

Despite the fact that microtubules (or actin filaments) are randomly moved on a micro-scale by an array of the surface-confined kinesin units, the appropriate design of micro-channels could result in the directional transport of the microtubules on a longer scale due to the re-direction of the moving microtubules by the sidewalls of the channel. Lithographically patterned channels (10 μm width and 1.5 μm depth) effectively guided the movements of microtubules over the kinesin-modified bottom along the channel (*24*). However, this micro-scale transportation is still not vectorial, since there is no preferential direction in the movement (right or left direction) along the channel. Rectification of the random movements, resulting in unidirectional transport of microtubules, could be achieved by the special design of the channel shape (*18*). Arrow-shaped micro-channels were made by photolithography, the bottom of the channel was functionalized by kinesin units that activated (in the presence of ATP) transport of microtubules along the channel. The sidewalls of the channel and particularly arrowheads bump and re-direct the microtubules upon collisions, thus resulting in the unidirectional rectified transport of the microtubules in the direction of the arrowhead of the channel (Figure 5). The arrow-shaped channels were assembled into complex systems and networks that allowed vectorial transport of microtubules over the kinesin-modified bottoms (*25*). For example, circular channels with the arrowheads allow the unidirectional movement of the microtubules upon activation of the kinesin motor units bound to the bottom by ATP (Figure 6). Similarly, the unidirectional transport of the microtubules from one micrometer-scale pool to another one

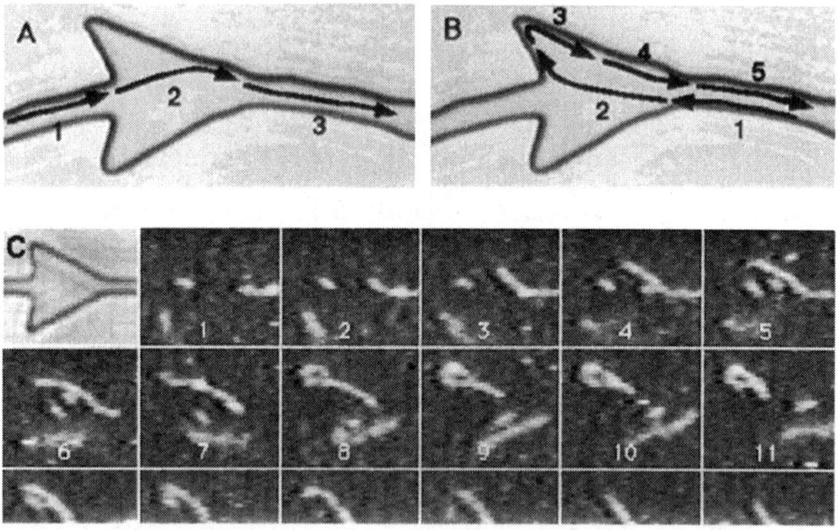

Figure 5. Arrowhead function as a rectifier of microtubule movements in the lithographically patterned channel over kinesin units at the bottom: (A) The scheme shows that a microtubule entering the arrowhead from the correct direction (from the left side entrance) passes through the arrowhead pattern. (B) The scheme demonstrates how a microtubule entering the arrowhead from the wrong direction (from the right side entrance) makes a 180° turn and moves out in the correct direction. (C) The set of images (8 sec per frame) shows turning of a microtubule at an arrowhead pattern. (Adapted with permission from reference 25. Copyright 2001 Biophysical Society.)

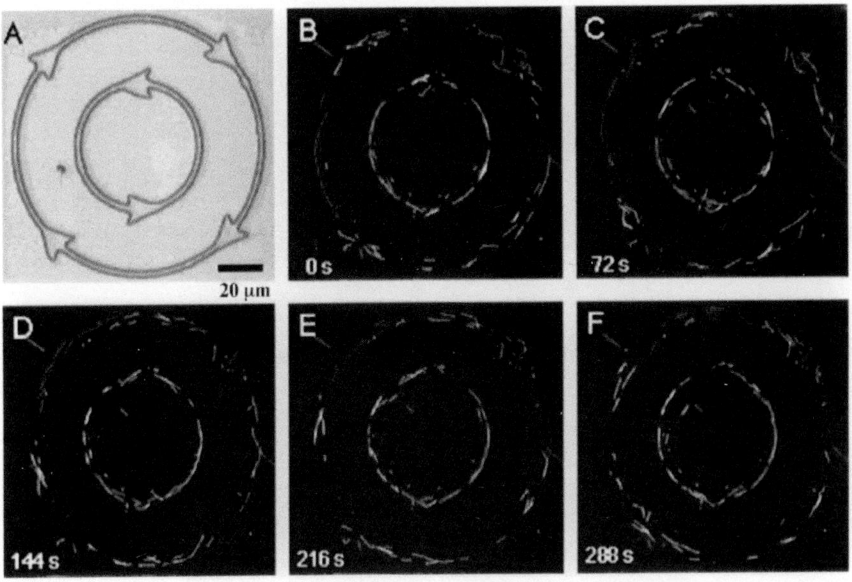

Figure 22.6. Unidirectional rotational movement of microtubules along circular tracks with the rectifying arrowheads: (A) The image of the transmission microscopy showing the patterned channels. (B-F) The snapshots of the movement of rhodamine-labeled microtubules taken at intervals of 72 s (20 min after the addition of ATP). Microtubules in the outer circle are moving clockwise (red), whereas those in the inner circle are moving counterclockwise (green). (Adapted with permission from reference 25. Copyright 2001 Biophysical Society.)

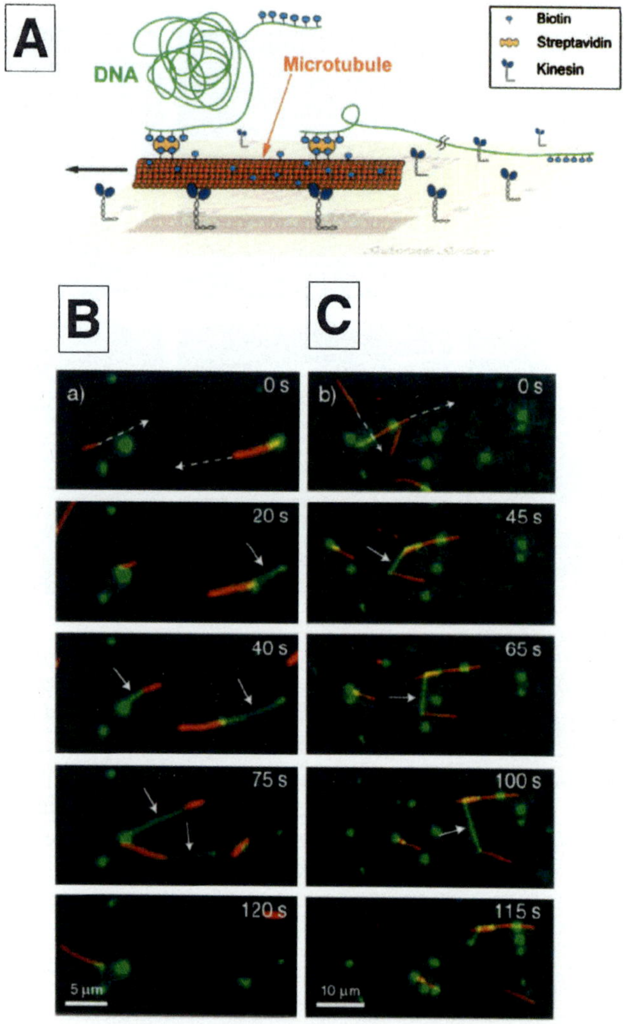

Figure 22.11. (A) DNA transporting and stretching using microtubules moving over kinesin-functionalized surface in the presence of ATP. (B) Stretching DNA between a surface and one motile microtubule: Two condensed DNA molecules attached to the substrate surface by one of their ends are grasped at their second end by motile microtubules (red, moving in the direction of the dotted white line) and are consequently stretched (green; white arrows). (C) Stretching DNA between two motile microtubules. (Adapted from reference 32, Figures 1 and 3, with permission. Copyright 2003 American Chemical Society.)

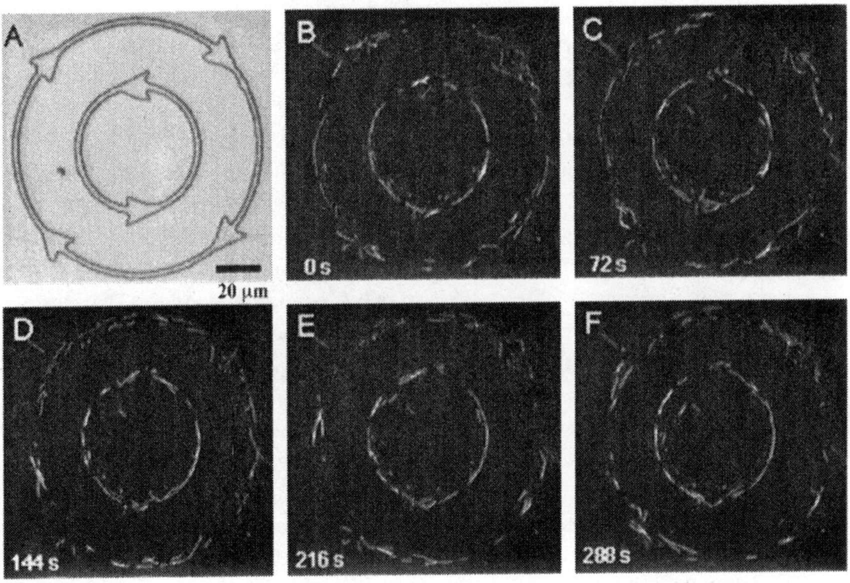

Figure 6. Unidirectional rotational movement of microtubules along circular tracks with the rectifying arrowheads: (A) The image of the transmission microscopy showing the patterned channels. (B-F) The snapshots of the movement of rhodamine-labeled microtubules taken at intervals of 72 s (20 min after the addition of ATP). Microtubules in the outer circle are moving clockwise (dark grey), whereas those in the inner circle are moving counterclockwise (light grey). (Adapted with permission from reference 25. Copyright 2001 Biophysical Society.) (See page 1 of color inserts.)

through three connecting channels with the rectifying arrowheads was demonstrated (Figure 7).

In another approach, the kinesin motor units were deposited on a surface functionalized with the aligned poly(tetrafluoroethylene) thin film (*26*). The aligned polymer film and the kinesin units associated with the polymer resulted in directional motion of microtubules along the film orientation axes.

Based on the polypeptide sequence, αβ-tubulin dimer carries 48 negative charges at physiological pH values that translate to ca. 84,000 negative charges per micron length of the microtubule (*18*). The experimentally measured charge of the microtubules is ca. 300-folds smaller (ca. 280 e^- per μm length) (*27*). This indicates that microtubule formation is accompanied by substantial changes of charge distribution within the tubulin subunits. However, even this negative charge is enough to direct the movement of the microtubules over kinesin motors assembled on a surface by application of an external electrical field (*18,27,28*). Upon application of an electric field, microtubules were moving to the positive electrode with the speed of ca. 1-8 μm s^{-1} (depending on the electric field strength and the ionic strength of the solution), indicating a negative effective charge. This allows directional transport of the microtubules controlled by the external electrical field.

Formation of multi-molecular assemblies, nanowires, and nanocircuitry is usually based on the self-assembly processes that proceed in the thermodynamically favorable direction. These processes are accompanied by the energy dissipation and result in equilibrium structures. Application of dynamic self-assembling, that requires an energy source to build up non-equilibrium systems, may open new possibilities of generating various systems, which cannot be created upon equilibrium. Active transport of nano-objects powered by motor proteins, with the energy source from ATP, could enable designing of self-assembling systems on the nanoscale, because it permits coupling the assembly process with an energy supplying process. This energy-driven assembling could yield the multi-molecular nano-structured systems that are not accessible via self-assembly driven by thermal forces. It was demonstrated that soluble microtubules mixed with kinesin motor units in the presence of ATP as an energy source could exert forces on the microtubules that form micro-scale patterns, such as asters, vortices, and interconnected poles, originating from the dynamic cross-linking of the microtubulues by the motor units (Figure 8) (*29*). Further development of the non-equilibrium assembling utilizing the energy-driven processes was directed to the formation of complex nanocircuitry on surfaces (*30*). Kinesin motor units were densely assembled on a surface (> 100 units per μm^{-2}) and microtubules functionalized with biotin and streptavidin units were moved over the surface in the presence of ATP. The coverage of the microtubules by streptavidin was ca. 50% of the amount of biotin units, thus upon collisions, the complementary streptavidin units and free biotin functions could couple resulting in the cross-linkage and aggregation of the functionalized

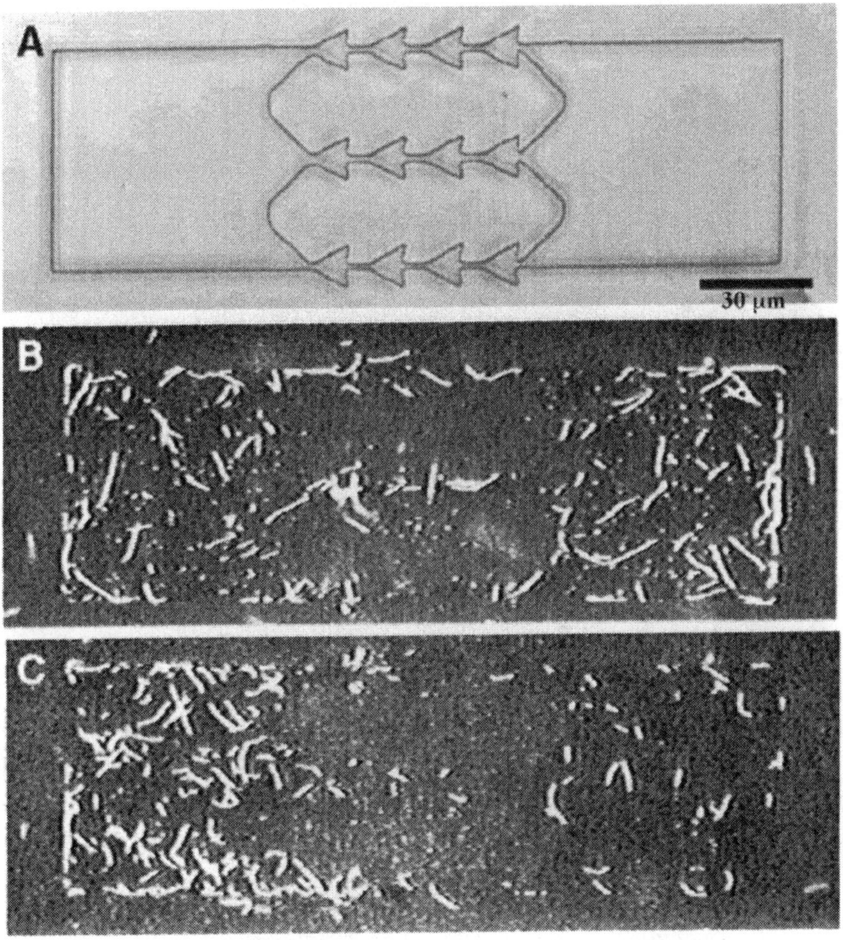

Figure 7. Active transport between two pools of micrometer scales: (A) Transmission micrograph of the patterned network. (B) Fluorescence image of rhodamine-labeled microtubules taken before ATP addition. (C) Fluorescence image of rhodamine-labeled microtubules taken 18 min after the ATP addition. (Adapted with permission from reference 25. Copyright 2001 Biophysical Society.)

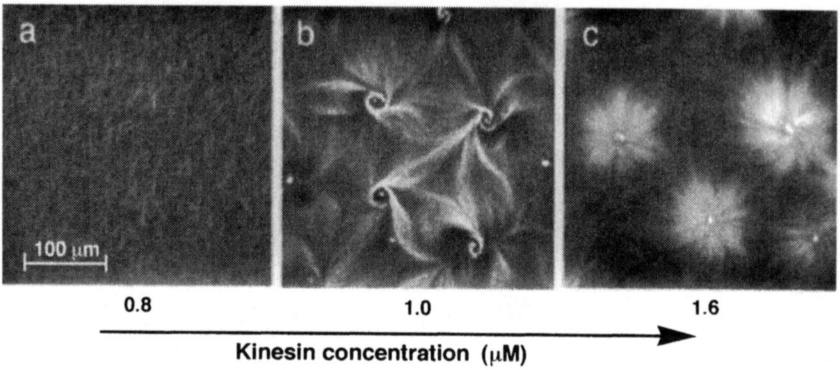

Figure 8. Dynamic assembly of microtubulin upon crosslinking by kinesin motor units in the presence of ATP as an energy source and in the presence of different concentrations of kinesin: a) 0.8 μM, b) 1.0 μM and c) 1.6 μM. (Figure 2A, Adapted with permission from reference 29. Copyright 2001 AAAS.)

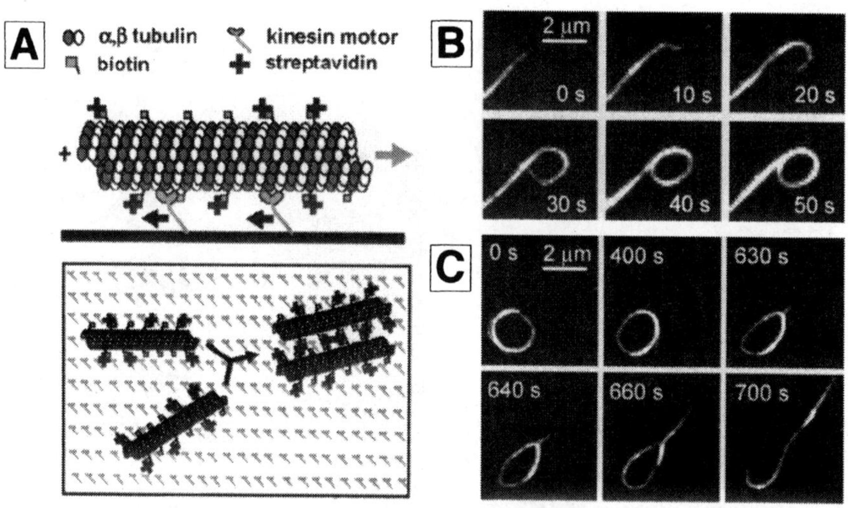

Figure 9. (A) Dynamic aggregation of biotin/streptavidin-functionalized microtubules upon the kinesin-powered translocation. (B) Formation of a "spool" structure upon random re-direction of the microtubule moving over the kinesin-functionalized surface. (C) Disassembling ("unspooling") of the dynamic structures formed by biotion/streptavidin-functionalized microtubules upon continuing movement over the kinesin-modified surface. (Adapted with permission from reference 30, Figures 1, 4 and 5. Copyright 2005 American Chemical Society.)

microtubules (Figure 9A). The binding energy between biotin and streptavidin units (particularly assuming the formation of multiple bonds) is high enough to preserve the aggregates under thermal equilibrium, however, upon application of moving forces exerted by the kinesin motors, these aggregates can dissociate. The dynamic formation and dissociation of the aggregates result in the formation of the patterns, which are not possible under thermal equilibrium. For example, random movements of the microtubules over the kinesin-functionalized surface often result in their re-direction and collisions of the free moving head with the side-domains of the long microtubules yielding "spool" structures (Figure 9B). The generated structures can be preserved upon inhibition of the active transport powered by kinesin motors (the generated structures are thermally stable). However, upon continuing active movements, the dynamic aggregates could be disassembled since the energy applied by the kinesin motor units is higher than the binding energy of the biotin-streptavidin couple (Figure 9C). The demonstrated examples clearly indicate the ability to generate dynamically labile assemblies of high complexity, however, it should be noted that the formation of these structures is not controllable yet and proceeds under random collisions of the units.

Application of dynamic assembling for the generation of complex meta-stable structures and also precise positioning and docking of the nanotransporters carrying a nano-cargo requires the fast activation and inhibition of the move. This task is not easy to achieve since the consumption of the ATP fuel by the kinesin motors is rather slow and they can be powered upon single addition of ATP for a long time. The addition of chemical inhibitors to the system can certainly solve the problem of interrupting the move, but the re-activation of the active transport requires the washing off of the inhibitor and change of the solution. The reversible activation/deactivation of the active transport by applying external stimuli signals (e.g. light signal) is a challenging goal. One step in this direction was made using a photo-labile light-activated ATP derivative (**3**) (Figure 10) (*31*). The light-induced deprotection of **3** yields the active ATP state that fuels the kinesin-powered active transport of microtubules over the surface modified by kinesin units. A biochemical system consisting of hexakinase and glucose was added to the solution in order to consume rapidly the excess of the light-generated ATP and to stop quickly the active transport. By applying light flashes, the active transport of microtubules was activated, and upon consuming the ATP fuel, it was rapidly stopped. This resulted in the controllable movements of the microtubules over the surface by a distance of 4-6 μm per each light flash. It should be noted, however, that within the short active periods the movements of the microtubules were random.

Controllable transportation of a nano-cargo could be also utilized to generate DNA-templates for nanocircuitry. The first example of the DNA transporting by microtubules moving over a kinesin-functionalized surface (still in the random, non-controllable mode) was reported recently (32). Microtubules

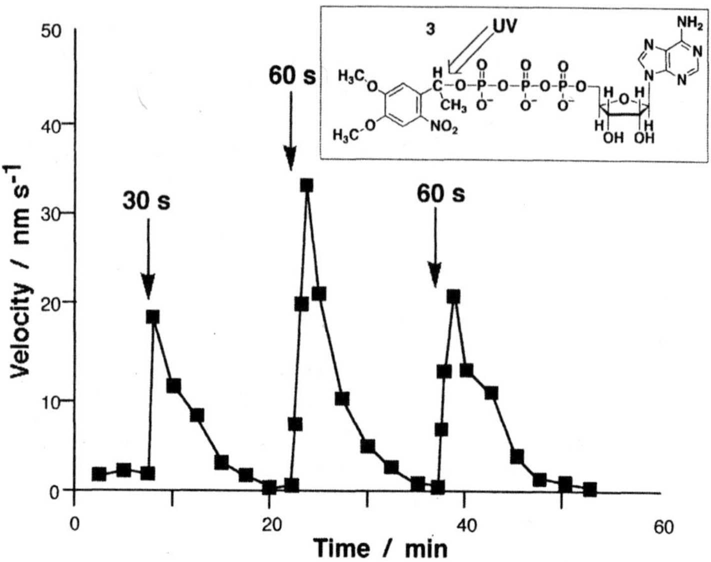

Figure 10. Average speed of microtubules after exposure of the photo-labile light-activated ATP derivative (3) to UV light for 30, 60, and 60 s converting 20%, 30%, and 20% of the initial 3 into free ATP. The presence of the ATP-consuming enzyme, hexokinase, leads to a rapid decline of the microtubule velocity. The inset shows the structure of 3. (Adapted with permission from reference 31, Figure 5. Copyright 2001 American Chemical Society.)

functionalized by biotin/streptavidin units were moved over a kinesin-modified surface in the presence of ATP. λ-Phage DNA with two biotinylated ends was added to the system. The microtubules upon moving were able to pick-up one end of the DNA due to the biotin-streptavidin coupling, while the second end of the same DNA might be linked to the surface (by electrostatic interactions or by biotin-streptavidin interactions, if the surface was also modified by streptavidin units) (Figure 11A). This could result in the stretching the DNA between the surface and the motile microtubule (Figure 11B). Upon coupling of both biotinylated ends of DNA with two different motile microtubules, the DNA chain could be stretched between the microtubules moving randomly in different directions (Figure 11C). However, to be useful for building nanostructures, the DNA transportation by the nanotransporting system should proceed in a controllable and directional mode.

This review has summarized recent advances in the rapidly developing area of bionanotransporters based on kinesin and related motor proteins coupled with functionalized surfaces and nano-objects. The developed systems allow the nano-cargo directional transport based on cooperative work of nano-designed

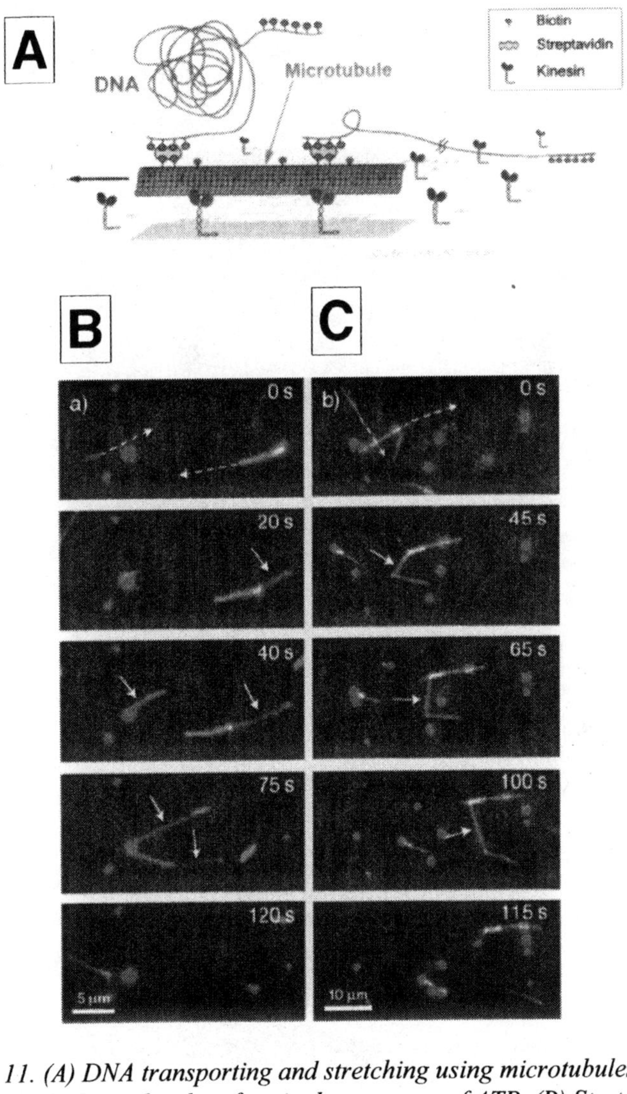

Figure 11. (A) DNA transporting and stretching using microtubules moving over kinesin-functionalized surface in the presence of ATP. (B) Stretching DNA between a surface and one motile microtubule: Two condensed DNA molecules attached to the substrate surface by one of their ends are grasped at their second end by motile microtubules (dark grey, moving in the direction of the dotted white line) and are consequently stretched (light grey; white arrows). (C) Stretching DNA between two motile microtubules. (Adapted with permission from reference 32, Figures 1 and 3. Copyright 2003 American Chemical Society.) (See page 2 of color insets.)

interfaces, functionalized nano-objects and biomotors. Future perspectives of this scientific field should include precise control of the speed and direction of the nano-transporting devices and their coupling with sophisticated bioenergy supplying systems and bioanalytical and logical elements providing ″smart″ control of the transporting units.

References

1. *Bioelectronics: From Theory to Applications*, Willner, I.; Katz, E., Eds.; Wiley-VCH: Weinheim, Germany, 2005.
2. Willner, I.; Katz, E.; Willner, in: *Biosensors and Their Applications*, Yang, V.C.; Ngo, T.T., Eds.; Kluwer Academic Publishers: New York, 2000, Chapter 4, pp. 47-98.
3. Katz, E. Biofuel cells, in: *2004 Yearbook of Science and Technology*, Blumel, D. Ed.; McGraw-Hill Professional: New York, 2004, pp. 33-37.
4. Katz, E.; Willner, I. *Angew. Chem. Int. Ed.* **2004**, *43*, 6042-6108.
5. Katz, E.; Willner, I. *ChemPhysChem* **2004**, *5*, 1084-1104.
6. Bao, G. *J. Mechanics Phys. Solids* **2002**, *50*, 2237-2274.
7. Mallik, R.; Gross, S.P. *Current Biology* **2004**, *14*, R971-R982.
8. Ishijima, A.; Yanagida, T. *Trends Biochem. Sci.* **2001**, *26*, 438-444.
9. Knoblauch, M.; Peters, W.S. *Cellular Molecular Life Sci.* **2004**, *61*, 2497-2509.
10. Ovechkina, Y.; Wordeman, L. *Traffic* **2003**, *4*, 367-375.
11. Howard, J. *Annu Rev. Physiol.* **1996**, *58*, 703-729.
12. Buss, F.; Spudich, G.; Kendrick-Jones, J. *Annu. Rev. Cell Dev. Biol.* **2004**, *20*, 649-676.
13. Svoboda, K.; Block, S.M. *Cell* **1994**, *77*, 773-784.
14. Hunt, A.J.; Gittes, F.; Howard, J. *Biophys. J.* **1994**, *67*, 766-781.
15. Svoboda, K.; Schmidt, C.F.; Schnapp, B.J.; Block, S.M. *Nature* **1993**, *365*, 721-727.
16. Muthukrishnan, G.; Roberts, C.A.; Chen, Y.-C.; Zahn, J.D.; Hancock, W.O. *Nano Lett.* **2004**, *4*, 2127-2132.
17. Vale, R.D.; Schnapp, B.J.; Reese, T.S.; Sheetz, M.P. *Cell* **1985**, *40*, 559-569.
18. Jia, L.; Moorjani, S.G.; Jackson, T.N.; Hancock, W.O. *Biomedical Microdevices* **2004**, *6*, 67-74.
19. Limberis, L.; Stewart, R.J. *Nanotechnology* **2000**, *11*, 47-51.
20. Limberis, L.; Magda, J.J.; Stewart, R.J. *Nano Lett.* **2001**, *1*, 277-280.
21. Howard, J.; Hudspeth, A.J.; Vale, R.D. *Nature* **1989**, *342*, 154-158.
22. Bachad, G.D.; Rivera, S.B.; Boal, A.K.; Gaudioso, J.; Liu, J.; Bunker, B.C. *Nano Lett.* **2004**, *4*, 817-821.

23. Patolsky, F.; Weizmann, Y.; Willner, I. *Nat. Mater.* **2004**, *3*, 692-695.
24. Moorjani, S.G.; Jia, L.; Jackson, T.N.; Hancock, W.O. *Nano Lett.* **2003**, *3*, 633-637.
25. Hiratsuka, Y.; Tada, T.; Oiwa, K.; Kanayama, T.; Uyeda, T.Q.P. *Biophys. J.* **2001**, *81*, 1555-1561.
26. Dennis, J.R.; Howard, J.; Vogel, V. *Nanotechnology* **1999**, *10*, 232-236.
27. Stracke, R.; Böohm, K.J.; Wollweber, L.; Tuszynski, J.A.; Unger, E. *Biochem. Biophys. Res. Commun.* **2002**, *293*, 602-609.
28. Van den Heuvel, M.G.L.; Butcher, C.T.; Lemay, S.G.; Diez, S.; Dekker, C. *Nano Lett.* **2005**, *5*, 235-241.
29. Surrey, T.; Nédélec, F.; Leibler, S.; Karsenti, E. *Science* **2001**, *292*, 1167-1171.
30. Hess, H.; Clemmens, J.; Brunner, C.; Doot, R.; Luna, S.; Ernst, K.-H.; Vogel, V. *Nano Lett.* **2005**, *5*, 629-633.
31. Hess, H.; Clemmens, J.; Qin, D.; Howard, J.; Vogel, V. *Nano Lett.* **2001**, *1*, 235-239.
32. Diez, S.; Reuther, C.; Dinu, C.; Seidel, R.; Mertig, M.; Pompe, W.; Howard, J. *Nano Lett.* **2003**, *3*, 1251-1254.

Indexes

Author Index

Subject Index

C

D

E

F

G

H

I

O

P

W

Y